Engineering Materials

More information about this series at http://www.springer.com/series/4288

P.M. Visakh · Yoshihiko Arao
Editors

Thermal Degradation of Polymer Blends, Composites and Nanocomposites

Springer

Editors
P.M. Visakh
Department of Ecology and Basic Safety
Tomsk Polytechnic University
Tomsk
Russia

Yoshihiko Arao
Doshisha University
Kyoto
Japan

ISSN 1612-1317 ISSN 1868-1212 (electronic)
Engineering Materials
ISBN 978-3-319-03463-8 ISBN 978-3-319-03464-5 (eBook)
DOI 10.1007/978-3-319-03464-5

Library of Congress Control Number: 2015934683

Springer Cham Heidelberg New York Dordrecht London

Printed on acid-free paper

Springer International Publishing AG Switzerland is part of Springer Science+Business Media (www.springer.com)

Preface

This book on "Thermal Degradation of Polymer Blends, Composites and Nanocomposites" summarizes many of the recent research accomplishments in the area of thermal stability of polymer blends, composites, and nanocomposites such as advances in thermal degradation of blend, composites and nanocomposites, thermal degradation of thermosetting blends, thermal degradation of thermosetting nanocomposites, effect of thermo-oxidation on the mechanical performance of polymer-based composites for high temperature applications, analysis for thermal degradation of a polymer by factor analysis, radiation effects on polymer-based systems, thermal degradation of synthetic rubber nanocomposites, outdoor exposure degradation of ethylene-vinyl-acetate, encapsulant material for photovoltaic application, thermal degradation of bio-nanocomposites, etc.

As the title indicates, the book emphasizes the various aspects of thermal stability of polymer blends, composites, and nanocomposites. This book is intended to serve as a "one stop" reference resource for important research accomplishments in the area of thermal stability-based nanocomposites book. This book will be a valuable reference source for university and college faculties, professionals, post-doctoral research fellows, senior graduate students, researchers from R&D laboratories working in the area of thermal stability-based nanocomposites. The various chapters in this book are contributed by prominent researchers from industry, academia, and government/private research laboratories across the globe. It covers an up-to-date record of the major findings and observations in the field of thermal stability-based nanocomposites. The first chapter on thermal stability-based nanocomposites gives an overview of the area of the state of the art, new challenges, and opportunities of thermal stability-based studies and research, chemistry of thermal degradation of polymers, chemistry of thermal degradation of nanocomposites, and future trends.

The following chapter provides a good structure of thermal degradation of thermosetting blends. In this chapter, the authors explain the literature studies on recent advances concerning the thermal behavior of different thermosetting blends. The introduction debates the general issue concerning polymer blends, that is, the occurrence of phase separation phenomena and lists a series of possibilities to overcome these undesired aspects. The introduction section also presents the most common polymers used as crosslinked scaffolds either individual or for different multicomponent polymeric materials. The subchapters that follow are focused on recent studies on the thermal stability and degradation of thermosetting blends, effect of reinforcement and nanofillers on the thermal stability of thermosetting blends, and applications and future trends of thermosetting blends, dealing with the latest issues and trying to reveal solutions.

The third chapter on thermal degradation of thermosetting nanocomposites discusses the rapidly-developing nanotechnology and nanoscience in recent years on thermosetting nanocomposites thermal degradation. The authors explain the effect of different nanoparticles, their dispersion, and use of modifiers on the polymer thermal stability. This chapter focuses on the thermal degradation study of thermosetting nanocomposites materials, evaluating their effect in thermal stability and in thermal degradation steps. The thermal applications of these nanocomposites are also evaluated and the challenges to the nanocomposites field in the following years are discussed. The next chapter manly concentrates on the effect of thermo-oxidation on the mechanical performance of polymer-based composites for high temperature applications. This chapter explains the effect of thermo-oxidation on the mechanical properties of polymer-based composites for high temperature applications. The polymer-based composites with high thermal stability and future trend towards modification of this type of composites have been discussed in this chapter.

The fifth chapter explains the analysis of thermal degradation of a polymer by factor analysis. In this chapter is introduced an application of multivariate curve resolution (MCR) technique based on factor analysis. The authors explain not only series of IR spectra but also two-dimensional data series of nuclear magnetic resonance (NMR), mass spectrometry (MS), and X-ray diffraction (XRD) that can deal in the same manner further two-dimensional data generated by hyphenated techniques such as gas chromatography/mass spectrometry (GC/MS) and liquid chromatography/ultraviolet (LC/UV) analysis, which combine two functions based on different principles, namely chromatography, which has a separating function, and spectrometry, which provides information related to molecular structure. Another chapter on radiation effects on polymer-based systems explains the improvements in thermal properties of polymeric materials/composites with effect of crosslinking and grafting on the thermal properties of polymer materials and composites. The authors also explain with accelerated degradation by radiation exposure.

The seventh chapter explains thermal degradation of synthetic rubber nanocomposites. The authors explain synthetic rubbers nanocomposites that have captured and held the attention of scientists as the materials of the future; these materials improve resistance to thermal degradation and stability of nanocomposites. They explain in this chapter these new materials that exhibit enhanced properties at very low filler level, usually ≤ 5 wt %. The properties of rubber nanocomposites strongly depend on the dispersion state of fillers and method of preparation. The effect of different nanoparticles on rubber properties is studied with thermal stability. This is mainly studied using TGA, TGA-MS TGA-FTIR and other techniques. Finally, the authors explain that rubber synthetic nanocomposites play an important role in engineering, automotive, aerospace, construction, packaging, and medical device applications due to the possibility to design new materials with unprecedented and improvements in their physical properties, particularly from the perspective of applications.

The eighth chapter discusses the outdoor exposure degradation of Ethylene-Vinyl-Acetate (EVA) encapsulant material for photovoltaic application. The authors explain the outdoor exposure of the materials leading to changes in polymer morphology. The main objective of this experimental investigation was to better understand the changes due to thermal transitions and the molecular organizations of the crosslinked ethylene-vinyl-acetate encapsulant material after aging in outdoor exposure. The authors discuss from the results, the significant decrease in most properties of EVA in natural field exposure due principally to the specificity of the exposure site. For aged EVA samples, the distinctive feature of these results is that there are two different endothermic processes due to the recrystallization phenomenon. Furthermore, the difference in the magnitude of peak current by TSC technique suggests increased crosslinking exposure occurring selectively in the high temperature phase as a result of outdoor exposure. The final chapter on thermal degradation of bio-nanocomposites explains the type of biomaterials and their nanocomposite thermal properties. This chapter reviews the recent developments in bio-nanocomposites where the related biodegradable polymers include Polylactic acid (PLA), polycaprolactone (PCL), polyhydroxyvalerate (PHV), polyhydroxyalkanoates (PHAs), polyhydroxybutyrate (PHB), poly(3-hydroxybutyrate-co-3-hydroxyvalerate) (PHBV), and poly(d,l-lactide) (PDLLA). A concise history outlining the development of bio-nanocomposites materials is explored, while the importance of environmental conditions and in particular the rate of biodegradability is highlighted. Furthermore, the authors discussed in this chapter the steps of thermal degradation and the systematic approaches used to overcome these concerns. It discusses the behavior of various nanoparticles on the thermal stability of biopolymers and other topics related to research challenges, future trends, and applications.

Finally, the editors would like to express their sincere gratitude to all the contributors of this book, who gave excellent support for the successful completion of this venture. We are grateful to them for the commitment and the sincerity they have shown toward their contribution in the book. Without their enthusiasm and support, the compilation of a book would not have not possible. We would like to thank all the reviewers who have taken their valuable time to give critical comments on each chapter. We also thank the publisher Springer for recognizing the demand for such a book, and for realizing the increasing importance of the area of "Thermal Degradation of Polymer Blends, Composites and Nanocomposites" and for starting such a new project, in which not many other publishers put their hands on.

P.M. Visakh

Contents

About the Editors

P.M. Visakh M.Sc., M.Phil. submitted his Ph.D. thesis to Mahatma Gandhi University, Kottayam, Kerala, India. He is working now at Tomsk Polytechnic University, Tomsk, Russia. He has edited 11 books from Scrivener (Wiley) and Springer and more than 14 books in press, (from Wiley, Springer, Royal Society of Chemistry and Elsevier). He has been invited as a visiting researcher in Portugal (2013, 2014), Czech Republic (2012, 2013), Italy (2009, 2012), Argentina (2010), Sweden (2010–2012), Switzerland (2010), Spain (2011, 2012), Slovenia (2011), France (2011), Belgium (2012), and Austria (2012) for his research work. He visited 15 universities in Europe. He published 5 publications, 3 reviews, and more than 15 book chapters. He has attended and presented more than 28 conferences, he has 136 citations, and his h-index is 6. His research interests include: polymer nanocomposites, bio-nanocomposites, and rubber based nanocomposites, fire retardant polymers and liquid crystalline polymers, and silicon sensors. Contact: visagam143@gmail.com

Yoshihiko Arao received his Ph.D. (2010) at the University of Waseda (Japan) under the supervision of Hiroyuki Kawada. There he spent 4 years as Assistant Professor at the Doshisha University. He published more than 30 papers, and his h-index is 6. He became involved from manufacturing to evaluation of polymer-based composite materials. Now he is focused on the mechanical and functional properties of nanocomposites. Contact: Yoshihiko.arao@gmail.com

Thermal Degradation of Polymer Blends, Composites and Nanocomposites

P.M. Visakh and Olga B. Nazarenko

Abstract This chapter deals with a brief account of thermal degradation of polymer-based blends, composites and nanocomposites. Different synthesising, preparation and characterisation methods of thermal degradation of polymer-based blends, composites and nanocomposites are discussed. Finally the applications, new challenges and opportunities for these thermal degradation of polymer-based blends, composites and nanocomposites are discussed.

1 Thermal Degradation of Thermosetting Blends

Outstanding properties of the epoxy resins, their great resistance to chemical agents and corrosion are noteworthy. Other properties include excellent electrical and mechanical behavior, flexibility and moderate toughness, great abrasion resistance, excellent adhesion to various building materials, low shrinkage during and after curing [1, 2]. Epoxy resins are also used in special application domains, such as encapsulating materials for miniature components and laminates in the aircraft industry. Carbon fiber reinforced epoxy resin composites are applied for structural modifications in aeroplanes, whilst their aramid fiber based composites are building materials for boats. Due to their properties, phenol-formaldehyde mouldings are used for producing telephones, handles, knobs, electrical iron parts, welding tongs and lamp housings, bottle caps and closures. Special applications include compression presses for injection moldings of thermoplastics and fuse-box covers and distributor heads in the automobile industry.

Urea-formaldehyde resins are applied in the field of electronics. They possess superior electrical insulation properties compared to phenolic resins and are used in the production of mainly sockets, plugs and switches. A special application of urea-formaldehyde resins consists in the obtaining of foams for placing on airport

P.M. Visakh (✉) · O.B. Nazarenko
Department of Ecology and Basic Safety, Tomsk Polytechnic University,
30 Lenin Avenue, Tomsk, Russia
e-mail: visagam143@gmail.com

P.M. Visakh and Y. Arao (eds.), *Thermal Degradation of Polymer Blends, Composites and Nanocomposites*, Engineering Materials,
DOI 10.1007/978-3-319-03464-5_1

runways for stopping overshooting aircrafts during emergency landings. Thermosetting polyimides exhibit superior mechanical strength, thermal and thermooxidative stability, are mostly chemically inert and inactive to high energy radiations, thus being suitable for aerospace applications, such as jet engines compressor seals, pressure discs, bearings, friction elements and sleeves. Shojaei and Faghihi [3] studied the effect of different concentrations of organoclay on styrene-butadiene rubber and phenolic resin blends prepared by two-roll mill. Thermogravimetric studies indicated that the organoclay enhanced the thermal stability of styrene-butadiene rubber vulcanizate, whilst manifesting a catalytic behavior in presence of the phenolic resin. Honmute et al. obtained IPNs based on polyaniline and poly(vinyl alcohol) as thin films with different concentrations of aniline through oxidative polymerization of aniline by using ammonium persulphate as oxidant [4]. All IPNs exhibited similar thermal behavior in nitrogen atmosphere, presenting the features of both comprising polymers.

Merlin and Sivasankar obtained SIPNs based on low (1000 Da) and high molecular weight (2000 Da) biocompatible polyurethane incorporated in poly (acrylamide) and studied their thermal decomposition characteristics in nitrogen atmosphere [5]. Alamri et al. investigated the effect of reinforcement with recycled cellulose fibres and nanoclay platelets (Cloisite 30B) on the thermal behavior of epoxy systems [6]. Cellulose nanocrystals were used as reinforcing filler for waterborne epoxy resin matrix [7]. In the case of nanocomposites, the onset temperature of thermal degradation decreased compared to neat epoxy. The same trend was obtained for the temperature of maximum weight loss for the first degradation step, while the temperature corresponding to the second step of degradation was slightly increased with increasing the cellulose nanocrystals content in the nanocomposites. Due to properties such as high tensile strength, high modulus and high chemical resistance, glass fibres are also used as reinforcement materials for thermosetting matrices. Epoxy resin modified with poly(styrene-co-acrylonitrile) was reinforced with glass fibres [8]. Amongst other plastics, thermosetting polymer blends continue playing a significant role in the development and application of polymer based materials because of their advantageous overall cost-performance relation and their capacity to improve the performance of single resins.

2 Thermal Degradation of Thermosetting Nanocomposites

Stability can be said as the protection of polymeric materials from which lead to deterioration of properties [9]. In literature, there are different and sometimes contradictory reported papers concerning the effect of the nanoparticles on polymer thermal stability. There are papers suggesting that nanoparticles have no obviously effect on thermal stability, some of them suggested a small to substantial enhancement and some others suggested acceleration of thermal decomposition. In a study performed by Ollier et al. [10], the author incorporated 5 % weight of bentonite in unsaturated polyester (UP) matrix. They noted that the addition of bentonite increases the thermal stability of the UP resin. Carrasco and Pagès [11] showed that, at low clay contents (up to 5 wt%) the addition of clay had no effect on the thermal stability of the epoxy matrix, whereas for higher concentration (10 wt%) a clear increase on this parameter was observed. In addition, Lakshimi et al. [12] reported an improvement in the thermal stability of epoxy resins with the incorporation of MMT with the same chemical treatments used in this work. Saitoh et al. [13] found that the phosphonium cations used to obtain the organoclays influenced the thermal resistance of the resulting epoxy/clay nanocomposites. The explanation for this behavior is that the dispersed MMT-Clay nanolayers can act as barrier protecting the epoxy polymer matrix degradation gaseous products from volatilizing.

In some reported studies it has been shown that the introduction of clays into polymeric matrices can accelerate the thermal decomposition of the polymer matrix, consisted by condensation polymers, due to the catalysis effect of water in MMT and hydroxyl groups on the clay platelets. The thermal stabilization effect on carbon nanotubes could be attributed to the increased interfacial interaction between the nanofibres and polymers which lead to an increase of the degradation's activation energy. Carrasco and Pagès [11] found that the thermal decomposition of cured materials was independent of cure temperature but from the content of clay. With the incorporation of crude clay, the initial temperature degradation and the maximum temperature degradation shifted towards higher temperature. The addition of unmodified MWCNTs has a negative effect on thermal stability of the epoxy nanocomposites. This is due to the poor affinity between as-received MWCNTs and epoxy resin matrix, which increases vacancies or voids in the nanocomposite. Organomodification of MMT have a negative effect in thermal degradation while the exfoliated structure due to the finer dispersion of the clay nanoparticles can lead to thermal stabilization than intercalated structure. Except montmorillonite, other clays like layered double hydroxide (LDH) can enhance thermal degradation of polymers. In such nanocomposites there are no reported for accelerating effect of LDH on polymer thermal degradation. This is because LDH can be more easily dispersed in intercalated or exfoliated structures, compared with MMT, into a polymer matrix and thus the stabilization effect is higher [14]. Carbon nanotubes (MWCNTs and SWCNTs) were also in all cases reported that can increase thermal stability of polymers. This was because the CNTs have excellent thermal conductivity. The improving thermal stability of the nanocomposites was also attributed to the formation and stabilization of MWCNTs-bonded macroradicals [15].

3 Thermal Degradation of Thermoplastic Nanocomposites

Reducing the flammability of polymers is an important issue. Different types of fillers are used to improve the polymer properties. Flame retardant additives include phosphorous-based compounds, nitrogen-containing, chlorinated and brominated compounds, antimony oxide, metal hydroxides, etc. [16–18]. The difficulty of reducing the polymer flammability is connected with finding effective fillers, addition of which to polymers would not reduce their physical, chemical and mechanical properties. Nanostructured materials are considered to be the promising additives to enhance mechanical properties as well as flame retardancy of polymeric materials [19–21]. The distinguishing characteristics of conventional fillers and nanoparticles are explained by low loading rate (~0.1–2 vol%), particle-particle correlation arising at low-volume fractions, large number density of particles per particle volume (10^6–10^8 particles/m^3), extensive interfacial area between the polymer and the nanoparticles, short distances between particles.

The research [22] revealed high activity of nanoscale powders obtained by electrical explosion of wires (EEW) to reduce polyolefin flammability. Aluminum hydroxide $Al(OH)_3$, bayerite β-$Al_2O_3 \cdot 3H_2O$, boehmiteγ-AlOOH, low-temperature modification of aluminum oxide γ-Al_2O_3 produced by the method of electrical explosion of wires (EEW) [23, 24] were used as fillers in polypropylene. All additives are resistant to oxidation under heating up to 400 °C, all of them release water in endothermic decomposition, except γ-Al_2O_3. The results of the study indicated that the oxidation rate decreases when polypropylene was filled with gibbsite and bayerite at concentration of 0.5–10 wt%.

The possibility of using aluminum nitride (AlN) nanopowders as flame retardant additive was studied [25]. Concentration of the nanopowder AlN incorporated in a polyethylene matrix was 0.1; 0.25; 0.75; 1.5 and 3 wt%. The incorporation of 1.5 wt% AlN in a polyethylene matrix caused the significant increase in the temperature of the beginning of oxidation of 33 °C (to 183 °C) in comparison with pure polyethylene (150 °C) and in the onset temperature of the intensive weight loss of 15 °C (to 375 °C) against 360 °C for pure polyethylene. The resulting effect is explained by the influence of nanoparticles on the microstructural characteristics of polyethylene. Nanoparticles are the crystallization centers and participate in the formation of fine-grained structure.

4 Effect of Thermo-Oxidation on the Mechanical Performance of Polymer Based Composites for High Temperature Applications

The properties of composite materials can be tailored to suit particular design requirements. Complex shapes can be manufactured with the composites easily leading to cheaper and better solutions [26–30]. The natural fibers offer a number of advantages over traditional synthetic fibers because of their superior corrosion resistance, excellent thermo-mechanical properties and high strength to weight ratio, which make them attractive as reinforcement in composite materials [31]. Hanu et al. [32] developed silicone polymer composites filled with mica, glass frit, ferric oxide and/or a combination of these for electrical power cables and other high temperature applications. The thermal stability of the polymer composites was determined by thermogravimetric techniques, thermal conductivity and heat release rate was measured by cone calorimetry.

The thermal energy storage phase change materials (PCM) based on paraffin/high density polyethylene (HDPE) composites were prepared by using twin-screw extruder technique. The morphology and properties of the PCM composites based on the flame retardant system with expanded graphite (EG) and ammonium polyphosphate (APP) were characterized by Scanning electron microscope (SEM), Differential scanning calorimeter (DSC), Thermogravimetric analyses (TGA) and Cone calorimeter tests [33]. The thermal degradation mechanism of these nanocomposites is related to the kind of used nanoparticles and its amount, the structure of the char formed during polymer degradation, the gas impermeability of inorganic nanoparticles, which inhibit the formation and escape of volatile byproducts during degradation and the interactions between polymer reactive groups and inorganic nanoparticles [34]. The enhanced thermal stability was confirmed by thermal analysis of the composite films, which showed an increase in the glass transition temperature of the polymer [35]. Synthesis of reduced silanized graphene oxide/epoxy-polyurethane (EPUAs/R-Si-GEO) composites with enhanced thermal and mechanical properties was reported by Lin et al. [36].

Continuous fiber-reinforced polymers are now widely used in many industries and provide properties that are superior to those of traditional ACSR (aluminum conductor steel reinforced) cables [37]. Effects of thermal treatment on the mechanical properties of poly(*p*-phenylenebenzobisoxazole) fiber reinforced phenolic resin composite materials was studied by Bian et al. [38]. Researchers are trying to improve the physical properties of the CNT/epoxy composites. Although carbon nanotubes (CNTs) have superior properties the interfacial bonding between the CNTs and the polymer matrix is weak. Kim et al. [39] treated CNTs by an acidic solution to remove impurities and subsequently by amine treatment or plasma oxidation for the improvement of interfacial bonding and dispersion of nanotubes in the epoxy matrix. The mechanical properties of the modified CNT/epoxy composites and rheological properties of nanotube containing epoxy resin were improved. This may be ascribed to the modification of CNTs, which improved dispersion and interaction between the CNT and the epoxy resin. The properties of carbon nanotubes such as high strength, high stiffness, high aspect ratio and low density have made them an excellent reinforcement for composite materials. Functionally graded carbon nanotube reinforced composite (FG-CNTRC) materials are being focused now to find out its prospect for future use [40]. Cellulose nanocrystals (CNCs) are reinforcing fillers of considerable interest for polymers due to their high modulus and potential for sustainable production. Xu et al. [41] prepared CNC-based composites with a waterborne epoxy resin matrix and characterized. The glass transition temperature (T_g) and modulus for the composites increased with increasing CNC content. The tensile strength increased from 40 to 60 MPa, suggesting good adhesion between epoxy and CNC surfaces exposed to the matrix.

5 Analysis for Thermal Degradation of a Polymer by Factor Analysis

In conventional analysis, one instance of analysis produces one data item, such as single-value data (scalars) obtained by titration or weighing, or qualitative data (category variables) obtained from a color reaction. The horizontal axis of a spectrum generally represents a specific range of frequencies or wavelengths. It is a collection of numeric absorbance data corresponding to that axis. The data can be saved in a spreadsheet as one row or column of spectrum absorbance data. Two-dimensional data can also be generated by hyphenated techniques such as gas chromatography/mass spectrometry (GC/MS) and liquid chromatography/ultraviolet (LC/UV) analysis, which combine two functions based on different principles, namely, chromatography, which has a separating function, and spectrometry, which provides information related to molecular structure. Researchers in the field of analysis generally refer to this as three-dimensional data, but more accurately, it is a three-dimensional representation of two-dimensional data. Because three-dimensional data result in a block data structure (suggestive of the Rubik's cube), it is difficult to represent these data in a single diagram, and thus, data compilation techniques such as slicing the data parallel to each facet are necessary. The first study on curve resolution, carried out by Kaiser [42] in 1958, proposed the varimax method, wherein factor rotation was used in factor analysis. Studies by Lawton and Sylvestre of Kodak clearly picked up on curve resolution technology as a means of reaction analysis in chemistry (1971, 1974) [43]. Window factor analysis (WFA), which was developed for data analysis of LC photo diode arrays by Malinowski [44] in 1992, is thought to share a common concept in that it uses eigenvalues of submatrices. SIMPLISMA, presented by Windig of Kodak in 1991, uses the concept of peak purity as the key to extracting components, which was a novel extension to the concepts used in the analysis methods proposed up to that point. The orthogonal projection approach (OPA) proposed by Sanchez et al. [45] of Bruges University (Belgium) in 1996, which uses the concept of spectral dissimilarity in peak extraction, is also an algorithm that has garnered attention.

6 Radiation Effects on Polymer-Based Systems

High energy radiation induces the "clean" modifications and efficient modification source of many applications. Along the time, several books and reviews presenting the effects induced during high energy irradiation were issued [46–56]. In the radiation field the polymer structures and reactivity are the features connected to initial molecular configuration. New formed radicals are involved in further complex processes, which define the final radiolysis products. The resistance of polymers against the action of ionizing radiation places them in a stability sequence, which depicts the material capacity to be modified. The successful irradiation processes present several advantages relative to thermal and chemical processes. The consequences of irradiation conditions related to the tremendous radiochemical yields place various macromolecular materials in different groups of stability: mainly degradable type and manly crosslinkable type [57, 58].

The great troublesome problem in the electron beam (EB) irradiation is the depth of penetration [59], which depends not only by the incident particle energy, but also on the material density. This problem can be solved by double face irradiation, when exposure dose can be considered uniform along profundity [60]. The main problem in the selection of the type of accelerator is the nominal power, which must be rigorously correlated with its applications [61].

Industrial scale of radiation processing is accomplished also on the sterilization of polymeric medical wear [62], food packaged in polymer bags [63] or electrical insulation of wires and cables [64, 65]. The radiation technologics arc based on the possibility of attaining improved properties and extension of service life is attained. The blends consisting of SBR and EPDM in different proportions are compatibilised by crosslinking under γ-irradiation. The viscosimetric measurements and the swelling investigations gave revealed the appearance of a new phase by the connections of molecular chains with chemical bridges, much stronger that physical interactions. The radiation technology applied for the modification of crystalline or semi-crystalline polymers allows manufacturing crosslink shape-memory products [66]. The exposure to high energy radiation promotes crosslinking of cylindrical tubes at certain dimensions. After heating the stain appears. This radiation processing can be applied successfully to natural rubber [67, 68], polyethylenes [69–71], ethylene vinyl acetate copolymer [72], poly(ε-caprolactone) [73].

Nanocomposites of two different kinds of rubber (acrylonitrile-butadiene rubber NBR and styrene butadiene rubber SBR)/organo-montmorillonite nanocomposites modified by hexadecyl trimethyl ammonium bromide have a remarkable resistance when they are radiochemically processes in the presence of TMPTMA [74]. Similar results were reported on epoxy resin modified with titania presents a very good resistance to the thermal oxidation after receiving 50 kGyγ-dose [75]. The radiation treatment of polymer materials, even they are as receives or modified with nanofillers converts them into engineering products with high economical values [76]. The applications of hydrogels in the production of medical items, resulting materials must have several features, which recommends them: non-toxicity, functionability, sterilizability, biocompatibility [77]. These characteristics are requires for wound dressings, drug delivery systems, transdermal systems, injectable polymers, implants, dental and ophthalmic applications, stimuli-responsive systems, hydrogel hybrid-type organs. The effects of an electron radiation dose (up to 300 kGy) and compatibilizer on the Charpy impact strength (σc) and tensile-impact strength (σt) of composites made of the following recycled polymers: low-density polyethylene (LDPE), high-density polyethylene (HDPE), polypropylene (PP), polystyrene (PS), and poly(ethylene terephthalate) (PET) indicate the processing potential of radiation treatment applied to polymers [78].

7 Thermal Degradation of Synthetic Rubber Nanocomposites

New generation nano scale fillers are challenging the domination of traditional fillers such a as carbon blacks and silica in the rubbery industry. Nanoscaled fillers such as layered silicates, carbon nanotubes, carbon nanofibers (CNFs), exfoliated graphite, spherical particles and Polyhedral oligomeric silsesquioxane (POSS), etc., dispersed as a reinforcing phase in a rubber matrix are emerging as a relatively new form of useful material.

Polymer nanocomposite having inorganic particles within nanoscale dimensions have received considerable attention because of their much improved unique properties and numerous potential applications as in automotive, aerospace and construction industry, manufacture of tires and inner tubes. Other industrial rubber goods include various belts, oil seals, gasket and food packing. The morphology-property correlations in nanocomposites the discussion now focuses on the different types of nanofillers and reinforcing in synthetic rubbers nanocomposites. In styrene-butadiene rubber (SBR) nanocomposites showed remarkable improvement in thermal stability their compared to that of the pure SBR. From another study also demonstrated that the increase in particle size is not beneficial in improving the thermal stability [79]. Xiong reported that both the thermal stability and the thermal conductivity of bromo-butyl rubber (BIIR) nanocomposites could be improved by incorporating the ionic liquids (ILs) modified graphene oxide (GO-Ils) using a solution compounding method [80]. Graphene has emerges as a subject of enormous scientific interest due to its exceptional electron transport, mechanical properties and high surface area.

Most of the reported literature on elastomer nanocomposites is based on solution mixing technique, where a polymer is dissolved in a suitable solvent along with nanofiller followed by evaporation of solvent to obtain the nanocomposite. Solution mixing can seldom be used for bulk production of nanocomposites as dissolution of elastomer in the solvent and subsequent removal of the solvent pose engineering difficulties and environmental problems. Ganter et al. [81] found that the inter-gallery distance of the organoclay increased more than two times when it was incorporated in styrene/butadiene rubber (SBR) using toluene solvent. The solution technique has been used in the past for SBR, ethylene/vinyl acetate [82], butadiene rubber (BR), polyepichlorohydrin [83], acrylonitrile/butadiene rubber (NBR) [84] and polyisoprene (IR) [85].

The addition of nanoparticles to synthetic rubber resulting in enhancement in thermal, stiffness and resistance to fracture is one of the most important phenomena in material science technology. Thermal and mechanical properties of clays multiwalled carbon nanotubes reinforced ethylene vinyl acetate (EVA) prepared through melt blending showed synergistic effect in properties [86]. Malas et al. reported (SBR/BR)/expanded graphite (EG) and black carbon (CB) nanocomposites by melt blending, this study demonstrated that the presence of EG improvement thermo-mechanical properties and the presence of CB are a factor important to improve such properties [87]. Recently carbon nanotubes are the organic nanoparticle more attractive to be incorporate in different polymeric matrix, due to their excellent electrical and conductive properties. The CNs can be surface modified using simple and economical methods [88], the modifications of the CNs are important for good dispersion in synthetic rubber nanocomposites. An important factor in study of thermal stability of synthetic rubber nanocomposites is the dispersion of the nanoparticle on the martix. The presence of nanoparticles exfolied influences the thermal or fire behavior of nanocomposite [89]. Ponomarenko et al. reported the melt-compounded composites of synthetic styrene-co-butadiene rubber (BUNA SL18) and silica particles (Silica VN3, Degussa). They showed that at low elongations, silica particles provided a considerably weaker reinforcement effect of the rubber matrix when compared to organoclay nanoparticles. Spherical and anisotropic rod-like particles, dispersed in the nanocomposites, formed a network of particles bridged by thin rubber layers throughout the SBR matrix [90].

8 Outdoor Exposure Degradation of Ethylene-Vinyl-Acetate Encapsulant Material for Photovoltaic Application

In several papers, the thermal behavior of EVA material is discussed in detail. The changes in thermal properties due to storage at room temperature and annealing at elevated temperatures are investigated and explained by characterizing the changes in polymer morphology. Many studies have been conducted, utilizing various technical analyses to assess the conditions of the EVA. These methods require the destructive extraction of samples of the polymer from the module [91, 92]. Extraction of EVA samples from field deployed PV module is very difficult and contributes to the whole destruction of the PV module. For this reason, the durability in outdoor exposure of EVA encapsulant under the conditions simulating those in PV module is often used to assess the weathering of polymeric materials [93]. In several papers, the thermal behavior of EVA material is discussed in detail. The changes in thermal properties due to storage at room temperature and annealing at elevated temperatures are investigated and explained by characterizing the changes in polymer morphology [94, 95]. TSC is based on the ability of polar molecules to be moved by an electrostatic field and also has been frequently employed to investigate the molecular motions in polymeric materials [96, 97]. Thermally Stimulated Current (TSC) TSC is a technique for detecting the transitions that depend on changes in the mobility of molecular scale dipolar structural units. TSC is based on the ability of polar molecules to be moved by an electrostatic field. Two types of current are generated: thermally stimulated polarization current (TSPC) and thermally stimulated depolarization current (TSDC). TSPC is generated when dipolar structures orient in a static electric field with increasing in the temperature.

9 Thermal Degradation of Bio-nanocomposites

These renewable biological resource polymers are usually called biopolymers, which are excellent vehicles for adding a wide range of additives such as antimicrobials, antioxidants, antifungal agents, colour and other nutrients [98, 99]. Biopolymers can be broken up into various categories which are based on the type of manufacturing process and the source of raw materials. These categories include natural biopolymers, synthetic biodegradable polymers and biopolymers developed by microbial fermentation [100]. The effectiveness of inorganic nanoparticles as additives such as nanoclays, cellulous nanowhiskers, carbon nanotubes and ultrafine layered titanate have been used to enhance the performance of many biopolymers [101–103]. By enhancing the thermomechanical properties via blending, copolymerisation and filling techniques, this could overcome such weaknesses [104]. Bio-nanocomposites represent an emerging group of hybrid materials that can provide excellent material characteristics through the incorporation of various nanofillers into the biopolymer matrix [105, 106].

Good thermal stability is one of many important properties that are essential for various industrial applications and this can be achieved by a low filler concentration such as clay. Clay nanofiller acts as a superior insulator and mass transport barrier to the volatile products generated during decomposition [107]. For this reason, TGA is employed to identify the clay content of bio-nanocomposites, as clay minerals such as montmorillonite (MMT) are thermally stable up to a temperature of 900 °C. The first work to report an enhancement in the thermal stability of bio-nanocomposites, combined PLA and organically modified fluorohectorite (FH) or MMT clay. Bandyopadhyay et al. [108] identified that the PLA was intercalated between layers of FH or MMT clay and this provided thermal degradation resistance towards conditions that would otherwise completely degrade pure PLA. Based on the work performed by previous authors [109–111], it was observed that nanofillers significantly increase the thermal stability of PLA in contrast to microfillers under a thermooxidative environment. Bafna et al. [112] reported that PCL/multi-walled carbon nanotube (MWCNT) composites prepared by ultrasonically mixing the PCL and as-fabricated MWCNT in a tetrahydrofuran solution lead to better thermal stability. Carrasco et al. [113] conducted a wide ranging study into the thermal stability of polylactic acid. The researchers studied the effects of different processing and thermal conditioning steps and of multiple polymer processing steps on thermal stability. The unprocessed PLA was more thermally stable than the processed and thermally conditioned PLA. For cellulose based nanoparticles there is a mixed effect on the thermal stability of biopolymers. A study by de Paula et al. [114] investigated the effects of incorporating cellulose nanowhiskers in poly(d,l-lactide). These nanoparticles significantly increased the thermal stability of biopolymer increasing the onset decomposition temperature by 12 and 26 °C for 1 and 5 % loadings, respectively. Cellulose nanowhiskers were shown to have negligible effect on the thermal stability of PLA [115].

References

1. Thomas, R., Vijayan, P., Thomas, S.: Recycling of thermosetting polymers. In: Fainleib, A., Grigoryeva, O. (eds.) Recent developments in polymer recycling, pp. 122–129. Transworld Research Network, Kerala (2011)
2. Irfan, M.H.: Chemistry and Technology of Thermosetting Polymers in Construction Applications, pp. 78–96, 230–239. Springer Science and Business Media, Dodrecht (1998)
3. Shojaei, A., Faghihi, M.: Physico-mechanical properties and thermal stability of thermoset nanocomposites based on styrene-butadiene rubber/phenolic resin blend. Mat. Sci. Eng. A. **527**, 917–926 (2010)
4. Honmute, S., Ganachari, S.V., Bhat, R., Naveen Kumar, H.M.P., Huh, D.S., Venkatarman, A.: Studies on polyaniline-polyvinyl alcohol (PANI-PVA) interpenetrating polymer network (IPN) thin films. Int. J. Sci. Res. **1**(2), 102–106 (2012)
5. Merlin, L.M., Sivasankar, B.: Synthesis and characterization of semi-interpenetrating polymer networks using biocompatible polyurethane and acrylamide monomer. Eur. Polym. J. **45**, 165–170 (2009)
6. Alamri, H., Low, I.M., Alothman, Z.: Mechanical, thermal and microstructural characteristics of cellulose fibre reinforced epoxy/organoclay nanocomposites. Compos. B Eng. **43**, 2762–2771 (2012)
7. Xu, S., Girouard, N., Schueneman, G., Shofner, M.L., Meredith, J.C.: Mechanical and thermal properties of waterborne epoxy composites containing cellulose nanocrystals. Polym. **54**, 6589–6598 (2013)
8. Hameed, N., Sreekumar, P.A., Francis, B., Yang, W., Thomas, S.: Morphology, dynamic mechanical and thermal studies on poly(styrene-co-acrylonitrile) modified epoxy resin/glass fibre composites. Compos. A Appl. Sci. Manuf. **38**, 2422–2432 (2007)
9. Pandey, J.K., Reddy, K.R., Kumar, A.P., Singh, R.P.: An overview on the degradability of polymer nanocomposites. Polym. Degrad. Stab. **88**, 234 (2005)
10. Ollier, R., Rodriguez, E., Alvarez, V.: Unsaturated polyester/bentonite nanocomposites: influence of clay modification on final performance. Compos. A Appl. Sci. Manuf. **48**, 137–143 (2013)
11. Carrasco, F., Pagès, P.: Thermal degradation and stability of epoxy nanocomposites: influence of montmorillonite content and cure temperature. Polym. Degrad. Stab. **93**, 1000 (2008)
12. Lakshmi, M.S., Narmadha, B., Reddy, B.S.R.: Enhanced thermal stability and structural characteristics of different MMT-Clay/epoxy-nanocomposite materials. Polym. Degrad. Stab. **93**, 20125–45213 (2008)
13. Saitoh, K., Ohashi, K., Oyama, T., Takahashi, A., Kadota, J., Hirano, H.: Development of high-performance epoxy/clay nanocomposites by incorporating novel phosphonium modified montmorillonite. J. Appl. Polym. Sci. **122**, 666 (2011)
14. Chrissafis, D.B.: Can nanoparticles really enhance thermal stability of polymers? Part I: an overview on thermal decomposition of addition polymers. Thermochim. Acta **523**, 1–24 (2011)
15. Sahoo, N.G., Rana, S., Cho, J.W., Li, L., Chan, S.H.: Polymer nanocomposites based on functionalized carbon nanotubes. Prog. Polym. Sci. **35**, 837 (2010)
16. Segev, O., Kushmaro, A., Brenner, A.: Environmental impact of flame retardants (persistence and biodegradability). Int. J. Environ. Res. Public Health **6**, 478–491 (2009)
17. Murphy, J.: Modifying specific properties: flammability-flame retardants. In: Additives for Plastics, Handbooks, pp. 115–140. Elsevier Science Ltd., New York (2001)
18. Kumara, A.P., Depana, D., Tomerb, N.S., Singha, R.P.: Nanoscale particles for polymer degradation and stabilization—Trends and future perspectives. Prog. Polym. Sci. **34**, 479–515 (2009)

19. Laoutid, F., Bonnaud, L., Alexandre, M., Lopez-Cuesta, J.-M., Dubois, Ph: New prospects in flameretardant polymer materials: from fundamentals to nanocomposites. Mater. Sci. Eng. R. **63**(3), 100–125 (2009)
20. Zhang, J., Ji, Q., Zhang, P., Xia, Y., Kong, Q.: Thermal stability and flame-retardancy mechanism of poly(ethyleneterephthalate)/boehmitena nocomposites. Polym. Degrad. Stab. **95**, 1211–1218 (2010)
21. Ke, Y.C., Wu, T.B., Xia, Y.F.: The nucleation, crystallization and dispersion behavior of PET with monodisperse SiO_2 composites. Polymer **11**, 3324–3336 (2007)
22. Ilyin, A.P., Nazarenko, O.B., Tikhonov, D.V., et al.: Hydroxide and oxide ultra fine powders —effective retardant additives in polymers. In: Abstract 10th Branch Meeting Problems and development prospects of the Tomsk Petrochemical Complex, Tomsk, Russia, p. 37 (1996) (In Russian)
23. Gromov, A.A., Nazarenko, O.B., Tikhonov, D.V., Iljin, A.P., Pautova, Y.I.: Electroex plosive Nanometals. In: Gromov, A., Teipel, U. (eds.) Metal Nanopowders Production, Characterization, and Energetic Applications, pp. 67–78. Wiley-VCH Verlag GmbH & Co. KGaA, Weinheim (2014)
24. Kwon, Y.-S., Kim, J.-C., Ilyin, A.P., Nazarenko, O.B., Tikhonov, D.V.: Electroexplosive technology of nano powders production: current status and future prospect. J. Korean Powder Metall. Inst. **19**(1), 40–48 (2012)
25. Nazarenko, O.B., Amelkovich, Y.A., Ilyin, A.P., Sechin, A.I.: Prospects of using nanopowders as flame retardant additives. Adv. Mater. Res. **872**, 123–127 (2014)
26. Haque, M.H., Upadhyaya, P., Roy, S., Ware, T., Voit, W., Lu, H.: The changes in flexural properties and microstructures of carbon fiber bismaleimide composite after exposure to a high temperature. Compos. Struct. **108**, 57–64 (2014)
27. La Mantia, F.P., Morreale, M.: Green composites: a brief review. Compos. A **42**, 579–588 (2011)
28. Salavatian, M., Smith, L.: An improved analytical model for shear modulus of fiber reinforced laminates with damage. Compos. Sci. Technol. **105**, 9–14 (2014)
29. Yu, T., Jiang, N., Li, Y.: Functionalized multi-walled carbon nanotube for improving the flame retardancy of ramie/poly(lactic acid) composite. Compos. Sci. Technol. **104**, 26–33 (2014)
30. Srikanth, I., Padmavathi, N., Kumar, S., Ghosal, P., Kumar, A., Subrahmanyam, Ch.: Mechanical, thermal and ablative properties of zirconia, CNT modified carbon/phenolic composites. Compos. Sci. Technol. **80**, 1–7 (2013)
31. Harle, S.M.: The performance of natural fiber reinforced polymer composites: review. Int. J. Civil. Eng. Res. **5**, 285–288 (2014)
32. Hanu, L.G., Simon, G.P., Cheng, Y.-B.: Thermal stability and flammability of silicone polymer composites. Polym. Degrad. Stab. **91**, 1373–1379 (2006)
33. Cai, Y., Wei, Q., Huang, F., Lin, S., Chen, F., Gao, W.: Thermal stability, latent heat and flame retardant properties of the thermal energy storage phase change materials based on paraffin/high density polyethylene composites. Renewable Energy **34**, 2117–2123 (2009)
34. Chrissafis, D.B.: Can nanoparticles really enhance thermal stability of polymers? Part I: an overview on thermal decomposition of addition polymers. Thermochim Acta **523**, 1–24 (2011)
35. Vadukumpully, S., Paul, J., Mahanta, N., Valiyaveettil, S.: Flexible conductive graphene/ poly(vinyl chloride) composite thin films with high mechanical strength and thermal stability. Carbon **49**, 198–205 (2011)
36. Lin, J., Zhang, P., Zheng, C., Wu, X., Mao, T., Zhu, M., Wang, H., Feng, D., Qian, S., Cai, X.: Reduced silanized graphene oxide/epoxy-polyurethane composites with enhanced thermal and mechanical properties. Appl. Surf. Sci. **316**, 114–123 (2014)
37. Santos, T.F.A., Vasconcelos, G.C., de Souza, W.A., Costa, M.L., Botelho, E.C.: Suitability of carbon fiber-reinforced polymers as power cable cores: galvanic corrosion and thermal stability evaluation. Mater. Des. **65**, 780–788 (2015)

38. Bian, L., Xiao, J., Zeng, J., Xing, S., Yin, C., Jia, A.: Effects of thermal treatment on the mechanical properties of poly(p-phenylenebenzobisoxazole) fiber reinforced phenolic resin composite materials. Mater. Des. **54**, 230–235 (2014)
39. Kim, J.A., Seong, D.G., Kang, T.J., Youn, J.R.: Effects of surface modification on rheological and mechanical properties of CNT/epoxy composites. Carbon **44**, 1898–1905 (2006)
40. Liew, K.M., Lei, Z.X., Zhang, L.W.: Mechanical analysis of functionally graded carbon nanotube reinforced composites: a review. Compos. Struct. **120**, 90–97 (2015)
41. Xu, S., Girouard, N., Schueneman, G., Shofner, M.L., Carson Meredith, J.: Mechanical and thermal properties of waterborne epoxy composites containing cellulose nanocrystals. Polymer **54**, 6589−6598 (2013)
42. Kaiser, H.F.: The varimax criterion for analytic rotation in factor analysis. Psychometrika **23**, 187−200 (1958)
43. Sylvestre, E.A., Lawton, W.H., Maggio, M.S.: Curve resolution using a postulated chemical reaction. Technometrics **16**(3), 353−368 (1974)
44. Malinowski, E.R.: Factor Analysis in Chemistry, 3rd edn. Wiley, New York (2002)
45. Sanchez, F.C., Toft, J., van den Bogaert, B. and Massart, D.L.: Orthogonal projection approach applied to peak purity assessment. Anal. Chem. Chem. **68**, 79 (1996)
46. Chapiro, A.: Radiation Chemistry of Polymer Materials. Wiley Interscience Publishers, New York (1962)
47. Clough, R.L.: Radiation-resistant polymers. In: Encyclopedia of Polymer Science and Engineering, 2nd edn. pp. 667–708. Wiley, New York (1988)
48. Bhattacharya, A.: Radiation and industrial polymers. Prog. Polym. Sci. **25**, 371–401 (2000)
49. Clegg, D.W., Collyer, A.A. (eds.): Irradiation Effects on Polymers. Elsevier Applied Science, London (1999)
50. Woods, R.J.: Applied Radiation Chemistry: Radiation Processing. Wiley Interscience Publishers, New York (1994)
51. Clough, R. L.: High-energy radiation and polymers. A review of commercial processes and emerging applications. Nucl. Instrum. Methods Phys. Res. B. **185**, pp. 8–33 (2001)
52. Spinks, J.W.T., Woods, R.J. (eds.): Introduction to Radiation Chemistry, 3rd edn. Wiley, New York (1990)
53. Dawes, K., Glover, L.C., Vroom, D.A.: The effects of electron beam and γ-irradiation on polymer materials. In: Mark, J.E. (ed.) Physical Properties of Polymer. Handbook, 2nd edn. Springer, New York (2007)
54. Makuuchi, K., Chang, S. (eds.): Radiation Processing of Polymer Materials and its Industrial Applications. Wiley, New York (2012)
55. Zaharescu, T., Jipa S.: Radiochemical modifications in polymers. In: Arndt, K.F., Lechner, M.D. (eds.), Landolt-Börnstein Series, Polymer Solids and Polymer Melts, vol. VIII/6 C2, pp. 95–184. Springer, Heidelberg (2013)
56. Drobny, J.G.: Ionizing radiation and polymers: principles, technology, and applications. PDL Handbook Series, Elsevier (2012)
57. Cleland, M.R., Park, L.A., Chang, S.: Applications for radiation processes of material. Nucl. Instrum. Meth. Phys. Res. B **208**, 66–73 (2003)
58. Gehring, J.: With radiation crosslinking of polyolefin engineering plastics into the next millennium. Radiat. Phys. Chem. **57**, 361–365 (2000)
59. Nablo, S.V., Chrusciel, J., Cleghorn, D.A., Rangwalla, I.: Factors influencing equipment selection in electron beam processing. Nucl. Instrum. Meth. Phys. Res. B **208**, 90–101 (2003)
60. Miller, A.: Approval and control of radiation processing, EB and gamma. Radiat. Phys. Chem. **31**, 385–393 (1988)
61. Cleland M.R., Park L.A.: Medium and high-energy electron beam radiation processing for commercial applications. Nucl. Instrum. Meth. Phys. Res. B **208**, 74–89 (2003)
62. Saylor, M.C., Parks, L.A., Herring, C.H.: Technical and regulatory for radiation sterilization facilities using electron beam accelerators. Nucl. Instrum. Meth. Phys. Res. B **79**, 875–878 (1993)

63. Pilette, L.: Effects of ionizing treatments on packaging—food simulant combinations. Packag. Technol. Sci. **3**, 17–20 (1990)
64. Zimek, Z., Przybytniak, G., Nowicki, A., Mirkowski, K., Roman, K.: Optimization of electron beam crosslinking for cables. Radiat. Phys. Chem. **94**, 161–165 (2014)
65. Bartoníček, B., Plaček, V., Hnát, V.: Comparison of degradation effects induced by gamma radiation and electron beam radiation in two cable jacketing materials. Radiat. Phys. Chem. **76**, 857–863 (2007)
66. Voit, W., Ware, T., Gall, K.: Radiation crosslinked shape-memory polymers. Polymer **51**, 3551–3559 (2010)
67. Banik, I., Bhowmick, A.K.: Effect of electron beam irradiation on the properties of crosslinked rubbers. Radiat. Phys. Chem. **58**, 293–298 (2000)
68. Haque, M.E., Dafader, N.C., Akhtar, F., Ahmad, M.U.: Radiation dose required for the vulcanization of narural rubber latex. Radiat. Phys. Chem. **48**, 505–510 (1996)
69. Kurtz, S.M., Muratoglu, O.K., Evans, M., Edidin, A.A.: Advances in the processing, sterilization and crosslinking of ultra-high molecular weight polyethylene for total joint arthroplast. Biomaterials **20**, 1659–1688 (1999)
70. Rezanejad, S., Kokab, M.: Shape memory and mechanical properties of cross-linked polyethylene/clay nanocomposites. Eur. Polym. J. **43**, 2856–2865 (2007)
71. Mahapram, S., Poompradub, S.: Preparation of natural rubber (NB) latex/low density polyethylene (LDPE) blown film and its properties. Polym. Test. **30**, 716–725 (2011)
72. Chattopadhyay, S., Chaki, T.K., Bhowmick, A.K.: Heat shrinkability of electron-beam-modified thermoplastic elastomeric films from blemds of ethylene vinylacetate copolymer and polyethylene. Radiat. Phys. Chem. **59**, 501–505 (2000)
73. Zhu, G., Liang, G., Xu, Q., Yu, Q.: Shape-memory effects of radiation crosslinked poly(ε-caprolactone). J. Appl. Polym. Sci. **90**, 1589–1595 (2003)
74. Mohamed, R.M.: Radiation induced modification of NBR and SBR montmorillonite nanocomposites. J. Ind. Eng. Chem. **19**, 80–86 (2013)
75. Crăciun, E., Jitaru, I., Zaharescu, T., Jipa, S.: Qualification of epoxy resin by radiochemical ageing. Optoelectr. Adv. Mater. Rapid Commun. **4**, 1819–1822 (2010)
76. Thomas, J.K.: Fundamental aspects of the radiolysis of solid polymers, crosslinking and degradation. Nucl. Instrum. Meth. Phys. Res. B **265**, 1–7 (2007)
77. Rosiak, J.M., Ulanski, I.P., Pajewski, L.A., Yoshii, F., Makuuchi, K.: Radiation formation of hydrogel for biomedical purposes. Some remarks and comments. Radiat. Phys. Chem. **46**, 161–168 (1995)
78. Żenkiewicz, M., Dzwonkowski, J.: Effects of electron radiation and compatibilizers on impact strength of composites of recycled polymers. Polym. Test. **26**, 903–907 (2007)
79. Zhang, Y., Liu, Q., Xiang, J., Frost, R.L.: Thermal stability and decomposition kinetics of styrene-butadiene rubber nanocomposites filled with different particles sized kaolinites. Appl. Clay Sci. **95**, 159–166 (2014)
80. Xiong, X., Wang, J., Jia, H., Fang, E., Ding, L.: Structure, thermal conductivity, and thermal stability of bromobutyl rubber nanocomposites with ionic liquid modified graphene oxide. Polym. Degrad. Stab. **98**, 2208–2214 (2013)
81. Ganter, M., Gronski, W., Semke, H., Zilg, T., Thomann, R., Mülhaupt, R.: Surface-compatibilized layered silicates—A novel class of nanofillers for rubbers with improved mechanical properties. Kautsch. Gummi Kunstst. **54**(4), 166–171 (2001)
82. Pramanik, M., Srivastava, S.K., Samantaray, B.K., Bhowmick, A.K.: Rubber-Clay nanocomposite by solution blending. J. Appl. Polym. Sci. **87**, 2216–2220 (2003)
83. Lim, S.K., Lim, S.T., Kim, H.B., Chin, I., Choi, H.J.: Preparation and physical characterization of polyepichlorohydrin elastomer/clay nanocomposites. J. Macromol. Sci. Part B Phys. B **42**(6), 1197–1199 (2003)
84. Wu, C.M., Hwang, W.G., Tien, K.C., Chang, Y.C., Fu, H.L.: In: 11th National Conference on Science and Technology of National Defense, Taipei, Taiwan (2003)
85. Jeon, H.S., Rameshwaram, J.K., Kim, G.: Structure-property relationships in exfoliated polyisoprene/clay nanocomposites. J. Polym. Sci. Part B Polym. Phys. **42**, 1000–1009 (2004)

86. Peeterbroeck, S., Lepoittevin, B., Pollet, E., Benali, S., Broekaert, C., Alexandre, M., Bonduel, D., Viville, P., Lazzaroni, R., Dubois, P.: Polymer layered silicate/carbon nanotube nanocomposites: The catalyzed polymerization approach. Polym. Eng. Sci. **46**, 1022–1030 (2006)
87. Malas, A., Pal, P., Das, Ch.K.: Effect of expanded graphite and modified graphite flakes on the physical and thermo-mechanical properties of styrene butadiene rubber/polybutadiene rubber (SBR/BR) blends. Mater. Des. **55**, 664−673 (2014)
88. Cabello, Ch., Saénz, A., López, L.I., Pérez, C., Barajas, L., Ávila, C.: Modificación superficial de (MWCNT) con H_2SO_4/HNO_3 mediante ultrasonido. Afinidad **68**, 370–374 (2012)
89. Cerin, O., Fontaine, G., Duquesne, S., Bourbigot, S.: Thermal stability of synthetic rubber nanocomposites. In: Mittal, V. (ed.) Recent Advance in Elastomeric Nanocomposites. Springer, Heidelberg (2011)
90. Scotti, R., Conzatti, L., D'Arienzo, M., Di Credico, B., Giannini, L., Hanel, T., Stagnaro, P., Susanna, A., Tadiello, L., Morazzoni, F.: Shape controlled spherical (0D) and rod-like (1D) silica nanoparticles in silica/styrene butadiene rubber nanocomposites: role of the particle morphology on the filler reinforcing effect. Polymer **55**, 1497–1506 (2014)
91. Dhere, N.G., Gadre, K.S.: Comparison of mechanical properties of EVA encapsulant in new and field-deployed PV modules. In: Proceedings of the 2nd World Photovoltaic Solar Energy Conference and Exhibition, Vienna, Austria, 6–10 July 1998
92. Dechthummarong, C., Wiengmoon, B., Chenvidhya, D., Jivacate, C., Kirtikara, K.: Physical deterioration of encapsulation and electrical insulation properties of PV modules after long-term operation in Thailand. Sol. Energy Mater. Sol. Cells **94**(9), 1437–1440 (2010)
93. ASTMD1435–05: Standard Practice for Outdoor Weathering of Plastics. ASTMD, Philadelphia (1985)
94. Stark, W., Jaunich, M.: Investigation of Ethylene Vinyl Acetate copolymer (EVA) by thermal analysis DSC and DMA. Polym. Test. **30**(2), 236–242 (2011)
95. Oreski, G., Wallner, G.M.: Damp heat induced physical aging of PV encapsulation materials. In: 12th IEEE Intersociety Conference on Thermal and Thermomechanical Phenomena in Electronic Systems (Itherm 2010), pp. 1−6. Las Vegas, NV, 2–5 June 2010
96. Collins, G., Yoo, S.U., Recber, A., Jaffe, M.: Thermal analysis of complex relaxation processes in Poly(Desaminotyrosyl-Tyrosine Arylates). Polymer **48**(4), 975–988 (2007)
97. Saffell, J.R., Matthiesen, A., McIntyre, R., Ibar, J.P.: Comparing thermal stimulated current (TSC) with other thermal analytical methods to characterize the amorphous phase of polymers. Thermochim Acta **192**, 243–264 (1991)
98. Kümmerer, K.: Sustainable from the very beginning: rational design of molecules by life cycle engineering as an important approach for green pharmacy and green chemistry. Green Chem. **9**, 899–907 (2007)
99. Clarinval, A.M., Halleux, J.: Classification of biodegradable polymers, pp. 3–31. CRC Press, Boca Raton (2005)
100. Rhim, J.W., Park, H.M., Ha, C.S.: Bio-nanocomposites for food packaging applications. Prog. Polym. Sci. **38**, 1629–1652 (2013)
101. Kumar, A.P., Depan, D., Singh Tomer, N., Singh, R.P.: Nanoscale particles for polymer degradation and stabilization-trends and future perspectives. Prog. Polym. Sci. (Oxford). **34**, 479–515 (2009)
102. Raquez, J.M., Habibi, Y., Murariu, M., Dubois, P.: Polylactide (PLA)-based nanocomposites. Prog. Polym. Sci. **38**, 1504–1542 (2013)
103. Sinha Ray, S., Bousmina, M.: Biodegradable polymers and their layered silicate nanocomposites: in greening the 21st century materials world. Prog. Mater. Sci. **50**, 962–1079 (2005)
104. Bikiaris, D.: Can nanoparticles really enhance thermal stability of polymers? Part II: an overview on thermal decomposition of polycondensation polymers. Thermochim. Acta **523**, 25–45 (2011)
105. Yang, K.K., Wang, X.L., Wang, Y.Z.: Progress in Nanocomposite of Biodegradable Polymer. J Ind Eng Chem. **13**, 485–500 (2007)

106. Mohanty, A.K., Wibowo, A., Misra, M., Drzal, L.T.: Development of renewable resource–based cellulose acetate bioplastic: Effect of process engineering on the performance of cellulosic plastics. Polym. Eng. Sci. **43**, 1151–1161 (2003)
107. Ray, S.S., Bousmina, M.: Biodegradable polymers and their layered silicate nanocomposites: in greening the 21st century materials world. Prog. Mater Sci. **50**, 962–1079 (2005)
108. Bandyopadhyay, S, Chen, R, Giannelis, E.P.: Biodegradable organic-inorganic hybrids based on poly(L-lactide). Polym. Mater. Sci. Eng. **81**, 159−160 (1999)
109. Pluta, M., Galeski, A., Alexandre, M., Paul, M.A., Dubois, P.: Polylactide/montmorillonite nanocomposites and microcomposites prepared by melt blending: Structure and some physical properties. J. Appl. Polym. Sci. **86**, 1497–1506 (2002)
110. Chen, C.X., Yoon, J.S.: Morphology and thermal properties of poly(L -lactide)/poly (butylene succinate-co-butylene adipate) compounded with twice functionalized clay. J. Polym. Sci. Part B Polym. Phys. **43**, 478–487 (2005)
111. Marras, S.I., Zuburtikudis, I., Panayiotou, C.: Nanostructure vs. microstructure: Morphological and thermomechanical characterization of poly(l-lactic acid)/layered silicate hybrids. Eur. Polymer J. **43**, 2191–2206 (2007)
112. Bafna, A., Beaucage, G., Mirabella, F., Mehta, S.: 3D Hierarchical orientation in polymer–clay nanocomposite films. Polymer **44**, 1103–1115 (2003)
113. Carrasco, F., Pagès, P., Gámez-Pérez, J., Santana, O.O., Maspoch, M.L.: Processing of poly (lactic acid): Characterization of chemical structure, thermal stability and mechanical properties. Polym. Degrad. Stab. **95**, 116–125 (2010)
114. de Paula, E.L., Mano, V., Pereira, F.V.: Influence of cellulose nanowhiskers on the hydrolytic degradation behavior of poly(d, l-lactide). Polym. Degrad. Stab. **96**, 1631–1638 (2011)
115. Hossain, K.Z., Ahmed, I., Parsons, A., Scotchford, C., Walker, G., Thielemans, W., et al.: Physico-chemical and mechanical properties of nanocomposites prepared using cellulose nanowhiskers and poly(lactic acid). J. Mater. Sci. **47**, 2675–2686 (2012)

Thermal Degradation of Thermosetting Blends

Dan Rosu, Cristian-Dragos Varganici, Liliana Rosu and Oana Maria Mocanu (Paduraru)

Abstract The chapter coalesces literature studies on recent advances concerning the thermal behavior of different thermosetting blends. The introduction debates the general issue concerning polymer blends, that being the occurrence of phase separation phenomena and lists a series of possibilities to overcome these undesired aspects. The introduction section also presents the most common polymers used as crosslinked scaffolds either individual or for different multicomponent polymeric materials. The subchapters that follow are focused on recent studies on the thermal stability and degradation of thermosetting blends, effect of reinforcement and nanofillers on the thermal stability of thermosetting blends and applications and future trends of thermosetting blends, dealing with the latest issues and trying to reveal solutions.

1 Introduction

The beginnings of the polymer industry date back to 1846 when the first polymer blend was patented by Parkes which co-dissolved natural rubber (amorphous *cis*-1,4-polyisoprene) and gutta-percha (semi-crystalline *trans*-1,4-polyisoprene) in CS_2. The blending process yielded a partially crosslinked material. The process was known as co-vulcanization and the interesting aspect of the final blend was that its rigidity could be controlled by composition. Due to this aspect, the blends were used in multiple application domains. What followed was the obtaining of synthetic

D. Rosu (✉) · C.-D. Varganici · L. Rosu · O.M. Mocanu (Paduraru)
Centre of Advanced Research in Bionanoconjugates and Biopolymers, "Petru Poni" Institute of Macromolecular Chemistry, 41A Grigore Ghica-Voda Alley, 700487 Iasi, Romania
e-mail: drosu@icmpp.ro; dan_rosu50@yahoo.com

C.-D. Varganici
e-mail: varganici.cristian@icmpp.ro; varganicicristian@yahoo.com

L. Rosu
e-mail: lrosu@icmpp.ro

O.M. Mocanu (Paduraru)
e-mail: paduraruo@icmpp.ro

P.M. Visakh and Y. Arao (eds.), *Thermal Degradation of Polymer Blends, Composites and Nanocomposites*, Engineering Materials,
DOI 10.1007/978-3-319-03464-5_2

rubbers, beginning with the first polymerization of isoprene in sealed bottles by Tilden in 1884, and modified natural polymers which included nitrocellulose, cellulose acetate (1926) and cellulose ethers and esters (1927).

Phenol-formaldehyde was reported as the first commercially synthetic polymer (1899) which was introduced as Bakelite™ by Baekeland in 1909. This was the period which marked the dawn for the production of commercial synthetic thermosetting polymers. Other advances in the field included the discovery of urea-formaldehyde resins in 1884 and the beginning of their commercialization as Beetle™ moldable resin in 1928, followed by thiourea-formaldehyde (1920), aniline-formaldehyde (*Cibatine*™ by Ciba, 1935) and melamine-formaldehyde (1937) moulding powders. The year 1909 marked the discovery of epoxy compounds by Prileschaiev, which were not used until World War 2. The first thermoset polyesters, invented by Ellis, date back to 1934 and in 1938 was reported their first use in the forms of glass-reinforced materials [1].

The typical polymers used as scaffolds in thermosetting blends are epoxy (Fig. 1), phenolic (Fig. 2), urea-formaldehyde (Fig. 3) and melamine-formaldehyde resins (Fig. 4), unsaturated polyesters and polyimides [2].

Fig. 1 Crosslinked epoxy resins

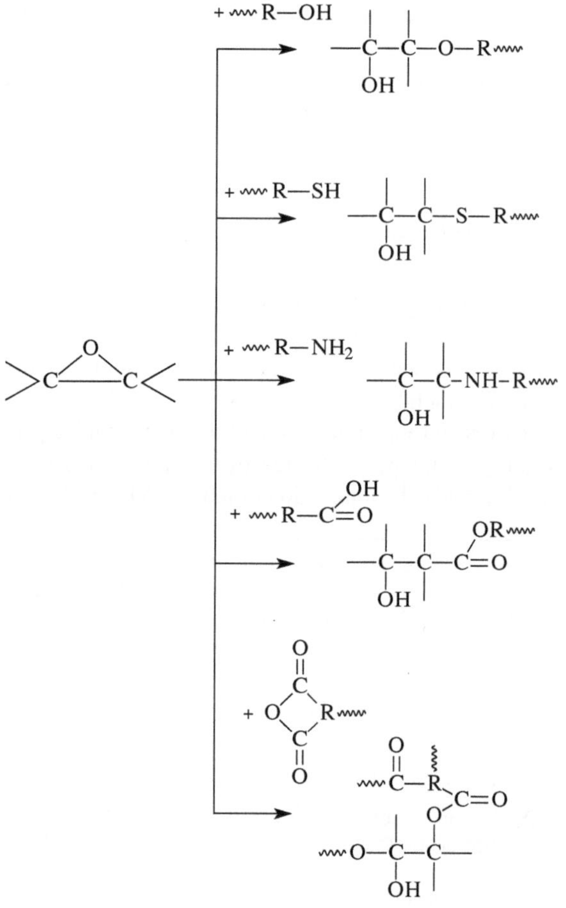

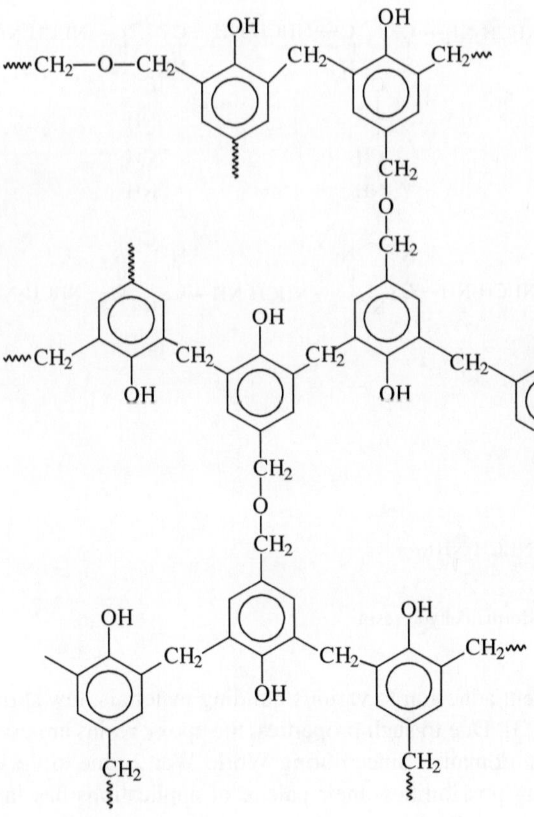

Fig. 2 Crosslinked phenolic resin

Fig. 3 Crosslinked urea-formaldehyde resin

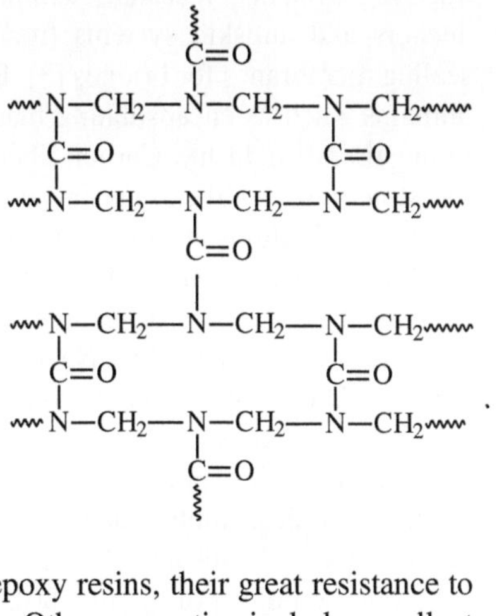

Amongst the outstanding properties of the epoxy resins, their great resistance to chemical agents and corrosion are noteworthy. Other properties include excellent electrical and mechanical behavior, flexibility and moderate toughness, great

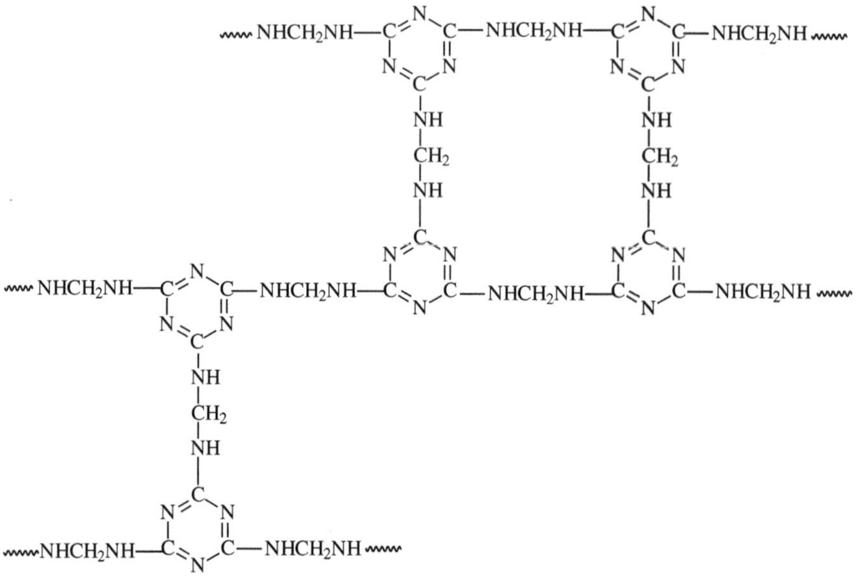

Fig. 4 Crosslinked melamine-formaldehyde resin

abrasion resistance, excellent adhesion to various building materials, low shrinkage during and after curing [2, 3]. Due to such properties, the epoxy resins are used in a wide variety of application domains. Since during World War 2, due to the epoxy resins multiple crosslinking possibilities, their palette of applications has fast and considerably broadened, especially in the civil engineering and building industry sectors were they are used as: structural, general purpose and tile adhesives, crack injecting systems, resistant, laminate and epoxy-tar coatings, floorings, grouts, anchors and antiskid systems (road resurfacing epoxies), mortars and interlayer sealing membranes for bridges [3]. Epoxy resins are also used in special application domains, such as encapsulating materials for miniature components and laminates in the aircraft industry. Carbon fiber reinforced epoxy resin composites are applied for structural modifications in aeroplanes, whilst their aramid fiber based composites are building materials for boats.

Unlike epoxy resins, the chemical resistance of phenolic resins is mostly dependent on the type of resin and filler. It has been reported that simple phenol-formaldehyde is affected by NaOH attack, whereas xylenol and cresol based resins are inert. Due to the high crosslinking degree and interlocking, phenolic mouldings are characterized as insoluble, hard and heat resistant (up to 200 °C) materials. Due to their properties, phenol-formaldehyde mouldings are used for producing telephones, handles, knobs, electrical iron parts, welding tongs and lamp housings, bottle caps and closures. Special applications include compression presses for injection mouldings of thermoplastics and fuse-box covers and distributor heads in the automobile industry.

Urea-formaldehyde resins are applied in the field of electronics. They possess superior electrical insulation properties compared to phenolic resins and are used in the production of mainly sockets, plugs and switches. Urea-formaldehyde resins may come in a wide color range and do not impart taste and odour, thus being also applied in the domestic and sanitary sectors in the making of pot and panhandles and toilet seats. Such resins also exhibit good adhesive properties in the furniture industry for plywood, chipboards and particleboards. A special application of urea-formaldehyde resins consists in the obtaining of foams for placing on airport runways for stopping overshooting aircrafts during emergency landings.

Same as urea-formaldehyde resins, melamine-formaldehyde resins present a wide color range, are track and scratch resistant, possess good electrical properties and superior heat resistance and hardness. The resins excellent adhesive properties make them useful in the furniture industry. Due to freedom from color and the above mentioned properties, urea-formaldehyde resins are employed for laminating. Other applications depend on the nature of different fillers. Electrical applications demand mineral-filled urea-formaldehyde resins, whilst the cellulose-filled resins are used for manufacturing of clock cases and trays. Special applications include flame retardant agents [2].

Unsaturated polyesters present excellent properties for applications in the construction sector, such as: excellent abrasion, chemical, corrosion and heat resistance, high impact and excellent compressive strength and rapid strength gain. Due to these properties, unsaturated polyesters are used as anchoring gouts, coatings, concretes and loop and lighting sealants, amongst other applications. This class of polymers is also used as fiber reinforced materials [3].

Thermosetting polyimides exhibit superior mechanical strength, thermal and thermooxidative stability, are mostly chemically inert and inactive to high energy radiations, thus being suitable for aerospace applications, such as jet engines compressor seals, pressure discs, bearings, friction elements and sleeves [2].

In order for thermosetting polymer blends to be used in different special applications, the thermal stability of such materials, amongst other properties, is a key factor which must be tested and, where the case, improved. For this purpose advanced knowledge on the different thermal decomposition mechanisms and their influence on the miscibility of the comprising polymers are crucial.

2 Recent Studies on the Thermal Stability and Degradation of Thermosetting Blends

The general issue concerning polymer blends consists in the fact that most polymers are highly immiscible and generate phase separation phenomena, this meaning that the free energy change of the mixture is not negative, i.e. the heat of mixing is too high, as previously discussed in this book. According to Hildebrand [4] the heat of mixing can be expressed as:

$$\Delta H = V(\phi_1 - \phi_2)^2 \upsilon_1 \upsilon_2$$

where V is the total volume, υ_1 and υ_2 are the volume fractions of the respective components, ϕ_1 and ϕ_2 are the solubility parameters.

$$\phi = \left(\frac{\Delta E}{V}\right)^{1/2}$$

where ΔE is the energy of vaporization (also cohesive-energy density or internal pressure) and V is the molar volume. Hildebrand indicated that a difference in solubility parameters values between the two components of the blend must ideally not exist or be minimal, in order not to increase the positive value of the heat of mixing.

In order to avoid the occurring of phase separation phenomena, a variety of compatibilization methods have been thoroughly discussed in the literature. Phase compatibilization may be achieved by in situ polymerization, introduction of specific groups, crosslinking between phases, polymer-polymer reactions, ternary polymer addition, reactive extrusion, block copolymer addition or obtaining of interpenetrating (IPNs) and semi-interpenetrating polymer networks (SIPNs) [5]. An ideal classification of thermosetting blends can be made as follows [1]:

- thermoset-thermoset blends and IPNs;
- thermoset-thermoplastic blends and SIPNs;
- rubber modified thermosets.

Amongst the above mentioned compatibilization methods, the obtaining of IPNs and SIPNs often proved to be a promising and very efficient route. An IPN is a polymer alloy comprised of two or more chemically crosslinked polymers. The difference between polymer blends and IPNs is that the latter ones swell instead of dissolving in solvents and do not creep or flow. Types of IPNs include sequential, simultaneous, latex and gradient IPNs and may also be thermoplastic (i.e. when physical crosslinks are implied). Thermoplastic IPNs behave as thermosets at ambient temperature, but usually flow when heated at certain temperatures, possess IPN properties and often exhibit dual phase behavior [1].

SIPNs are polymer alloys comprised of one or more linear or branched polymers entangled in one or more crosslinked polymers. The great advantage of obtaining SIPNs is that during such synthesis there occurs a synergism of the individual properties of the components due to a forced compatibilization of the phases. Of course, in not all cases the compatibilization occurs, but rather in limited ones. Whether the occurrence of phase separation phenomena is desired or not, depends on the application domain(s) sought for the future material. However, in most cases phase separation proves to lead to a decline in polymer blends performances, by especially altering their thermal and mechanical properties and thus the occurrence of such phenomena is usually undesired. Another aspect regarding the miscibility of a SIPN is that, in addition to close solubility parameters of the comprising polymers, the system is also characterized by the existence of a single glass transition temperature domain (T_g) value, which is dependent on the T_g values of the individual components and their weight fractions.

Epoxy resins are excellent coating materials widely used in constructions, due to their high mechanical strength and good adhesion to metals, amongst other properties, as mentioned earlier in this chapter. However, epoxy resins stress crack under impact forces, possess brittleness and usually exhibit stress after curing. One route to resolve this impediment consists in the obtaining of SIPNs with elastomeric polyurethanes which have excellent elasticity, damping properties and abrasive resistance, but in turn do not have the superior properties of the epoxy resins.

Rosu et al. [6] obtained six SIPNs based on a linear aromatic polyurethane and increasing crosslinked epoxy resin (CER) content of 5, 10, 15, 20, 30 and 40 %. Varganici et al. [7] conducted miscibility studies, mostly based on observations of thermal investigations of the six SIPNs, and tried to determine if phase separation phenomenon occurred and at what epoxy resin content value. DSC heating curves showed a single T_g value which increased in the range −33 to 16 °C for the SIPNs containing up to an including 30 % CER content, indicating a good miscibility between the two comprising polymers. For the SIPN containing 40 % CER, the system generated two T_g values, indicating the occurrence of phase separation phenomenon, which was confirmed by scanning electron microscopy and optical microscopy techniques. The evidence of the formation of the SIPNs was made by the reducing in intensities of the melting and crystallization profiles of the polyurethane until their total disappearance with CER content increase. By applying the Fox [8] and Gordon–Taylor [9] equations, authors demonstrated the existence of some specific and strong interactions between the two comprising polymers.

Thermogravimetric (TG) studies indicated that samples thermal stability of the SIPNs depended on the CER concentration. SIPNs containing up to a 10 % CER started decomposing at intermediate temperatures between the ones of the pure comprising polymers. This aspect indicated the existence of some specific weak interactions between the crude comprising polymers. These interactions were also evidenced by thermo-mechanical measurements [10] and miscibility studies [7]. The thermograms and the first derivative curves (DTGs) of the SIPNs and the pure comprising polymers, recorded at a heating rate of 10 °C/min, are given in Fig. 5. The SIPNs containing a higher than 10 % CER content started decomposing at temperatures below that of the polyurethane. The increasing of CER content led to the increase of the strength of interactions of the comprising polymers. The cause may consist in disruption of hydrogen bonding in the pure comprising polymers during the obtaining of the SIPNs. These interactions led to a slight decrease in thermal stability, the findings being in good agreement with the miscibility studies.

At least three stages of thermal decomposition were evidenced for the SIPNs, from the DTGs. The polyurethane and CER thermally decomposed in four and one stage, respectively [11, 12]. The DSC thermograms, recorded simultaneously with the TG ones, were in good agreement with the DSC thermograms recorded for the miscibility studies [7], indicating a melting profile around 200 °C for the polyurethane, with

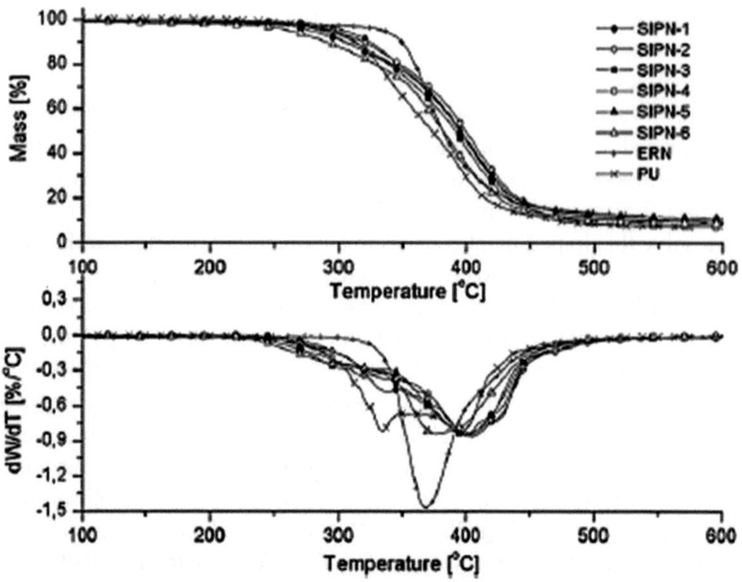

Fig. 5 TG and DTG thermograms recorded at 10 °C/min (reproduced with kind permission from Elsevier—License no. 3280641074690)

reducing intensity until disappearance, and also three endothermic processes, corresponding to each individual thermal decomposition stage.

The global kinetic parameters values of the thermal decomposition process were determined for the structure containing 20 % CER, since the other SIPNs presented similar thermal behavior, by applying the differential and integral isoconversional methods of Friedman and Ozawa, Flynn and Wall (OFW) [13, 14]. In both cases activation energy (E_a) values increased with the conversion degree (α), suggesting a complex thermal decomposition mechanism through successive and/or parallel reactions [15]. A three consecutive stage thermal decomposition mechanism was proposed. The OFW imposes a first order reaction model (i.e. $1 - \alpha$) in describing the thermal decomposition process. If the process is indeed characterized by a first order reaction model, then the straight lines of the OFW graph should respect the same parallelism, however, in this case, this condition was not met. Authors tested 14 kinetic models, detailed in the literature [16], by multivariate non-linear regression method, in order to find the real form of the conversion function which best characterized the global thermal decomposition process and to determine the values of the kinetic parameters (E_a, pre-exponential factor (A) and reaction order (n)) characteristic to each individual stage of the thermal degradation, corresponding to each DTG curve peak. A good correlation between simulated and experimental data was found for an n-th order reaction model (i.e. $(1 - \alpha)^n$) for all studied SIPNs. The first stage of thermal decomposition of SIPNs containing 5 and 10 % CER content were characterized by values of $n > 2$, indicating that mass loss occurred through scission and intermolecular transfer phenomena [17]. For higher

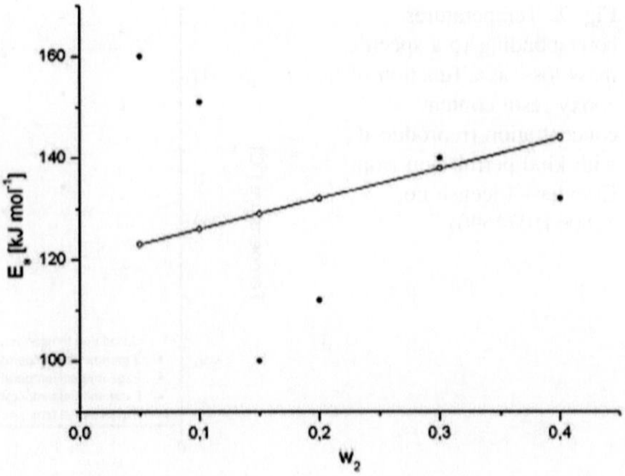

Fig. 6 E_a variation as function of CER content for the first stage of thermal decomposition min (reproduced with kind permission from Elsevier—License no. 3280641074690)

than 10 % CER content $n \approx 1$, suggesting that mass loss occurred by random scission of the polyurethane main chain [17].

Two instances can be observed from Fig. 6, which shows a comparison between E_a experimental values and those calculated with the additivity rule in Eq. (1), for the first stage of thermal decomposition of the SIPNs, where E_{a1} and E_{a2} are the E_a values of the polyurethane and CER, and y_1 and y_2 are their corresponding weight fractions:

$$E_a = y_1 E_{a1} + y_2 E_{a2} \tag{1}$$

In the first instance, the SIPNs containing 5 and 10 % CER were characterized by E_a values higher than the calculated ones, aspect attributed to the high mobility of the polyurethane chains in the melt states, thus increasing the probability of recombination reactions occurrence between macroradicals generated during thermal decomposition [18].

In the second instance, the experimental E_a values of structures containing 15 and 20 % CER content were lower than the calculated ones, due to the enhanced stiffness of the SIPNs induced by the increasing CER content which reduced the macroradicals deactivation probability by diffusion processes and increased the thermal decomposition rate [18]. For the structure containing 30 % CER content, the experimental E_a value exceeds the calculated one due to a maximum crosslinking density and afterwards phase separation phenomenon occurred and the E_a value of the SIPN containing 40 % CER becomes lower than the calculated value [19].

A comparative study between experimental and calculated temperatures values corresponding to 10, 25, 50 and 75 % weight loss is shown in Fig. 7, where ERN is the epoxy resin content.

The occurrence of specific interactions between the two comprising polymers is confirmed by the experimental values which are lower than the calculated ones for a 10 % mass loss and which decrease with CER concentration. The existing specific interactions are lower for a CER concentration up to 10 % and become more intense

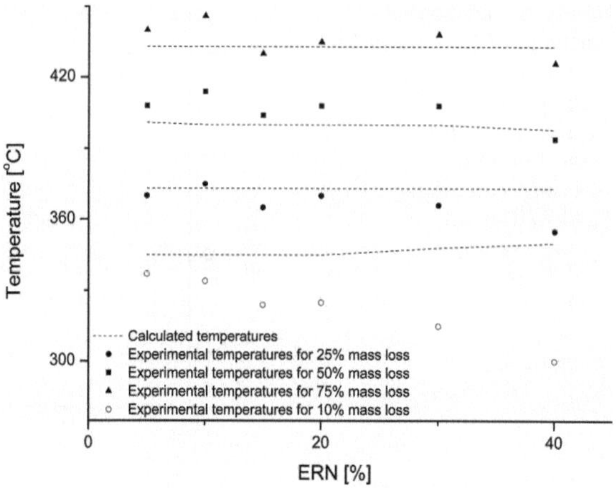

Fig. 7 Temperatures corresponding to a specific mass loss, as a function of epoxy resin content concentration (reproduced with kind permission from Elsevier—License no. 3280641074690)

with CER concentration increase up to 40 %. It can also be observed that experimental and calculated values start approaching at mass loss values higher than 25 %.

The evolved gas analyses were conducted on FTIR and mass spectroscopy (MS) devices, coupled to the TG apparatus. The main volatile compounds released below 350 °C were carbon dioxide and water vapours, which increased during the whole thermal decomposition process. Alcohol traces and aromatic structured compounds were also identified. Volatile structures containing carbonyl groups were found in the gaseous mixture at temperature values above 350 °C. Ammonia evolvement was also found in the gaseous mixture. The MS spectra were in good correlation with the findings from the FTIR spectra data.

Shojaei and Faghihi [20] studied the effect of different concentrations of organoclay on styrene-butadiene rubber and phenolic resin blends prepared by two-roll mill. Thermogravimetric studies indicated that the organoclay enhanced the thermal stability of styrene-butadiene rubber vulcanizate, whilst manifesting a catalytic behavior in presence of the phenolic resin.

Varshney et al. [21] prepared oxazolidone modified epoxy blends by cycloaddition reactions of diglycidil ether of bisphenol A and tolylene 2,6-diisocyanate, forming linear thermoplastic polymers. The obtained polymers were cured with benzophenone tetracarboxylic dianhydride, leading to the obtaining of thermosets by the crosslinking reaction of the epoxy chain ends. The oxazolidone concentrations varied in the range 5–50 %. All blends exhibited high thermal stability in nitrogen atmosphere and in the onset temperature range 373–384 °C. Char yields increased with the increasing number of heterocyclic oxazolidone moieties in the blends. It was concluded that 15 % oxadiazole in the blend was an optimum concentration value in yielding the highest thermal stability.

Phenolic-acrylic IPNs have been extensively studied due to their excellent mechanical and thermal properties [22–26]. It was reported that such IPNs exhibited lowering of T_g values owed to an enhancement of toughness, but weak thermal stability in comparison to pure phenolic resin. This aspect was explained by the occurring of certain chemical reactions between the IPNs components at certain temperature values [27]. The decreasing in T_g values related to the toughening of phenolic–acrylic IPNs was attributed to novolac resin possible curing reaction occurrence in the presence of the acrylic polymer. In order to correlate this possible reaction with the thermal behavior, a quantitative evaluation of both global thermal decomposition parameters and T_gs of phenolic–acrylic IPNs is necessary. For this purpose, Goswami and Kiran [28] obtained sequential IPNs and SIPNs based on novolac resin and poly(2-ethyl hexyl acrylate) in varying ratios and studied their thermal decomposition behavior and kinetics in nitrogen atmosphere. All structures showed a good miscibility by exhibiting a single T_g value which shifted to lower temperature domains with the increase of acrylate concentration. Activation energy values of the T_gs decreased for all samples, and were lower for the SIPNs compared to IPNs which had both components crosslinked. All structures thermally decomposed in two stages with the SIPNs exhibiting lower thermal stability than the IPNs. A thermal decomposition mechanism of the phenolic resin in the absence and presence of the acrylate network interpenetration was reported by Chakrabarty and Goswami [29] (Scheme 1).

It was demonstrated that the thermal decomposition of phenolic resin followed a first order kinetic model due to only one reactant evolving during the rate determining step, whilst the thermal decomposition of phenolic-acrylic IPNs was described by a second order reaction model due to parallel evolvement of formaldehyde and poly(2-ethyl hexyl acrylate) in the second thermal decomposition stage. Global thermal decomposition activation energy values decreased with the increasing of acrylic concentration in the IPNs and reaction order values remained unmodified, indicating that the 2-ethyl hexyl acrylate molecules facilitate phenolic resin decomposition.

Honmute et al. obtained IPNs based on polyaniline and poly(vinyl alcohol) as thin films with different concentrations of aniline through oxidative polymerization of aniline by using ammonium persulphate as oxidant [30]. All IPNs exhibited similar thermal behavior in nitrogen atmosphere, presenting the features of both comprising polymers. The structures thermally decomposed in three stages, exhibiting dopant release in the first stage (30–100 °C), covalently linked water in

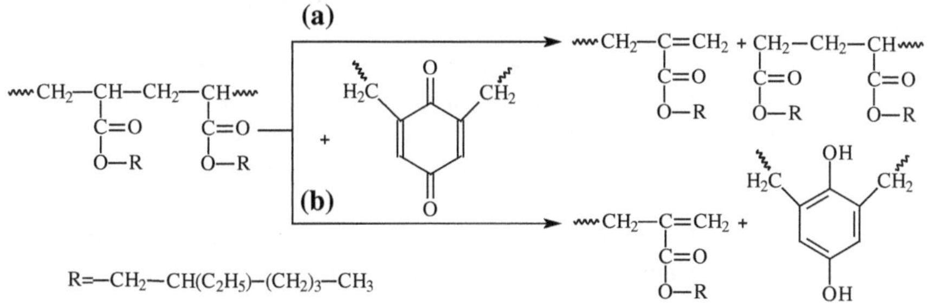

Scheme 1 Thermal decomposition of polyacrylate: **a** in absence and **b** in presence of novolac resin

the second stage (180–240 °C) and decomposition of both comprising polymers leading to polyene entities and volatile compounds in the temperature range 410–450 °C. The structures yielded char residue values of 26, 25 and 21 % at the end of the thermal decomposition process.

IPN membranes based on poly(dimethyl siloxane) with vinyl terminations and increasing concentrations of aromatic polyimide (5, 10, 15 %) for methanol and toluene azeotrope separation were obtained and their thermal stability was studied in both nitrogen and air atmospheres [31]. The blends were first cured at 300 °C, which is the temperature for optimum imidization process. In both atmospheres the structures exhibited thermal stability higher than 400 °C which increased with the polyimide content due to electrostatic attraction between silicone moieties and polyimide molecules. The blend components each decomposed following identical pathways. The thermal stability of all structures slightly decreased in air compared to nitrogen atmosphere and in both situations it was higher than that of neat poly (dimethyl siloxane). The thermal decomposition processes of the IPNs and the pure comprising polymers followed similar first order kinetics. Activation energy values increased in both atmospheres with the polyimide content due to its shielding mechanism which protected the poly(dimethyl siloxane) during thermal decomposition. Char yields values at 800 °C varied in the range 46–56 % in nitrogen atmosphere and in the range 0–49 % in air atmosphere.

Pielichowski and Janowski obtained a series of SIPNs based on polyurethane and poly(vinyl chloride) by prepolymer method using partially polymerized diphenylmethane-4,4′-diisocyanate, polyoxypropylenediol and 1,4-butanediol as chain extender. The networks were thermally characterized in order to gain new insights in further understanding of the flame retardant effect of poly(vinyl chloride) in different polymer based blends [32]. All structures thermally decomposed in three stages in argon atmosphere. The first stage ranged between 313 and 503 °C with a mass loss value of 12 % and was explained by decomposition initiation. The second stage was found in the interval 503–693 °C with a mass loss value of 50 % and was attributed to main thermal degradation during which the decomposition rate was the fastest. The third thermal decomposition stage occurred in the range 693–863 °C with a mass loss value of 14 % and was attributed to the end of the decomposition process leading to the formation of solid residue due to crosslinking processes [33]. Authors reported the occurrence of a decrease of maximum thermal decomposition rate (i.e. DTG peak maximum) for poly(vinyl chloride) in comparison to the crude polyurethane because of hydrogen chloride evolvement which retards the burning process by playing a scavenger role for free radicals [34]. Furthermore, the amount of solid char remained after thermal decomposition process exhibited an increase with molecular mass of polyoxypropylenediol increase, due to the multiphase morphology of the polyurethane. The polyoxypropylenediol comprises the soft segments creating a flexible matrix scattered between hard segments domains which physically crosslink to form a complex temperature dependent morphology with phase separations.

Merlin and Sivasankar obtained SIPNs based on low (1000 Da) and high molecular weight (2000 Da) biocompatible polyurethane incorporated in poly (acrylamide) and studied their thermal decomposition characteristics in nitrogen atmosphere [35]. Both SIPNs thermally decomposed in two stages in comparison to pure poly(acrylamide) which decomposed in three stages. The first stage of thermal decomposition occurred in the range 180–210 °C for pure poly(acrylamide), 120–257 °C for the low molecular weight based SIPN and 250–380 °C for the high molecular weight based SIPN and was attributed to the decomposition of poly (acrylamide) by ammonia loss to form imide group via cyclization reactions. The second stage of thermal decomposition occurred in the range 250–450 °C for the low molecular weight based SIPN and 390–490 °C for the high molecular weight based SIPN and may describe the decrosslinking of the network structure. The second stage of thermal decomposition of pure poly(acrylamide) ranged between 284 and 314 °C and was due to loss of bonded water and ammonia, whilst the third occurred in the range 381–425 °C when the degradation of the cyclized products was observed.

An earlier study by Mathew et al. reports the obtaining and thermal stability studies of natural rubber and polystyrene SIPNs and IPNs with varying components ratios (30/70, 50/50, 70/30), initiators and crosslinking agents types and concentrations for both components [36]. The IPNs were obtained sequentially by first crosslinking the natural rubber with dicumyl peroxide followed by the polymerization and crosslinking with divinyl benzene of styrene. Authors investigated the effects of the different initiating system (1 % benzoyl peroxide, 1 % dicumyl peroxide or 0.5 % azo-bis-isobutyronitrile), composition and crosslinking density on the thermal stability of the obtained IPNs and SIPNs. The uncrosslinked polystyrene and vulcanized natural rubber presented similar thermal stability up to 280 °C in nitrogen atmosphere and decomposing until 400 °C. Both homopolymers thermally decomposed in one single stage. Natural rubber decomposed by yielding a mixture of isoprene, dipentene and p-menthene, whilst polystyrene decomposed by yielding a mixture of styrene, toluene and methyl styrene. The IPNs and SIPNs exhibited superior thermal stability compared to the homopolymers. All network structures started decomposing after 300 °C and exhibited approximately 90 % mass loss around 450 °C. It was observed that the increase of natural rubber concentration led to a decrease in thermal decomposition temperatures. The IPNs showed higher thermal stability than the SIPNs, presenting a 30–50 % mass loss at 400 °C compared to a mass loss of 60–80 % at the same temperature, due to the higher crosslinking density reducing the bond cleavage in the IPNs. Up to a 4 % of polystyrene crosslinking agent content proved to contribute to the enhancement of all the networks thermal stability. The change in the initiating system had no influence on the thermal stability of the samples.

Boonpoo-nga et al. obtained SIPNs based on poly(4-styrenesulfonic acid) and crosslinked poly(acrylic acid) for proton-conducting membranes in fuel cell applications [37] and studied the influence of the crosslinking degree on the membranes water and thermal stability and hydrophilicity properties. The membranes were obtained by mixing the comprising components solutions in different ratio and crosslinked afterwards by maintaining those 2 h at 120 °C. All structures were thermally stable up to 200 °C in nitrogen atmosphere and afterwards decomposing in three stages. The first mass loss occurred below 200 °C and was attributed to absorbed water removal. The second mass loss occurred in the temperature range 200–340 °C and was due to the thermal decomposition of polymeric side chains. The last thermal decomposition stage occurred in the temperature range 340–480 °C and was attributed to main chains degradation. Authors observed a slight increase in the thermal decomposition rate with lower crosslinking degree values. By comparison to the membranes with the same crosslinking degree, the membrane with higher content of poly(4-styrenesulfonic acid) exhibited a larger weight loss in the first thermal decomposition stage. This aspect was explained by the strong hygroscopic character of poly(4-styrenesulfonic acid).

Huang et al. synthesized a class of thermosetting poly(2,6-dimethyl-1,4-phenylene oxide)s containing pendant epoxide groups [38]. The procedure was conducted by bromination reaction of poly(2,6-dimethyl-1,4-phenylene oxide) in halogenated aromatic hydrocarbons to yield polymers with different bromination values (50, 35 and 26 %). The obtained compounds were afterwards submitted to a Witting reaction to obtain vinyl-substituted polymer derivatives. The obtained vinyl-substituted polymers, with vinyl ratios of 38, 31 and 23 %, were treated with *m*-chlorobenzoic acid to yield epoxidized poly(2,6-dimethyl-1,4-phenylene oxide) with different pendant epoxide molar ratios (35, 28 and 20 %). The thermal stability of the brominated polymers in nitrogen atmosphere decreased slightly with the increase of bromo substituent quantity and the char yield increased up to 45 % at the end of the measurements (800 °C). The structures thermally decomposed in two stages. The first stage, in the range 250–340 °C with a mass losses of 2–20 %, was attributed to side chain decomposition due to weaker C–Br bond dissociation energy compared to that of C–H benzylic bond. The second stage of thermal decomposition occurred in the range 420–500 °C and was attributed to backbone scission [39]. A similar thermal decomposition trend was followed by the vinyl-substituted polymers. The increase of the degree of functionalization led to a slight decrease in thermal stability, due to the greater disorder in the structural conformation introduced by the vinyl group. The char yield increased in the range 33–50 %. The epoxidized compounds exhibited an intense exothermic peak in the range 257–267 °C attributed to the curing reaction of the epoxy groups during heating. The curing process led to a significant increase in the thermal stability of the cured compounds in comparison to the epoxy analogues. The thermal decomposition temperature corresponding to 5 % mass loss shifted from within the range 250–256 to 340–355 °C, with tolerable increase in char yield.

3 Effect of Reinforcement and Nanofillers on the Thermal Stability of Thermosetting Blends

Reinforced thermosets are used in a wide variety of applications, some of which involve the exposure to high temperatures. In these cases thermal stability of such thermosetting materials is investigated by thermogravimetric analysis. The char yield of these materials is correlated to their flame retardancy; the increased char yield can limit the production of combustible gases and inhibit the thermal conductivity of the material [40].

In the last years, natural fibres were investigated as reinforcement materials in polymer composites instead of synthetic fibres, due to their properties such as low cost, high specific strength, low weight, good thermal and acoustic insulation properties or lack of toxicity for human health. Alamri et al. investigated the effect of reinforcement with recycled cellulose fibres and nanoclay platelets (Cloisite 30B) on the thermal behavior of epoxy systems [41]. The presence of cellulose fibres and nanoclay slightly increased the rate of nanocomposites thermal degradation at low temperatures. However, at higher temperatures the char yield of the nanocomposites was significantly higher than that of the pure epoxy and epoxy reinforced only with cellulose fibres. This means that nanoclay acted as a barrier and hindered the diffusion of volatile decomposition products from the nanocomposites.

Shih studied the thermal behavior of an epoxy resin reinforced with water bamboo husks fibres and powders [42]. The char yield increased from 8.9 % for the epoxy resin to 10.1–13.6 % for the composites containing 10 % bamboo fibres or powder. The results showed that the addition of bamboo powder or fibres to epoxy systems would raise the char yield of the sample, therefore could improve the flame retardancy of these materials. Similar results were obtained in the case of *Phormium tenax* fibres reinforced epoxy composites, containing 20 % fibres. The presence of plant fibres determined an increase of the composites thermal stability, due to the improved fibre-matrix interactions [43].

Kenaf fibres were used as reinforcing materials for an epoxy resin [44]. The maximum degradation temperature was shifted to higher values in the case of composites. The remaining residue at 500 °C, after thermal degradation, was the lowest in the case of neat epoxy and increased for the composites containing modified fibres with NaOH and for the composites with unmodified fibres. The composites containing modified fibres had a lower char yield than those containing unmodified fibres due to the removal of lignin by alkalization treatment, lignin being responsible for the char production in natural fibres. It was concluded that the thermal stability of kenaf fibres/epoxy composites depends on the composition of the fibres.

The thermal stability of composites containing natural fibres may also depend on the nature of the matrix. In the case of composites containing unsaturated polyester or unsaturated polyester modified with acrylic acid reinforced with jute fibres it was shown that the composites with modified polyester matrix were more resistant to temperature than the ones with unmodified polyester matrix. The reason for this enhanced thermal stability was the presence of acrylic acid as modifier [45].

Cellulose nanocrystals were used as reinforcing filler for waterborne epoxy resin matrix [46]. In the case of nanocomposites, the onset temperature of thermal degradation decreased compared to neat epoxy. The same trend was obtained for the temperature of maximum weight loss for the first degradation step, while the temperature corresponding to the second step of degradation was slightly increased with increasing the cellulose nanocrystals content in the nanocomposites. The results of this study revealed that the addition of cellulose nanocrystals to the epoxy matrix improved the mechanical properties without compromising the neat matrix properties.

The thermal behavior of the composites based on raw and surface modified *Grewia optiva* particle fibres reinforced unsaturated polyester matrix was studied [47]. The fibres were modified by mercerization, silanation, benzoylation and graft copolymerization. The thermal stability of the composites containing modified fibres was improved compared to the unsaturated polyester matrix and to the composites containing untreated fibres. This was a consequence of the additional intermolecular bonding between fibres and matrix induced by the surface modification. The highest thermal stability was obtained by the composites based on silanated fibres due to the formation of strong covalent bonds between silanes and cellulose.

Due to properties such as high tensile strength, high modulus and high chemical resistance, glass fibres are also used as reinforcement materials for thermosetting matrices. Epoxy resin modified with poly(styrene-co-acrylonitrile) was reinforced with glass fibres [48]. The thermal stability of these composites was improved, their decomposition starting at higher temperature than that of the neat epoxy. Also the temperature at maximum weight loss was increased in the case of composites. Moreover the char residue was around 60 % for composites, while for the neat epoxy was only 6.6 %. This behavior was a consequence of the presence of thermally stable glass fibres. Glass and aramid (aromatic polyamides) fibres were used as reinforcement for epoxy foams [49]. All composites exhibited improved thermal stability compared to the neat epoxy. However, samples reinforced with glass fibres exhibited higher thermal stability than those reinforced with aramid fibres. The ash content was of 95 % for the samples reinforced with glass fibres, while the ash content of the samples containing aramid fibres was 45 %. This proved that the addition of glass fibre produced epoxy foams with improved thermal stability.

In the last years the thermal behavior of carbon fibre reinforced epoxy resins has been investigated by different research groups [50]. Régnier et al. performed a kinetic study on the thermal degradation of carbon fibre/epoxy composites, both in air and in inert atmosphere. The thermal degradation of the composites occurred in three stages [51]. The presence of vapour-grown carbon nanofibres into the epoxy resin matrix did not influence the thermal stability of the resin. The decomposition temperatures in the case of composites were almost the same with the decomposition temperature of epoxy resin [52].

Huntite ($Mg_3Ca(CO_3)_4$) was used as reinforcement for an unsaturated polyester resin [53]. A content of 3 % huntite had a reinforcing effect for the resin, which caused an improvement of the mechanical and thermal properties. Regarding the thermal behavior, the onset temperature of composites thermal degradation was increased with approximately 50 °C, while the maximum degradation temperature was improved with 16 °C, compared to the polyester matrix.

The properties of a composite material can be changed with the particle size or distribution, dispersion state and geometric shape. The incorporation of nanometer sized fillers has become an important strategy for the enhancement of polymers physical and mechanical properties. The obtained nanocomposites have properties different than the pure polymers due to the small size of the filler and increased surface area [54, 55]. The literature data showed that the performances of different nanocomposites are difficult to compare because the same filler can produce different effects, depending on the matrix, or in the same matrix can induce different effects, depending on the processing conditions [56]. For example, different results were reported regarding the effect of nanoclay incorporation on the thermal properties of thermosetting matrices.

The incorporation of montmorillonite (MMT) nanoparticles in different epoxy resins caused an enhancement of the nanocomposites thermal stability, which was more significant in the first stage of degradation. In this case, the MMT nanolayers acted as a barrier and prevented the volatilization of the epoxy polymer matrix. Moreover, the segmental motion of the polymer network was restricted in the presence of clay layers, which was reflected in an increased thermal stability, compared to the systems without MMT [57]. Similar results were obtained when commercial Cloisite 25A organoclay was used [58]. Both degradation steps were shifted to higher temperatures in the samples filled with nano-organoclay, compared to neat epoxy resin. Also, the activation energy for the first step of degradation was higher for the nanocomposite system because of the inorganic clay thermal resistance, which increased the energy demand for thermal degradation of the resin.

The thermal properties of the nanocomposites of diglycidyl ether of bisphenol A epoxy resin containing different montmorillonite nanoclays depends on the surface modifications of the nanoclays. Moreover, the addition of nanoclays improved the thermal properties of the nanocomposites, compared to the neat epoxy [59].

Brnardic et al. studied the thermal stability of nanocomposites based on organically modified MMT and an epoxy resin [60]. Compared to the neat resin, small changes in the thermal stability were observed in the case of nanocomposites. The interlayer organic ions in organically modified MMT were more prone to degradation than the neat epoxy resin, causing a slight decrease in the onset temperatures of nanocomposites thermal degradation. The nanocomposites containing 5 % nanoclay had a thermal stability equal to that of the epoxy resin in the absence of nanoclay. An increase of the nanoclay content, up to 10 % determined a slight decrease in the thermal stability of the nanocomposites [61].

Phenolic resin/montmorillonite nanocomposites were prepared by intercalative polymerization of phenol and formaldehyde in the presence of organo-modified montmorillonites [62]. The nanocomposites thermal stabilities were higher than that of the neat phenolic resin. The highest thermal stability was observed for the nanocomposites containing MMT treated with quaternary ammonium salts containing benzene ring. This effect was attributed to the enhanced chemical affinity between the resin and the modifiers containing benzene groups. The favorable interactions promote the exfoliation and homogenous distribution of the MMT into the phenolic resin matrix.

Polyfurfuryl alcohol (PFA) nanocomposites were obtained by in situ polymerization of furfuryl alcohol in the presence of MMT (natural and organomodified) or cellulose nanowhiskers [63]. Below 400 °C, the temperatures at the onset of degradation were increased for the nanocomposites, compared to pure PFA. The highest increase was obtained for nanocomposites containing cellulose whiskers, due to a stronger matrix-nanoparticle interaction as compared to the nanocomposites containing MMT. Above 400 °C the nanocomposites containing organo modified MMT exhibited superior weight retention, compared to the nanocomposites containing cellulose nanowhiskers or to pure PFA. This increase was attributed to the complete exfoliation of the clay in the nanocomposites.

In the case of Novolac phenolic resin, the incorporation of highly dispersed organoclays containing cationic pillaring agents, did not improved the thermal stability of the resin [64]. Upon reaching the resins decomposition temperatures, the clay surfaces may provide some catalytic activity which promotes decomposition. The thermal stabilities of the nanocomposites, expressed by their 5 and 10 % weight loss temperatures, were established for the initially cured samples (80 °C/2 h) and for the samples which were additionally cured at 100 °C for 0.5 h. These temperatures were higher for the samples cured than for the nanocomposites partially cured. Moreover, the char yield at 900 °C was higher for the totally cured nanocomposites.

The effects of Cloisite loading on the thermal decomposition behavior of an epoxy resin were reported by Ingram et al. [65]. It was shown that the addition of Cloisite in the systems improved the thermal stability of the epoxy resins which underwent an initial cure at 180 °C. However, when the nanocomposites were post cured at 220 °C, the nanoclay incorporation induced a decrease in the thermal stability of the systems. This behavior may be attributed to the dissociated alkyl chains, which destabilize the thermal properties after being subjected to high temperatures [66]. Therefore, a careful selection of cure and post cure temperatures must be made in order to obtain nancomposites with improved physical properties and with enhanced thermal stability.

The thermal stability of some thermosetting nanocomposites obtained by thermal cationic cure of diglycidilether of bisphenol A and γ-valerolactone using rare earth metal triflates (trifluoromethanesulfonate) as initiators and containing different types of Cloisite was studied. The addition of clay into the polymeric matrix was found to increase the thermal stability, acting as a superior insulator and mass transport barrier to the volatile products evolved during thermal decomposition [67].

In the case of phenolic resin/organic expanded vermiculite nanocomposites (a mica-type silicate) the thermal behavior was investigated in both air and nitrogen atmosphere [68]. The onset temperature of thermal degradation, as well as the temperature of 15 % weight loss showed a significant increase in the presence of vermiculite. Also, the mass loss rate of the resin was higher than that of the nanocomposites. These results were obtained for the experiments carried out in both atmospheres. The addition of 1.5 % vermiculite in the nanocomposites determined an increase in the temperature of 15 % weight loss from 476 to 511 °C, in nitrogen atmosphere and from 422 to 467 °C in oxidative conditions. The improvement of thermal stability in the case of nanocomposites was attributed to a uniform dispersion of the exfoliated platelets of vermiculite into the polymer matrix.

These platelets acted as superior insulator and as barrier for oxygen transport and volatilization of the products resulted after the thermal degradation.

The incorporation of 5 % organically modified sepiolite, which is a microcrystalline-hydrated magnesium silicate, in a bisphenol A-based epoxy resin has no significant effect over the thermal stability of the epoxy resin, due to the poor dispersion of the clay and poor diffusion of the resin between fibres [69]. The effect of attapulgite (magnesium aluminium phyllosilicate) over the thermal properties of hyperbranched polyimides was studied. The presence of this silicate in the nanocomposites significantly improved the thermal stability of the neat polyimide [70].

Ollier et al. reported the effects of modified and unmodified bentonite loading on the decomposition behavior of unsaturated polyester thermosets [71]. It was shown that the addition of bentonite increased the thermal stability of the resin. Furthermore, the clay modification did not significantly influence the degradation temperatures of the nanocomposites.

The incorporation of other type of nanoparticles, such as CuO, TiO_2, silsequioxanes was found to affect the thermal degradation of the thermosetting materials. The thermal stability of the epoxy resin/cupric oxide (CuO) nanocomposites in air atmosphere was improved by the presence of CuO nanoparticles. Furthermore, the presence of these nanoparticles affected the degradation mechanism of the epoxy resin [72].

Cupric oxide unmodified and functionalized with methacryloxypropyl trimethoxysilane was used to obtain vinyl-ester resin polymeric nanocomposites [73]. The resistance to thermal degradation was improved for the nanocomposites with and without postcuring, compared to the pure cured resin. The thermal stability of the nanocomposites containing functionalized CuO was higher than that of the nanocomposites containing the unmodified CuO nanoparticles. The presence of the coupling agent prevented the intimate contact between CuO and vinyl-ester resin, passivating the particle surface, and thus increasing the thermal stability of the nanocomposites with functionalized CuO nanoparticles. Similar results were obtained in the case of vinyl-ester resin nanocomposites with iron oxide nanoparticles (un-modified and functionalized with the same coupling agent) [74, 75].

In the case of Ni–La–Fe–O/epoxy nanocomposites, the thermal degradation showed a more complicated behavior than the neat epoxy resin, with two peaks in the 300–475 °C temperature range. Moreover, the thermal stability of the resin decreased in the presence of Ni–La–Fe–O nanoparticles, due to the fact that these nanoparticles may act as catalysts to degrade the epoxy matrix [76].

The thermal stability of epoxy resin/TiO_2 nanocomposites was found to be dependent on the nanoparticles loading, as well as on their dispersion state [77]. At a very low TiO_2 loading into the matrix, the nanoparticles were dispersed uniformly and formed a barrier to heat and oxygen, due to their ceramic nature. When the content of the nanoparticles was increased, they tended to agglomerate into lumps, which were non-uniformly distributed into the matrix. In this case the nanoparticles were less effective in blocking the heat and oxygen, but still more effective than the neat epoxy resin.

The incorporation of the hybrid TiO_2–SiO_2 nanofillers into an epoxy resin increased the thermal stability of the neat resin [78]. Also the char yield increased from 0 % for the neat resin to 25 % for the nanocomposites. These phenomena are a consequence of the hybrid nanoparticles which acted as thermal stabilizers for the epoxy resin.

When SiO_2 was incorporated into poly(furfuryl alcohol) thermosetting matrix, the thermal stability of the furanic matrix was enhanced [79]. The thermal decomposition temperature corresponding to 10 % weight loss increased with 30 °C in the presence of SiO_2. Also in the early stage of degradation the rate of thermal degradation decreases significantly. These results were a consequence of the interconnections which appeared between the furanic end groups and silica, reducing the possibility of unzipping the furanic chain. This meant that the chain scission predominated and high molecular weight compounds with low volatility were formed. Above 400 °C the rate of degradation was the same for the systems with or without silica, suggesting that the silica network had no influence on the latter stages of poly(furfuryl alcohol) thermal degradation.

Silsesquioxanes are Si–O–Si networks with organic substituents attached to each Si atom, which can improve the compatibility with the polymer matrix. The materials containing silsesquioxanes have improved thermal stability because char formation is promoted and the diffusion of gases into the material is retarded [80]. Nagendiran et al. studied the thermal degradation behavior of epoxy/polyhedral oligomeric silsesquioxane (POSS) nanocomposites [81]. The thermal stability of the nanocomposites containing octa-aminophenyl silsesquioxane (OAPS) was improved compared to the pure epoxy resin, with a retarded mass loss rate and an enhanced char yield. The increased OAPS concentration and the nanoscale dispersions of POSS into the epoxy matrices contributed to the enhanced thermal stability. The high char yield implied that fewer volatiles were released during nanocomposites thermal degradation, suggesting that the flame retardancy of the materials was increased.

Epoxy acrylate/vinyl POSS nanocomposites were obtained by in situ polymerization and UV-curing method [82]. The thermal degradation mechanism of epoxy acrylate was not modified by the presence of vinyl-POSS and the thermal stability of the nanocomposites was improved with the increasing of POSS loading. The temperatures corresponding to 50 % mass loss increased in the case of nanocomposites, compared to the neat epoxy acrylate. Also the char yield was increased because of SiO_2 yield which was produced by POSS thermal degradation.

Schutz et al. studied the thermal properties of a phenolic resin nanocomposite containing silsesquioxanes [80]. The thermal oxidative stability of the nanocomposites was improved, as compared to that of the pure resin. This effect may be a consequence of the formation of a protective layer of SiO_2 during silsesquioxanes pyrolysis at the surface of nanocomposites, which retarded the thermal oxidative degradation. The temperatures characteristic to thermal degradation processes were higher in the case of nanocomposites, compared to the pure resin matrix.

In the case of a resole phenolic resin, the presence of trisilanolphenyl POSS markedly increased the thermal stability of the phenolic nanocomposites. The temperatures corresponding to 5 and 50 % mass loss increased with 34 and 90 °C, respectively, for the maximum amount of POSS used (10.4 %) [83].

Aflori et al. synthesized POSS-based hybrid nanocomposites with methacrylate units containing titania and/or silver nanoparticles (POSS-AgTi and POSS-Ag) and tested these materials as antibacterial/antifungal coatings for monumental stones [84]. The thermal decomposition of the nanocomposites presented different stages (93–260, 260–285, 384–456, 456–550 °C for POSS-Ag; 78–156, 156–230, 300–435 °C for POSS-AgTi) and residues of 51.22 % for POSS-Ag and 27.7 % for POSS-AgTi, respectively, at 600 °C in nitrogen atmosphere. The first decomposition stage of POSS-Ag corresponded to residual isopropyl loss and dehydration, with a secondary condensation process between Si–OH groups in the range 85–175 °C, loss of AgO and decomposition of dodecylamine surfactant (around 220 °C). A similar thermal behavior was observed for POSS-AgTi up to 230 °C, the degradation processes occurring during the succession of two decomposition stages and with similar mass loss (6 %). POSS-AgTi exhibited a lower initial temperature corresponding to solvent loss compared to POSS-Ag, probably due to the rigid network structure which did not trap the solvent molecules inside the polymer chains. The second thermal decomposition stage (260–285 °C) of POSS-Ag occurred with a small mass loss (3.35 %) due to chain-end initiation phenomena from end double bonds around 277 °C. For POSS-AgTi, the presence of titania delayed the degradation of the end double bonds, up to 300 °C. Within the temperature range 300–550 °C, the mass loss corresponding to random polymeric chain scission was lower for POSS-AgTi (21.64 %), compared to POSS-Ag (45.68 %). The improvement of the thermal stability of POSS-AgTi was correlated with the fact that this hybrid nanocomposite with silsesquioxane and titania content might act as a cage, trapping free radicals generated during thermal degradation. Furthermore, the influence of titania-catalyzed oxidative decomposition of the polymer matrix was negligible, thus it was concluded that titania nanoparticles were wrapped within the silica components. This observation led to the state that the silsesquioxane-based coating protected the polymer scaffold from catalytic degradation.

The thermal behavior of nanocomposites films based on poly(triazole-imide) reinforced with neat and modified SiC nanoparticles was investigated [85]. The nanocomposites decomposition temperatures increased with SiC content, indicating an enhancement of the thermal stability in the presence of SiC. Moreover, the char yield increased from 58 % for the poly(triazole-imide) to 85 % for the nanocomposites containing 10 % SiC. The nanocomposites containing SiC nanoparticles functionalized with epoxide-end groups showed a lower thermal stability compared to the nanocomposites containing neat SiC.

Carbon nanotubes (CNTs) can be used as nano-modifiers in the thermosetting matrices due to their attractive mechanical, thermal and electrical properties. Nevertheless, the incorporation of CNTs into these matrices has not had remarkable effect on the polymers thermal properties, with some exceptions [86]. The effect of CNTs over the thermal stability of thermosetting polymers depends on some factors such as CNTs type and content, aspect ratio and dispersion, the interactions between CNTs and the polymer matrix. The addition of multi-walled carbon nanotubes (MWCNTs) leads to a decrease of the thermal stability of epoxy matrices [87, 88]. This effect is caused by the increase of the polymer thermal conductivity as a consequence of MWCNTs addition. It was also reported that the nanotubes length and modification with amino groups has no significant influence on the thermal stability of MWCNTs/epoxy nanocomposites. The activation energies of the thermal degradation reactions were calculated using the method proposed by Flynn-Wall-Ozawa. For all studied nanocomposites, the activation energy had lower values, for all conversion degrees, compared to the neat epoxy. This proved that the addition of MWCNTs to the epoxy matrix increased the materials degradation efficiency. MWCNTs have high thermal conductivity properties, facilitating heat transportation in nanocomposites and implicitly decreasing the activation energy [89]. Similar results were reported in the case of single-walled CNTs [90].

Nanocomposites containing functionalized CNTs exhibited increased thermal stability, compared to the neat epoxy resin. In this case the functionalized CNTs had a better affinity for the polymeric matrix than the un-functionalized CNTs [91]. Kuan et al. reported that the incorporation of the MWCNTs functionalized with vinyltriethoxysylane into an epoxy resin increased its thermal stability [92]. The same effects were obtained in the case of MWCNTs grafted with triethylenetetramine [91] and MWCNTs functionalized with silane [88].

The thermal stability of nanocomposites containing cyanate ester/MWCNTs was found to be dependent on the surface modification of CNTs [93]. The results showed that the presence of unmodified MWCNTs in the nanocomposites produced a decrease of the initial decomposition temperature and a char yield similar to the neat resin. On the contrary, the initial decomposition temperature increased with 20 °C in the presence of modified MWCNTs. Moreover, the char yield of the nanocomposites containing modified MWCNTs was much higher than that of the cyanate ester resin. These results proved that the nature of carbon nanofiller is important for developing new nanocomposites with enhanced thermal properties.

The nanocomposites obtained from phenolic resin and carboxylated MWNTs showed an improvement of the thermal stability than the neat phenolic resin. The highest thermal stability was obtained in the case of the nanocomposites obtained by in situ polymerization, due to the quality of dispersion of the functionalized MWCNTs [94]. An enhancement of the thermal stability was also obtained in the case of nanocomposites containing boron phenolic resin and MWCNTs modified with nitric acid, 4,4′-diaminodiphenyl methane and boric acid. This effect was ascribed to better interfacial interactions between modified MWCNTs and the resin matrix [95].

The effect of carbon nanofibres on the thermal behavior of phenolic resins was studied by Bafekrpour et al. [96]. The presence of carbon nanofibres produced an increase in the thermal stability of the nanocomposites, compared to the neat phenolic resin. Thus, the decomposition temperature was shifted to higher values with an increase of the nanofibres content. The same trend was obtained for the nanocomposites char yield.

Faraz et al. obtained new composites from carbon nanofibres (CNFs) and bismaleimide, using a thermokinetic mixing method [97]. The addition of CNFs slightly improved the thermal stability of the nanocomposites. The presence of CNFs caused the formation of more char residue, but this char was thermally unstable and it degraded faster compared to the char formed in the case of pure resin.

Graphene is a new type of nanofiller, with two-dimensional structure and with excellent mechanical, thermal and electrical properties [98]. The addition of two kinds of nanofillers: graphene and magnetic graphene (graphene nanosheets coated with iron core iron oxide shell nanocomposites, $Fe@Fe_2O_3$) to an epoxy resin matrix caused a decrease of the thermal decomposition temperatures. This effect was attributed to the spatial obstruction of the nanoparticles on the formation of highly cross-linked epoxy structure. The nanocomposites containing magnetic graphene showed a much lower weight loss compared to the pure epoxy and epoxy matrix containing graphene, due to protruding $Fe@Fe_2O_3$ nanoparticles on the surface of graphene, which favored the char formation during the thermal degradation process [99]. Wang et al. reported an improvement in the thermal stability of epoxy nanocomposites containing 0.1 % graphene oxide nanoparticles [100].

The presence of graphene oxide slightly changed the thermal stability of the polyurethane/graphene oxide/epoxy nanocomposites. In the case of these nanocomposites, a content of 0.033 % graphene oxide casued a slight increase of the degradation temperatures [101]. Isocyanate-treated graphene oxide and cyanate ester resin were used to prepare nanocomposites by a solution intercalated method. The composites presented a higher char residue than the cyanate ester resin, as a result of the interactions between graphene oxide and the resin matrix. Also the degradation temperatures slightly increased with increasing the grapheme oxide content and the decomposition rates significantly decreased [102].

New materials can be developed by using interpenetrating polymer networks (IPN) as matrices for various nano-reinforcing agents. Polyhedral oligomeric silsequioxanes (POSS) was used for the reinforcement of dimethacrylic/epoxy IPN. The addition of POSS to the neat resin caused a slightly decrease of the onset temperature of degradation and no change in the maximum weight loss rate. Moreover, the presence of POSS did not influence the thermal stability of the IPN [103].

The thermal decomposition of polyurethane/epoxy resin interpenetrating polymer network (IPN) nanocomposites containing organophilic montmorillonite was found to be a complex, two-stage process. The thermal stability of these nanocomposites was improved by the presence of nanoclay [104]. Potassium titanate whiskers were used as reinforcement of polyurethane/epoxy resin IPN [105]. The incorporation of the whiskers enhanced the thermal stability of the polyurethane/epoxy resin IPN. The temperatures for 20 and 80 % mass loss were increased compared to the IPN in the absence of potassium titanate whiskers. Moreover the residual weight at 600 °C increased from 11.13 % for the IPN to 19.27 % for the IPN containing 5 % potassium titanate whiskers. When carbon nanotubes were used as reinforcement for these polyurethane/epoxy resin IPNs, the thermal stability decreased slightly. This was caused by the CNTs high thermal conductivity [106].

4 Applications and Future Trends of Thermosetting Blends

There are several emerging domains incorporating a great deal of research from multidisciplinary fields and from which great expectations arise for the future. Such continuous evolving domains include the obtaining of new construction materials, drug delivery systems, biomaterials, fuel cells [107], cells for solar energy captation, renewable resources for biofuels, biodegradable polymer systems and smart materials which are covered in a series of recent reviews [108–110].

Paduraru et al. obtained cryogels based on crosslinked poly(vinyl alcohol) and increasing quantities of microcristalline cellulose (10, 30, 50 %) and conducted in vitro tests for the release of vanillin (4-hydroxy-3-methoxybenzaldehyde), an antimicrobial and antioxidant agent, from poly(vinyl alcohol)/cellulose matrices [111]. The tests yielded good results due to the increased release percent of bioactive agent with cellulose content increase in the cryogels and also shortening of the half and maximum release time. Prior to the in vitro tests authors conducted thermal stability studies of the cryogel supports. Varganici et al. conducted thermal decomposition kinetics and evolved gas analysis of the cryogels in nitrogen atmosphere as background to industrial processing of the future bioactive formulations for wound dressings [112]. According to the two latter literature reports, pure microcrystalline cellulose exhibited one thermal decomposition stage in the range 30–124 °C with a 2.1 % mass loss, corresponding to physical dehydration. This was followed by a main and major decomposition stage with a maximum temperature at 345 °C, characterized by the highest mass loss (86.09 %), due to the unzipping of cellulose chains and levoglucosan formation with char and volatile products yield. Pure poly(vinyl alcohol) thermally decomposed in four stages, generating water loss, partial dehydration of poly(vinyl alcohol) together with polyene formation, decomposition of the polyenes formed in the second stage together with destruction of macroradicals followed by advanced degradation of the polymeric backbone and scission reactions in the last stage. The highest mass loss occurred in the range 306–419 °C for poly(vinyl alcohol). These temperature values shifted to lower domains for the cryogels and remained situated between those of the pure components. The second thermal decomposition stage (217–306 °C) characteristic to poly(vinyl alcohol) disappeared due to the cellulose influence over the cryogels. An increase of the residual mass with cellulose content increase was also observed, from 3.4 % for the poly(vinyl alcohol) to 11.18 %. The evolved gas analysis indicated that the major volatile thermal decomposition products were water, carbon dioxide and carbonyl and olefin structures.

Other excellent wound healing accelerators consists in various materials based on modified chitosan. The modification of chitosan is necessary, due to its poor solubility in common organic solvents and water, thus inhibiting its utilization. Various modification methods, such as chemical grafting, physical blending or crosslinking have been reported in the literature [113–116]. Rodkate et al. obtained SIPN hydrogels based on polydimethylsiloxane (PDMS)/polyethylene glycol (PEG) modified chitosan by interpenetrating 20 wt% of PDMS and PEG into the chitosan hydrogels crosslinked with 10 wt% hexamethylene-1,6-di-(aminocarboxylsulfonate) (HDA) [117]. Thermogravimetry in inert atmosphere was used to investigate the effect of PDMS and PEG on the thermal stability of the hydrogels. For PDMS-chitosan SIPNs the mass loss became retarded in comparison to HDA crosslinked chitosan without PDMS, indicating an enhanced thermal stability after addition of PDMS. A slight enhancement of thermal stability was also observed in PEG-chitosan SIPNs. In both cases, the PDMS/PEG-chitosan SIPNs decomposed in two stages, ranging between 230–300 and 380–410 °C, as opposed to HDA-crosslinked chitosan (without PDMS or PEG) which thermally decomposed in a single step.

Another example of biocompatible polymer hydrogels for controlled drug release, tissue engineering, artificial muscles, biosensors, etc. consists of dextran based materials. Dinu et al. obtained hydrogels based on polyacrylamide and dextran, tuned by the synthesis temperature, via free radical crosslinking copolymerization in aqueous solution at three temperatures (−18, 5 and 22 °C) [118]. Dextran decreased the hydrogels thermal stability compared to the corresponding polyacrilamide gels, indicating that the dextran chains interdiffuse and are physically entangled in the crosslinked polyacrilamide. The hydrogels thermal stability did not depend on preparation temperature. Dextran thermally decomposed in two stages, whilst the hydrogels and polyacrilamide gels in three stages, irrespective of the preparation temperature. The first mass loss was attributed to evolvement of water and other volatiles. The dextran chain degradation also explained the lower second stage onset temperatures of the hydrogels and polyacrilamide gels. The hydrogels underwent a higher mass loss in the third stage compared to the crosslinked polyacrilamide matrix which decomposes mainly at higher temperatures. The presence of dextran in the gels led to an increase in char quantity.

Another expanding research field consists in the obtaining of polymers from renewable resources for reducing global warming by replacing of petroleum-based resins. For this purpose, vegetable oils represent very promising raw materials. A favorite candidate is soybean oil because it may be 'genetically engineered' for such applications. 'Functionalization' of vegetable oils is necessary because double bonds from fatty acids in triglycerides are not very reactive in free radical crosslinking compared to those situated in terminal positions.

Grishchuk and Karger-Kocsis obtained a series of hybrid thermosets from vinyl ester resin and acrylated epoxidized soybean oil by free radical-induced crosslinking, with ratios varying from 75/25 to 25/75 % and the 50/50 % hybrid was modified with different amounts of phthalic anhydride (1, 5 and 10 %) [119]. The thermal stability of the hybrid materials in nitrogen atmosphere increased in comparison to the individual comprising polymers. This aspect was explained by the synergistic effects within the IPN structures, slowing decomposition and volatilization and diffusion of the degradation products. The incorporation of phthalic anhydride was expected to yield even better results, due to a higher molecular confinement induced by the grafts. However, the incorporation of phthalic anhydride led to a decrease in thermal stability. An explanation may be correlated with a surplus of uncoupled phthalic anhydride in the synthesis or phthalic anhydride acting as initiating site during thermal degradation.

Wang and Schuman [120] synthesized glycidyl esters of epoxidized fatty acids derived from soybean (EGS) and linseed oil to gain higher oxirane content for the purpose of enhancing reactivity and lowering the viscosity compared to epoxidized soybean (ESO) or linseed oil. The EGS and ESO were characterized individually and in blends of different ratios with diglycidyl ether of bisphenol A (DGEBA) and obtaining of thermosetting resins with epoxy monomers and 1,2-cyclohexanedicarboxylic anhydride or BF_3 as catalyst. ESO-DGEBA and the EGS-DGEBA blends presented similar thermal stability profiles. Thermogravimetric analysis results in inert atmosphere showed that al cured resins were stable up to 300 °C. EGS-DGEBA blends led to a slight lowering in thermal stability, compared to ESO-DGEBA blends. All structures thermally decomposed in two stages. The first stage occurred in the range 300–450 °C and was attributed to unreacted 1,2-cyclohexanedicarboxylic anhydride decomposition, hydroxyl groups dehydration and crosslinked epoxy resin network degradation. The second stage ranged between 450 and 600 °C and was correlated to the complete degradation of smaller fragments such as aromatic or cyclized decomposition byproducts as shown by a decrease in char residue with EGS component increase.

Mustata et al. investigated the influences of different amounts of ESO on the curing kinetics and thermal properties of DGEBA cured with p-aminobenzoic acid (p-ABA) by dynamic thermogravimetry in inert atmosphere, amongst other methods [121]. All resins, except crosslinked ESO, exhibited similar thermal behavior and decomposed in a single stage, unaffected by the ESO content, due to the networks structural similarity. All structures started decomposing at approximately 370 °C, with a major mass loss (around 50 %) in the range 370–440 °C. Sample ESO/p-ABA decomposed at a lower temperature in comparison to DGEBA/p-ABA, due to the presence of aromatic ring in the DGEBA molecular structure, which cannot be easily broken. Activation energy values of thermal decomposition decreased with ESO content increase for the DGEBA/p-ABA/ESO samples, due to their chemical structures containing several ester groups, introduced by ESO, which decompose more facile.

The majority of thermal applications of different thermoset blends related to the construction domains consist of thermosetting adhesives. Such adhesives may also be used for circuit boards manufacturing. This may be achieved by the dispersion of a powder of cured heat resistant resin soluble in an acid into a matrix of heat resistant resin which is uncured and hardly soluble in the same acid. One of the phases is cured whilst the other is not, this way forming an adhesive layer. For example, the obtaining of marine adhesives implies the using of polyaromatics which contain reactive end groups and a thermoset resin. This is done by using a reactive catalyst which contains a Lewis acid bearing amine functionality. After the curing process, the amine reacts with the polymer generating an adhesive which is curable at high temperatures and with environmental resistance and excellent mechanical properties. Cocontinuous structures may also be used in combination with curable thermosets. Such structures give great heat resistance and toughness to circuit boards, good adhesion to copper wiring and little thermal deformation. They also make fine patterns formation possible [122].

5 Conclusions

Most developments in the field of polymer blends have aimed at improving materials final properties. Amongst other plastics, thermosetting polymer blends continue playing a significant role in the development and application of polymer based materials because of their advantageous overall cost-performance relation and their capacity to improve the performance of single resins. Although in thc majority of cases thermosetting polymer blends provided great combination of properties to single resins, blending techniques require a rich interdisciplinary knowledge, including surface and phase phenomena characteristics, thermodynamic principles of miscibility, morphology, processing, and performance. It is a known fact that the vast majority of commercial blends are not in a thermodynamic equilibrium state, thus occurring the possibility of generating phase separation phenomena. The selection of adequate processing method determines the product performance. With the continuous improvement of fundamental knowledge and research grows the demand for materials with enhanced performances, making the role of blending more and more important. Processability is also improved by blending. Blending may lead to a reduction in viscosity which is of crucial importance to processing. Since the final properties of the blends are dictated by the properties, composition and morphology of the individual components, processing must be very carefully examined.

Different special applications of thermosetting blends require different processing techniques of such materials, thus their thermal stability, along with other properties, is very important in proving the miscibility of the phases and settling the threshold temperatures for the processing of the future materials. It is for such purposes that fundamental knowledge must be gained on the different thermal decomposition mechanisms and their influence on the miscibility and morphology of the comprising polymers.

Acknowledgments Authors of this chapter acknowledge a grant of the Romanian National Authority for Scientific Research, CNCS—UEFISCDI, project number PN-II-ID-PCE-2011-3-0187.

References

1. Utracki, L.A.: In: Utracki, L.A. (ed.) Polymer Blends Handbook, vol. 1. Kluwer Academic Publishers, Dordrecht (2002)
2. Thomas, R., Vijayan, P., Thomas, S.: Recycling of thermosetting polymers. In: Fainleib, A., Grigoryeva, O. (eds.) Recent Developments in Polymer Recycling, pp. 122–129. Transworld Research Network, Kerala (2011)

3. Irfan, M.H.: Chemistry and Technology of Thermosetting Polymers in Construction Applications, pp. 78–96, 230–239. Springer Science and Business Media, Dodrecht (1998)
4. Skiest, I.: Handbook of Adhesives, 2nd edn, pp. 12–17. Litton Educational Publishing, New York (1977)
5. Benson, L.M.: Polymer Blends: A Comprehensive Review, pp. 65–108. Carl Hanser Verlag, Munich (2007)
6. Rosu, D., Rosu, L., Varganici, C.-D.: The thermal stability of some semi-interpenetrated polymer networks based on epoxy resin and aromatic polyurethane. J. Anal. Appl. Pyrol. **100**, 103–110 (2013)
7. Varganici, C.-D., Rosu, L., Rosu, D., Simionescu, B.C.: Miscibility studies of some semi-interpenetrating polymer networks based on an aromatic polyurethane and epoxy resin. Compos. B Eng. **50**, 273–278 (2013)
8. Fox, T.G.: Influence of diluent and of copolymer composition on the glass temperature of a polymer system. Bull. Am. Phys. Soc. **1**, 123 (1956)
9. Gordon, M., Taylor, J.: Ideal copolymers and the second-order transitions of synthetic rubbers. I. Non-crystalline copolymers. J. Appl. Chem. **2**, 493–500 (1952)
10. Cristea, M., Ibanescu, S., Cascaval, C.N., Rosu, D.: Dynamic mechanic analysis of polyurethane-epoxy interpenetrating polymer networks. High Perform. Polym. **21**, 608–621 (2009)
11. Rosu, D., Tudorachi, N., Rosu, L.: Investigations on the thermal degradation of a MDI based polyurethane elastomer. J. Anal. Appl. Pyrol. **89**, 152–158 (2010)
12. Rosu, D., Rosu, L., Brebu, M.: Thermal stability of silver sulfathiazole-epoxy resin network. J. Anal. Appl. Pyrol. **92**, 10–18 (2011)
13. Friedman, H.L.: Kinetic of thermal degradation of char forming plastics from thermogravimetry-application of phenolic plastics. J. Polym. Sci. **C6**, 183–195 (1965)
14. Ozawa, T.: A new method of analysing thermogravimetric data. Bull. Chem. Soc. Jpn. **38**, 1866–1881 (1965)
15. Opferman, J.: Kinetic analysis using multivariate non-linear regression. J. Therm. Anal. Calorim. **60**, 641–658 (2000)
16. Galwey, A.K., Brown, M.E.: Kinetic background to thermal analysis and calorimetry. In: Brown, M.E. (ed.) Handbook of Thermal Analysis and Calorimetry, pp. 169–171. Elsevier, Amsterdam (1998)
17. Denq, B.L., Chin, W.Y., Lin, K.F.: Kinetic model of thermal degradation of polymers from nonisothermal process. J. Appl. Polym. Sci. **66**, 1855–1867 (1997)
18. Tiptipakorn, S., Damrongsakkul, S., Ando, S., Hemvichian, K., Rimdusit, S.: Thermal degradation behaviours of polybenzoxazine and silicone polyimide blends. Polym. Degrad. Stab. **92**, 1265–1278 (2007)
19. Rosu, D., Rosu, L., Mustata, F., Varganici, C.-D.: Effect of UV radiation on some semi-interpenetrating polymer networks basedon polyurethane and epoxy resin. Polym. Degrad. Stab. **97**, 1261–1269 (2012)
20. Shojaei, A., Faghihi, M.: Physico-mechanical properties and thermal stability of thermoset nanocomposites based on styrene-butadiene rubber/phenolic resin blend. Mater. Sci. Eng. A **527**, 917–926 (2010)
21. Varshney, A., Mathur, R.M., Prajapati, K.: Thermal characteristics of oxazolidone modified epoxy anhydride blends. Int. J. Chem. **4**(3), 113–120 (2012)
22. Knop, A., Scheib, W.: Chemistry and Application of Phenolic Resin. Springer, New York (1979)
23. Gardziella, A., Knop, A., Pilato, L.A.: Phenolic Resins: Chemistry, Applications, Standardization, Safety and Ecology, 2nd edn. Springer, Germany (2000)
24. Wolfgang, H.: 'Phenolic Resins' in Ullmann's Encyclopedia of Industrial Chemistry. Wiley-VCH, Weinheim (2002)
25. Wang, H., Yan, Y., Yu, Y., Zhao, T., Zhi, L.: Synthesis of novolac/layered silicate nanocomposites by reaction exfoliation using acid-modified montmorillonite. Macromol. Rapid Commun. **23**, 44–48 (2002)

26. Bandyopadhyay, D., Chakrabarty, D., Mandal, P.K., Goswami, S.: Novolac resin-poly(ethyl methacrylate) interpenetrating polymer networks: morphology and mechanical and thermal properties. J. Appl. Polym. Sci. **90**, 412–420 (2003)
27. Goswami, S., Nad, S., Chakrabarty, D.: Modification of novolac resin by interpenetrating network formation with poly(butyl acrylate). J. Appl. Polym. Sci. **97**, 2407–2417 (2005)
28. Goswami, S., Kiran, K.: Application of Kissinger analysis to glass transition and study of thermal degradation kinetics of phenolic-acrylic IPNs. Bull. Mater. Sci. **35**(4), 657–664 (2012)
29. Goswami, S., Chakrabarty, D.: Synthesis and characterization of sequential interpenetrating polymer networks of novolac resin and poly(ethyl acrylate). J. Appl. Polym. Sci. **99**, 2857–2867 (2006)
30. Honmute, S., Ganachari, S.V., Bhat, R., Naveen, H.M.P., Kumar, D.S., Venkatarman, H.A.: Studies on polyaniline-polyvinyl alcohol (PANI-PVA) interpenetrating polymer network (IPN) thin films. Int. J. Sci. Res. **1**(2), 102–106 (2012)
31. Garg, P., Singh, R.P., Choudhary, V.: Selective polydimethylsiloxane/polyimide blended IPN pervaporation membrane for methanol/toluene azeotrope separation. Sep. Purif. Technol. **76**, 407–418 (2011)
32. Pielichowski, K., Janowski, B.: Semi-interpenetrating polymer networks of polyurethane and poly(vinyl alcohol). Thermal stability assessment. J. Therm. Anal. Calorim. **80**, 147–151 (2005)
33. Vieira, E.F.S., Cestari, A.R., Zawadzki, S.F., Rocha, S.M.: Evaluation of tg data of htpb-based polyurethanes. J. Therm. Anal. Calorim. **75**(2), 501–506 (2004)
34. Starnes Jr., W.H.: Structural and mechanistic aspects of the thermal degradation of poly(vinyl chloride). Progr. Polym. Sci. **27**, 2133–2170 (2002)
35. Merlin, L.M., Sivasankar, B.: Synthesis and characterization of semi-interpenetrating polymer networks using biocompatible polyurethane and acrylamide monomer. Eur. Polym. J. **45**, 165–170 (2009)
36. Mathew, A.P., Packirisamy, S., Thomas, S.: Studies on the thermal stability of natural rubber/polystyrene interpenetrating polymer networks: thermogravimetric analysis. Polym. Degrad. Stab. **72**, 423–439 (2001)
37. Boonpoo-nga, R., Sriring, M., Nijpanich, S., Wongbuth, L., Martwiset, S.: Semi-interpenetrating polymer networks of poly(4-styrenesulfonic acid) and poly(acrylic acid) for fuel cell applications. KKU Res. J. **16**(7), 757–763 (2011)
38. Huang, C.-C., Yang, M.-S., Liang, M.: Synthesis of new thermosetting poly(2,6-dimethyl-1,4-phenylene oxide)s containing epoxide pendant groups. J. Polym. Sci. A Polym. Chem. **44**, 5875–5886 (2006)
39. Takayama, S., Mathubara, T., Arai, T., Takedo, K.: Rearrangement of the main-chain and subsequent thermal degradation of polyphenylene-ether. Polym. Degrad. Stab. **50**(3), 277–284 (1995)
40. Pearce, E.M., Liepins, R.: Flame retardants. Environ. Health Perspect. **11**, 59–70 (1975)
41. Alamri, H., Low, I.M., Alothman, Z.: Mechanical, thermal and microstructural characteristics of cellulose fibre reinforced epoxy/organoclay nanocomposites. Compos. B Eng. **43**, 2762–2771 (2012)
42. Shih, Y.F.: Mechanical and thermal properties of waste water bamboo husk fiber reinforced epoxy composites. Mater. Sci. Eng. A **445–446**, 289–295 (2007)
43. De Rosa, I.M., Santulli, C., Sarasini, F.: Mechanical and thermal characterization of epoxy composites reinforced with random and quasi-unidirectional untreated *Phormium tenax* leaf fibers. Mater. Des. **31**, 2397–2405 (2010)
44. Azwa, Z.N., Yousif, B.F.: Characteristics of kenaf/epoxy composites subjected to thermal degradation. Polym. Degrad. Stab. **98**, 2752–2759 (2013)
45. Manfredi, L.B., Rodríguez, E.S., Przybylak, M.W., Vázquez, A.: Thermal degradation and fire resistance of unsaturated polyester, modified acrylic resins and their composites with natural fibres. Polym. Degrad. Stab. **91**, 255–261 (2006)

46. Xu, S., Girouard, N., Schueneman, G., Shofner, M.L., Meredith, J.C.: Mechanical and thermal properties of waterborne epoxy composites containing cellulose nanocrystals. Polymer **54**, 6589–6598 (2013)
47. Singha, A.S., Rana, A.K., Jarial, R.K.: Mechanical, dielectric and thermal properties of *Grewia optiva* fibers reinforced unsaturated polyester matrix based composites. Mater. Des. **51**, 924–934 (2013)
48. Hameed, N., Sreekumar, P.A., Francis, B., Yang, W., Thomas, S.: Morphology, dynamic mechanical and thermal studies on poly(styrene-co-acrylonitrile) modified epoxy resin/glass fibre composites. Compos. A Appl. Sci. Manuf. **38**, 2422–2432 (2007)
49. Alonso, M.V., Auad, M.L., Nutt, S.: Short-fiber-reinforced epoxy foams. Compos. A Appl. Sci. Manuf. **37**, 1952–1960 (2006)
50. Daoa, D.Q., Luche, J., Richard, F., Rogaume, T., Bourhy-Weber, C., Ruban, S.: Determination of characteristic parameters for the thermal decomposition of epoxy resin/carbon fibre composites in cone calorimeter. Int. J. Hydrogen Energy **38**, 8167–8178 (2013)
51. Régnier, N., Fontaine, S.: Determination of the thermal degradation kinetic parameters of carbon fibre reinforced epoxy using TG. J. Therm. Anal. Calorim. **64**, 789–799 (2001)
52. Pervin, F., Zhou, Y., Rangari, V.K., Jeelani, S.: Testing and evaluation on the thermal and mechanical properties of carbon nano fiber reinforced SC-15 epoxy. Mater. Sci. Eng. A **405**, 246–253 (2005)
53. Seki, Y., Sever, K., Sarikanat, M., Sakarya, A., Elik, E.: Effect of huntite mineral on mechanical, thermal and morphological properties of polyester matrix. Compos. B Eng. **45**, 1534–1540 (2013)
54. Sun, Y., Zhang, Z., Moon, K.S., Wong, C.P.: Glass transition and relaxation behavior of epoxy nanocomposites. J. Polym. Sci. B Polym. Phys. **42**, 3849–3858 (2004)
55. Bikiaris, D.: Can nanoparticles really enhance thermal stability of polymers? Part II: an overview on thermal decomposition of polycondensation polymers. Thermochim. Acta **523**, 25–45 (2011)
56. Preghenella, M., Pegoretti, A., Migliaresi, C.: Thermo-mechanical characterization of fumed silica-epoxy nanocomposites. Polymer **46**, 12065–12072 (2005)
57. Lakshmi, M.S., Narmadha, B., Reddy, B.S.R.: Enhanced thermal stability and structural characteristics of different MMT-Clay/epoxy-nanocomposite materials. Polym. Degrad. Stab. **93**, 20125–45213 (2008)
58. Saad, G.R., Elhamid, E.E.A., Elmenyawy, S.A.: Dynamic cure kinetics and thermal degradation of brominated epoxy resin-organoclay based nanocomposites. Thermochim. Acta **524**, 186–193 (2011)
59. Narteh, A.T., Hosur, M., Triggs, E., Jeelani, S.: Thermal stability and degradation of diglycidyl ether of bisphenol A epoxy modified with different nanoclays exposed to UV radiation. Polym. Degrad. Stab. **98**, 759–770 (2013)
60. Brnardic, I., Macan, J., Ivankovic, H., Ivankovic, M.: Thermal degradation kinetics of epoxy/organically modified montmorillonite nanocomposites. J. Appl. Polym. Sci. **107**, 1932–1938 (2008)
61. Carrasco, F., Pages, P.: Thermal degradation and stability of epoxy nanocomposites: influence of montmorillonite content and cure temperature. Polym. Degrad. Stab. **93**, 1000–1007 (2008)
62. Jiang, W., Chen, S.H., Chen, Y.: Nanocomposites from phenolic resin and various organo-modified montmorillonites: preparation and thermal stability. J. Appl. Polym. Sci. **102**, 5336–5343 (2006)
63. Pranger, L.A., Nunnery, G.A., Tannenbaum, R.: Mechanism of the nanoparticle-catalyzed polymerization of furfuryl alcohol and the thermal and mechanical properties of the resulting nanocomposites. Compos. B Eng. **43**, 1139–1146 (2012)
64. Zhang, Z., Ye, G., Toghiani, H., Pittman Jr, C.U.: Morphology and thermal stability of novolac phenolic resin/clay nanocomposites prepared via solution high-shear mixing. Macromol. Mater. Eng. **295**, 923–933 (2010)

65. Ingram, S.E., Liggat, J.J., Pethrick, R.A.: Properties of epoxy nanoclay system based on diaminodiphenyl sulfone and diglycidyl ether of bisphenol F: influence of post cure and structure of amine and epoxy. Polym. Int. **56**, 1029–1034 (2007)
66. Park, J., Jana, S.C.: Adverse effects of thermal dissociation of alkyl ammonium ions on nanoclay exfoliation in epoxy-clay systems. Polymer **45**, 7673–7679 (2004)
67. Arasa, M., Pethrick, R.A., Mantecón, A., Serra, A.: New thermosetting nanocomposites prepared from diglycidyl ether of bisphenol and γ-valerolactone initiated by rare earth triflate initiators. Eur. Polym. J. **46**, 5–13 (2010)
68. Chongqing, Y., Shunping, L., Jianying, Y., Yong, N., Congcong, F., Hua, W., Yufeng, C.: Preparation and thermal properties of phenolic resin/organic expanded vermiculite nanocomposites. Adv. Chem. Lett. **1**, 51–55 (2013)
69. Nohales, A., Solar, L., Porcar, I., Vallo, C.I., Gómez, C.M.: Morphology, flexural, and thermal properties of sepiolite modified epoxy resins with different curing agents. Eur. Polym. J. **42**, 3093–3101 (2006)
70. Zhang, Y., Shen, J., Li, Q., Pang, L., Zhang, Q., Xu, Z., Yeung, K.W.K., Yi, C.: Synthesis and characterization of novel hyperbranched polyimides/attapulgite nanocomposites. Compos. A Appl. Sci. Manuf. **55**, 161–168 (2013)
71. Ollier, R., Rodriguez, E., Alvarez, V.: Unsaturated polyester/bentonite nanocomposites: influence of clay modification on final performance. Compos. A Appl. Sci. Manuf. **48**, 137–143 (2013)
72. Zabihi, O., Ghasemlou, S.: Nano-CuO/epoxy composites: thermal characterization and thermo-oxidative degradation. Int. J. Polym. Anal. Charact. **17**, 108–121 (2012)
73. Guo, Z., Liang, X., Pereira, T., Scaffaro, R., Hahn, H.T.: CuO nanoparticle filled vinyl-ester resin nanocomposites: fabrication, characterization and property analysis. Compos. Sci. Technol. **67**, 2036–2044 (2007)
74. Guo, Z., Lei, K., Li, Y., Ng, H.W., Prikhodko, S., Hahn, H.T.: Fabrication and characterization of iron oxide nanoparticles reinforced vinyl-ester resin nanocomposites. Compos. Sci. Technol. **68**, 1513–1520 (2008)
75. Zabihi, O., Hooshafza, A., Moztarzadeh, F., Payravand, H., Afshar, A., Alizadeh, R.: Isothermal curing behavior and thermo-physical properties of epoxy-based thermoset nanocomposites reinforced with Fe_2O_3 nanoparticles. Thermochim. Acta **527**, 190–198 (2012)
76. Asiri, A.M., Hussein, M.A., Abu-Zied, B.M., Hermas, A.E.A.: Effect of $NiLa_xFe_{2-x}O_4$ nanoparticles on the thermal and coating properties of epoxy resin composites. Compos. B Eng. **51**, 11–18 (2013)
77. Chatterjee, A., Islam, M.S.: Fabrication and characterization of TiO_2–epoxy nanocomposite. Mater. Sci. Eng. A **487**, 574–585 (2008)
78. Omrani, A., Afsar, S., Safarpour, M.A.: Thermoset nanocomposites using hybrid nano TiO_2–SiO_2. Mater. Chem. Phys. **122**, 343–349 (2010)
79. Guigo, N., Mija, A., Zavaglia, R., Vincent, L., Sbirrazzuoli, N.: New insights on the thermal degradation pathways of neat poly(furfuryl alcohol) and poly(furfuryl alcohol)/SiO_2 hybrid materials. Polym. Degrad. Stab. **94**, 908–913 (2009)
80. Schutz, M.R., Sattler, K., Deeken, S., Klein, O., Adasch, V., Liebscher, C.H., Glatzel, U., Senker, J., Breu, J.: Improvement of thermal and mechanical properties of a phenolic resin nanocomposite by in situ formation of silsesquioxanes from a molecular precursor. J. Appl. Polym. Sci. **117**, 2272–2277 (2010)
81. Nagendiran, S., Alagar, M., Hamerton, I.: Octasilsesquioxane-reinforced DGEBA and TGDDM epoxy nanocomposites: Characterization of thermal, dielectric and morphological properties. Acta Mater. **58**, 3345–3356 (2010)
82. Wang, Y., Liu, F., Xue, X.: Synthesis and characterization of UV-cured epoxy acrylate/POSS nanocomposites. Prog. Org. Coat. **76**, 863–869 (2013)
83. Zhang, Y., Lee, S., Yoonessi, M., Liang, K., Pittman, C.U.: Phenolic resin-trisilanolphenyl polyhedral oligomeric silsesquioxane (POSS) hybrid nanocomposites: structure and properties. Polymer **47**, 2984–2996 (2006)

84. Aflori, M., Simionescu, B., Bordianu, I.-E., Sacarescu, L., Varganici, C.-D., Doroftei, F., Nicolescu, A., Olaru, M.: Silsesquioxane-based hybrid nanocomposites with methacrylate units containing titania and/or silver nanoparticles as antibacterial/antifungal coatings for monumental stones. Mater. Sci. Eng. B Solid–State Mater. Adv. Technol. **178**(19), 1339–1346 (2013)
85. Bazzar, M., Ghaemy, M.: 1,2,4-Triazole and quinoxaline based polyimide reinforced with neat and epoxide-end capped modified SiC nanoparticles: Study thermal, mechanical and photophysical properties. Compos. Sci. Technol. **86**, 101–108 (2013)
86. Ma, P.C., Siddiqui, N.A., Marom, G., Kim, J.K.: Dispersion and functionalization of carbon nanotubes for polymer-based nanocomposites: a review. Compos. A Appl. Sci. Manuf. **41**, 1345–1367 (2010)
87. Zhou, Y., Pervin, F., Lewis, L., Jeelani, S.: Experimental study on the thermal and mechanical properties of multi-walled carbon nanotube-reinforced epoxy. Mater. Sci. Eng. A **452–453**, 657–664 (2007)
88. Ma, P.C., Kim, J.K., Tang, B.Z.: Effects of silane functionalization on the properties of carbon nanotube/epoxy nanocomposites. Compos. Sci. Technol. **67**, 2965–2972 (2007)
89. Ciecierska, E., Boczkowska, A., Kurzydlowski, K.J., Rosca, I.D., Hoa, S.V.: The effect of carbon nanotubes on epoxy matrix nanocomposites. J. Therm. Anal. Calorim. **111**, 1019–1024 (2013)
90. Loosa, M.R., Coelhoa, L.A.F., Pezzina, S.H., Amicob, S.C.: Effect of carbon nanotubes addition on the mechanical and thermal properties of epoxy matrices. Mater. Res. **11**, 347–352 (2008)
91. Yang, K., Gu, M.: The Effects of triethylenetetramine grafting of multi-walled carbon nanotubes on its dispersion, filler-matrix interfacial interaction and the thermal properties of epoxy nanocomposites. Polym. Eng. Sci. **49**, 2158–2167 (2009)
92. Kuan, C.F., Chen, W.J., Li, Y.L., Chen, C.H., Kuan, H.C., Chiang, C.L.: Flame retardance and thermal stability of carbon nanotube epoxy composite prepared from sol–gel method. J. Phys. Chem. Sol. **71**, 539–543 (2010)
93. Han, C., Gu, A., Liang, G., Yuan, L.: Carbon nanotubes/cyanate ester composites with low percolation threshold, high dielectric constant and outstanding thermal property. Compos. A Appl. Sci. Manuf. **41**, 1321–1328 (2010)
94. Cui, J., Yan, Y., Liu, J., Wu, Q.: Phenolic resin-MWNT nanocomposites prepared through an in situ polymerization method. Polym. J. **40**, 1067–1073 (2008)
95. Liu, L., Ye, Z.: Effects of modified multi-walled carbon nanotubes on the curing behavior and thermal stability of boron phenolic resin. Polym. Degrad. Stab. **94**, 1972–1978 (2009)
96. Bafekrpour, E., Simon, G.P., Naebe, M., Habsuda, J., Yang, C., Fox, B.: Preparation and properties of composition-controlled carbon nanofiber/phenolic nanocomposites. Compos. B Eng. **52**, 120–126 (2013)
97. Faraz, M.I., Bhowmik, S., De Ruijter, C., Laoutid, F., Benedictus, R., Dubois, Ph, Page, J.V.S., Jeson, S.: Thermal, morphological, and mechanical characterization of novel carbon nanofiber-filled bismaleimide composites. J. Appl. Polym. Sci. **117**, 2159–2167 (2010)
98. Potts, J.R., Dreyer, D.R., Bielawski, C.W., Ruoff, R.S.: Graphene-based polymer nanocomposites. Polymer **52**, 5–25 (2011)
99. Zhang, X., Alloul, O., He, Q., Zhu, J., Verde, M.J., Li, Y., Wei, S., Guo, Z.: Strengthened magnetic epoxy nanocomposites with protruding nanoparticles on the graphene nanosheets. Polymer **54**, 3594–3604 (2013)
100. Wang, X., Jin, J., Song, M.: An investigation of the mechanism of graphene toughening epoxy. Carbon **65**, 324–333 (2013)
101. Li, Y., Pan, D., Chen, S., Wang, Q., Pan, G., Wang, T.: In situ polymerization and mechanical, thermal properties of polyurethane/graphene oxide/epoxy nanocomposites. Mater. Des. **47**, 850–856 (2013)
102. Lin, Q., Qu, L., Lü, Q., Fang, C.: Preparation and properties of graphene oxide nanosheets/cyanate ester resin composites. Polym. Test. **32**, 330–337 (2013)

103. Lungu, A., Florea, N.M., Iovu, H.: Dimethacrylic/epoxy interpenetrating polymer networks including octafunctional POSS. Polymer **53**, 300–307 (2012)
104. Jia, Q.M., Zheng, M.S., Chen, H.X., Shen, R.J.: Morphologies and properties of polyurethane/epoxy resin interpenetrating network nanocomposites modified with organoclay. Mater. Lett. **60**, 1306–1309 (2006)
105. Chen, S., Wang, Q., Wang, T., Pei, X.: Preparation, damping and thermal properties of potassium titanate whiskers filled castor oil-based polyurethane/epoxy interpenetrating polymer network composites. Mater. Des. **32**, 803–807 (2011)
106. Chen, S., Wang, Q., Wang, T.: Damping, thermal, and mechanical properties of carbon nanotubes modified castor oil-based polyurethane/epoxy interpenetrating polymer network composites. Mater. Des. **38**, 47–52 (2012)
107. Wu, X., He, G., Gu, S., Hu, Z., Yao, P.: Novel interpenetrating polymer network sulfonated poly(phthalazinone ether sulfone ketone)/polyacrylic acid proton exchange membranes for fuel cell. J. Membr. Sci. **295**, 80–87 (2007)
108. Banerjee, S., Ray, S., Maiti, S., Sen, K.K., Bhattacharyya, U.K., Kaity, S., Ghosh, A.: Interpenetrating polymer network (IPN): a novel biomaterial. Int. J. Appl. Pharm. **2**(1), 28–34 (2010)
109. Patel, J.M., Savani, H.D., Turakhiya, J.M., Akbari, B.V., Goyani, M., Raj, H.A.: Interpenetrating polymer network: a novel approach for controlled drug delivery. UJP **1**(1), 1–11 (2012)
110. Shivashankar, M., Mandal, B.K.: A review on interpenetrating polymer network. Int. J. Phram. Phram. Sci. **4**(5), 1–7 (2012)
111. Paduraru, O.M., Ciolacu, D., Darie, R.N., Vasile, C.: Synthesis and characterization of polyvinyl alcohol/cellulose cryogels and their testing as carriers for a bioactive component. Mater. Sci. Eng. C **32**, 2508–2515 (2012)
112. Varganici, C.-D., Paduraru, O.M., Rosu, L., Rosu, D., Simionescu, B.C.: Thermal stability of some cryogels based on poly(vinyl alcohol) and cellulose. J. Anal. Appl. Pyrol. **104**, 77–83 (2013)
113. Gibson, S.L., Walls, H.J., Kennedy, S.B., Welsh, E.R.: Reaction kinetics and gel properties of blocked diisocyanate crosslinked chitosan hydrogels. Carbohydr. Polym. **54**, 193–199 (2003)
114. Zeng, M., Fang, Z., Xu, C.: Effect of compatibility on the structure of the microporous membrane prepared by selective dissolution of chitosan/synthetic polymer blend membrane. J. Membr. Sci. **230**, 175–181 (2004)
115. Zeng, M., Fang, Z.: Preparation of sub-micrometer porous membrane from chitosan/polyethylene glycol semi-IPN. J. Membr. Sci. **245**, 95–102 (2004)
116. Welsh, E.R., Schauer, C.L., Qadri, S.B., Price, R.R.: Chitosan crosslinking with a water-soluble, blocked diisocyanate. 1. Solid state. Biomacromolecules **3**, 1370–1374 (2002)
117. Rodkate, N., Wichai, U., Boontha, B., Rutnakornpituk, M.: Semi-interpenetrating polymer network hydrogels between polydimethylsiloxane/polyethylene glycol and chitosan. Carbohydr. Polym. **81**, 617–625 (2010)
118. Dinu, M.V., Cazacu, M., Dragan, E.S.: Mechanical, thermal and surface properties of polyacrylamide/dextran semi-interpenetrating network hydrogels tuned by the synthesis temperature. Cent. Eur. J. Chem. **11**(2), 248–258 (2013)
119. Grishchuk, S., Karger-Kocsis, J.: Hybrid thermosets from vinyl ester resin and acrylated epoxidized soybean oil (AESO). Express Polym. Lett. **5**(1), 2–11 (2011)
120. Wang, R., Schuman, T.P.: Vegetable oil-derived epoxy monomers and polymer blends: a comparative study with review. Express Polym. Lett. **7**(3), 272–292 (2013)
121. Mustata, F., Tudorachi, N., Rosu, D.: Curing and thermal behavior of resin matrix for composites based on epoxidized soybean oil/diglycidyl ether of bisphenol A. Compos. B Eng. **42**, 1803–1812 (2011)
122. Harrats, C., Mekhilef, N.: Cocontinuous phase morphologies: predictions, generation and practical applications. In: Harrats, C., Thomas, S., Groeninckx, G. (eds.) Micro- and Nanostructured Multiphase Polymer Blend Systems: Phase Morphology and Interfaces, p. 124. Taylor & Francis Group, USA (2006)

Thermal Degradation of Thermosetting Nanocomposites

Matheus Poletto, Heitor L. Ornaghi Júnior and Ademir J. Zattera

Abstract The rapidly-developing nanotechnology and nanoscience in recent years, led to an increase of polymer-based nanocomposites focused on the enhancement of specific properties. Among these, the effect of different nanoparticles, their dispersion and the use of modifiers on the polymer thermal stability are studied. New materials with pre-tailored properties can be manufactured by the incorporation nanoreinforced materials and by the knowledge of the steps involved in their obtention. So, this work focuses on the thermal degradation study of thermosetting nanocomposites materials, evaluating their effect in thermal stability and in thermal degradation steps. Also, the thermal applications of these nanocomposites were also evaluated and the challenges to the nanocomposites field in the following years were also discussed.

1 Introduction

The use of organic and inorganic reinforcements as vegetal fibers, clays and inorganic powder (as barium sulfate) has become very common in polymeric systems [1–13]. This can be mainly attributed to reduce the cost of the end-products. Polymer composites are manufactured commercially for many diverse applications such as sport engineering [14] aerospace components [15] and automobile industry [7, 16, 17]. These applications are due among other characteristics, intrinsic lightweight and

M. Poletto (✉)
Center of Exact Sciences, Nature and Technology (CENT),
Caxias do Sul University (UCS), Rua Francisco Getúlio Vargas,
1130 Caxias do Sul/RS, Brazil
e-mail: matheus.poletto@hotmail.com; mpolett1@ucs.br

H.L. Ornaghi Júnior
Engineering and Materials Science Graduate Program (PGMAT),
Caxias do Sul University (UCS), 95070-490 Caxias do Sul/RS, Brazil

A.J. Zattera
Laboratory of Polymers (LPOL), Caxias do Sul University (UCS),
Rua Francisco Getúlio Vargas, 1130 Caxias do Sul/RS, Brazil

P.M. Visakh and Y. Arao (eds.), *Thermal Degradation of Polymer Blends, Composites and Nanocomposites*, Engineering Materials,
DOI 10.1007/978-3-319-03464-5_3

tailor-ability of composites that clearly promoted their rapid utilization [18, 19]. However, there are concerns related to their overall durability particularly when under load and exposed to harsh and changing environmental conditions [20, 21]. Besides, some properties cannot be reached in using reinforcement in the micro/macro scale.

With the rapid development of nanotechnologies and nanomaterials since 1990s, the studies on polymer-based nanocomposites have been extensively focused on the enhancement of their properties. Among these, it is focused on highlighting the effect of different nanoparticles, their dispersion and the used modifiers on polymer thermal stability. The whole range of polycondensation polymer matrices has reactive end groups which can interact with inorganic nanoparticles surface. Hydrogen or covalent bonds can be formed, which can increase the adhesion of nanoparticles with the polymer matrix, resulting in higher dispersion degrees. This, in most cases, leads to substantial enhancement in thermal decomposition properties. Only in nanocomposites containing montmorillonite there are conflicting results and accelerated degradation process were also reported. Organoclays also have similar effect on polymers thermal stability and in this case the achieved clay dispersion (intercalated-exfoliated), as well the used modifier, can alter the thermal decomposition of polymers. The used amount of nanoparticles plays an important role on the thermal stability of nanocomposites. In most cases thermal stability enhancement takes place at low loading (4–5 wt%) of nanoparticles, while at higher contents the stabilization becomes progressively smaller [22].

Nanocomposites materials can be defined as composite materials, that combine one or more separate components in order to improve performance properties, for which at least one dimension of the dispersed particles is in the nanometer range [23]. In polymer clay nanocomposites, nanoscale particles have typically 10–100 nm in size [24]. Depending on the reinforcement, remarkable improvement in material properties when compared with neat polymer or conventional micro and macro-composites can be obtained. These improvements can include high moduli, increased strength and heat resistance, decreased gas permeability and flammability, and increased biodegradability of biodegradable polymers [25].

Among all the nanocomposite precursors, those based on clay and layered silicates have been most widely investigated, probably because the starting clay materials are easily available and because their intercalation chemistry has been studied for a long time [26]. Besides clay [2, 9], it can be cited other examples of nanocomposite as reinforcements (some of them previously cited) such as polyhedral oligomeric silsesquioxanes (POSS) [10], nanodiamonds [27] and carbon nanotubes [28]. Simultaneously, the rapid growth of the computer technology has made it easier to characterize and predict the properties at the nanoscale via modeling and simulation [29]. In general, the unique combination of the nanomaterial's characteristics, such as size, mechanical properties, and low concentration are necessary to promote changes in a polymer matrix, coupled with the advanced characterization and simulation techniques now available have generated much interest in the field of nanocomposites. In addition, many polymer nanocomposites can be fabricated and processed in ways similar to that of conventional polymer composites, making them particularly attractive from a manufacturing point of view [30].

Two terms are used to describe the two general classes of nano-morphology that can be prepared: intercalated and delaminated (exfoliated). Intercalated structures are well ordered multi-layered structures where the extended polymer chains are inserted into the gallery space between the individual silicate layers. The delaminated (exfoliated) structures result when the individual silicate layers are no longer close enough to interact with the adjacent layers' gallery cations. In the latter, the interlayer spacing can be on the order of the radius of gyration of the polymer; therefore, the silicate layers may be considered well dispersed in the organic polymer. Furthermore, the silicate layers in a delaminated structure may not be as well ordered as in an intercalated structure. X-ray diffraction measurements are used to characterize the nanostructures. Reflections in the low angle region indicate the d-spacing (basal spacing) of ordered intercalated and ordered delaminated nanocomposites; disordered delaminated nanocomposites show no peaks in this region due to the loss of structural registry of the layers and the large d-spacings (>10 nm) [30].

This chapter has as objective to evaluate the thermal degradation of thermosetting nanocomposite and the effect of the reinforcement on the thermal stability of the nanocomposites.

2 Thermal Degradation of Thermosetting Nanocomposites

Degradation is a process where the deterioration in the properties of the polymer takes place due to different factors like ultraviolet (UV) radiation, thermal energy or mechanical loading. As a consequence of degradation, the resulting smaller fragments do not contribute effectively to the mechanical properties, the component becomes brittle and the life of the material becomes limited. Thus, any polymer or its composite, which is to be used in outdoor applications, for example, must be highly resistant to all environmental conditions [31]. There is a well-established mechanism of polymer degradation as well as their stabilization. The study of degradation and stabilization of polymers is an extremely important area from a scientific and industrial point of view and a better understanding of polymer degradation will ensure the long life of the product.

More specifically, Fig. 1 shows some properties commonly used for evaluating the polymer degradation. Several techniques were used for characterization of the polymer degradation. Among the mechanical properties characterization that can be obtained, it can be cited dynamic mechanical analysis, where the influence of the mechanical properties, such as the flexural and impact strengths can be directly related with the storage modulus and the tan delta dissipation energy [4, 8]. The percentage degradation can be estimated by the weight loss (from TGA) [5], for example. Scanning electron microscopy (SEM) or transmission electron microscopy (TEM) [9] and X-ray diffraction [2] can be used when a physical property is required. Finally, in chemical properties, functional changes can be measured by FTIR or UV spectroscopy [1, 13] and molecular weight changes by GPC and viscosity [25].

There is a well-established mechanism of polymer degradation as well as their stabilization, as can be seen in Fig. 2. The study of degradation and stabilization of

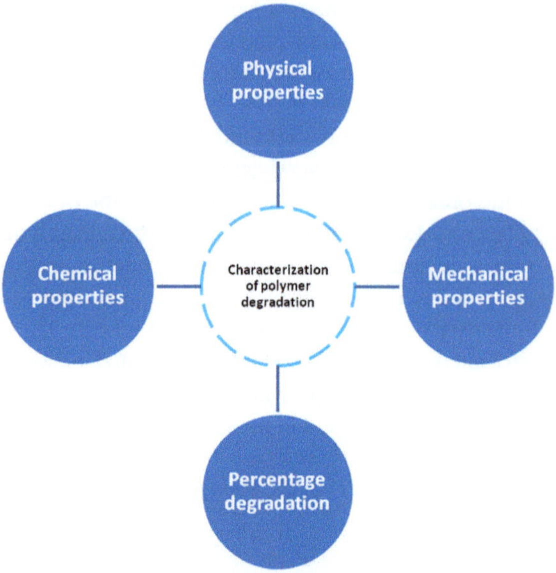

Fig. 1 Different properties commonly used for evaluate the polymer degradation. Adapted from [31]

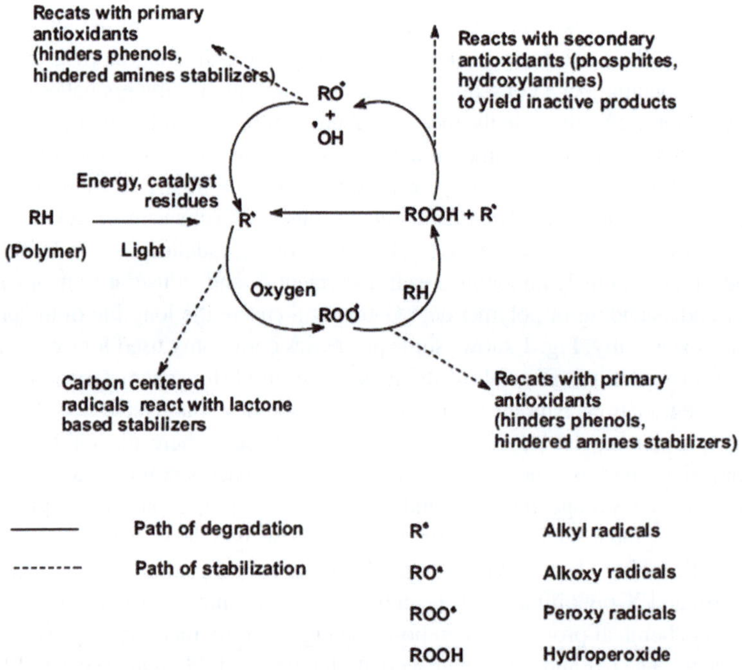

Fig. 2 Different pathways of degradation and stabilization in polymeric materials. Adapted from [31]

polymers is an extremely important area from the scientific and industrial point of view and a better understanding of polymer degradation will ensure the long life of the product and leads to the development of more efficient stabilizers.

The scheme above is based on an original scheme for autocatalytic oxidation of hydrocarbons. The oxidation of polymers begins during processing (mechano-oxidation), and the hydroperoxides formation during fabrication affects further the rate of thermal/photo-oxidation during subsequent use (aging and weathering). So, degradation of polymers is like a double-edged sword: it has harmful aspects as well as beneficial aspects. If unchecked it can play havoc with a polymers performance, if uncontrolled it can lead to safety hazards of fire and toxicity, but if properly harnessed it can be used for producing new and better materials.

For various applications, the service life of polymeric materials can be tuned by incorporating additives externally. In order to induce and promote oxidation of polymers, the sensitizers, which transfer the energy to polymer or decompose itself to give single oxygen molecule which can initiate the oxidation are added. Certain organic and inorganic compounds are excited in presence of light, heat, high-energy radiations, and transfer energy to polymers or free radicals [31].

3 Thermal Stability of Thermosetting Nanocomposites

Stability can be said as the protection of polymeric materials from which lead to deterioration of properties [31]. In literature, there are different and sometimes contradictory reported papers [31–38] concerning the effect of the nanoparticles on polymer thermal stability. There are papers suggesting that nanoparticles have no obviously effect on thermal stability, some of them suggested a small to substantial enhancement and some others suggested acceleration of thermal decomposition. Nanocomposites based on clay or with other nanoparticles with a larger amount of hydroxyl groups can exhibited a much more pronounced degradation because the hydroxyl groups acted as Bronzed acidic sites and accelerated the polymer's degradation. It was suggested that during thermal decomposition in nitrogen the clay can slow down degradation of polymer as a mass transport protective barrier, but the catalytic effect of the metal derivatives in the clays could accelerate the decomposition behavior of a polymer. The combination of these two effects determined the final thermal stability of the nanocomposite. Furthermore, some nanoparticles have reactive groups such as −COOH and −OH, which can also accelerate the decomposition of polymers or to form covalent bonds leading to their thermal degradation enhancement [32]. Thus, the main question is what is the really effect of nanoparticles on polymer thermal stability?

In a study performed by Ollier et al. [33], the author incorporated 5 % weight of bentonite in unsaturated polyester (UP) matrix. They noted that the addition of bentonite increases the thermal stability of the UP resin. Bharadwaj et al. [34] showed that the incorporation of commercially modified MMT to UP was detrimental for the thermal stability of the nanocomposites. The most important finding is that there was the formation of a nanocomposite with decreasing of the tensile modulus, storage and loss moduli as clay content was incorporated. In addition, the rate of thermal degradation was slightly but progressively hastened upon formation of a nanocomposite. These trends have been explained on the basis of a progressive decrease in the degree of crosslinking with increasing clay concentration. Becker et al. [35] also observed a slight decreased in the onset temperature (in the order of 5–10 °C) of the degradation process of epoxy resin by increasing organoclay content (at a clay concentration of 10 %). The changes in thermal stability are of very low significance and it is unlikely that they would be considered as a drawback to any possible industrial application. Carrasco and Pagès [36] showed that, at low clay contents (up to 5 wt%) the addition of clay had no effect on the thermal stability of the epoxy matrix, whereas for higher concentration (10 wt%) a clear increase on this parameter was observed. Other authors demonstrated that the thermal stability of epoxy based nanocomposite is dependent upon the dispersion of the organoclay in the epoxy matrix but that all the epoxy nanocomposites had enhanced thermal stability compared with the neat epoxy resin. In addition, Lakshimi et al. [37] reported an improvement in the thermal stability of epoxy resins with the incorporation of MMT with the same chemical treatments used in this work. The author incorporated two different modified clays with four different epoxies using 5 wt% modified clay. The SAXS and SEM characterization reveals that the modified clay is highly exfoliated and uniformly dispersed in the matrix. There was an increase in the thermal stability observed in all modified systems. This was attributed to the nanolayer acted as a barrier for volatile degradation of polymer matrix. Saitoh et al. [38] found that the phosphonium cations used to obtain the organoclays influenced the thermal resistance of the resulting epoxy/clay nanocomposites. The explanation for this behavior is that the dispersed MMT-Clay nanolayers can act as barrier protecting the epoxy polymer matrix degradation gaseous products from volatilizing.

For the majority of polymers, due to their hydrophobic character, the clay must be modified with a surfactant in order to make the gallery space sufficiently organophilic to permit it to interact with the polymer. In fact, several factors were found to govern the thermal stability of nanocomposite materials, such as intrinsic thermal resistance of polymer matrix, nanofiller content, chemical constitution of organic modifier and chemical character of polar compatibilizers as well as an access of oxygen to composite material during heating. For surface modification of clay, the surfactant is usually described as an 'onium' salt. The quaternary ammonium ion is nominally chosen to compatibilize the layered silicate with a given polymer resin. However, the molecular structure (length and number of alkyl chains and unsaturations) is also the determining factor of the thermal stability of the polymer/MMT nanocomposites [39]. The amount of surfactants with multiple alkyl tails has greater thermal stability than those with a single alkyl tail. It has been proposed that the organic modifiers starting at a temperature around 200 °C, and the small molecular weight organics are released first while the high-molecular weight organic species are still trapped by organically modified layered silicates (OLS) matrix. With the increase of temperature, the high-molecular organic polymer chains may still exist between the interlayers until the temperature is high enough to lead to their further decomposition. The incorporation of silicate layers with high-aspect ratio decomposed/charred material on the clay surface act as carbonaceous insulators. The silicate has an excellent barrier property that prevents against permeation of various degraded gaseous products. The addition of clay enhanced the performed by acting as a superior insulator and mass transport barrier to the volatile products generated during decomposition. The clay acts as a heat barrier, which could enhance the overall thermal stability of the system, as well as assisting in the formation of char during thermal decomposition. In the case of the nanocomposite, the temperature at which volatilization occurs, increases as compared of the micro-composite. Moreover, the thermal oxidation process of the polymer is strongly slowed down in the nanocomposite with high-char yield both by a physical barrier effect, enhanced by ablative reassembling of the silicate, and by a chemical catalytic action due to the silicate and to the strongly acid sites created by thermal decomposition of the protonated amine silicate modifier. The polymers that show good fire retardancy upon nanocomposite formation exhibit significant intermolecular reactions, such as inter-chain aminolysis/acidolysis, radical recombination and hydrogen abstraction. In the case of the polymers that degrade through a radical pathway, the relative stability of the radical is the most important for the prediction of the effect that nanocomposite formation has on the reduction in the peak heat release rate. The more stable is the radical produced by the polymer; the better is the fire retardancy, as measured by the reduction in the peak release rate, of the polymer/clay nanocomposite [40, 41].

4 Different Step for Thermal Degradation of Thermosetting Nanocomposites

The particles can behave in a different way on the polymers' thermal stability. In some reported studies it has been shown that the introduction of clays into polymeric matrices can accelerate the thermal decomposition of the polymer matrix, consisted by condensation polymers, due to the catalysis effect of water in MMT and hydroxyl groups on the clay platelets. Meanwhile, some metallic derivatives that can be found in clays interspaces have significant catalytic activity on thermal degradation, and reduce the thermal stability of nanocomposites. Furthermore, most of the used organic modifiers for clays have low thermal stability that condensation polymers, like alipharomatic polyesters or polyamides, and this can form byproducts that can accelerate thermal decomposition [42, 43]. Ammonium surfactants experience Hofmann elimination reaction, producing acidic sites which can accelerate thermal degradation of polycondensation polymers. Also, the architecture (trimethyl or dimethyl), chain length, surfactant mixture, exchanged ratio, or preconditioning (washing) does not alter the initial onset temperatures. However, these factors do affect the initial mass loss [44].

Except for these studies there are also some others, but limited which report that the addition of clays can enhance thermal stability. In these papers the improved thermal stability is probably due to the high thermal stability of clay and due to the interactions between the clay particles and the polymer matrices. Furthermore, the hindrance effect of clays was due to the formation of char layers obtained via collapsing of the nanocomposite structure and reformulation of silicate layers. This carbon silicate structure on the surface may act as a mass transport barrier that hinders volatilization of decomposed gaseous products via the degradation. However, the most probably reason to achieve a thermal stability is due to homogeneous dispersion of nanoclay platelets within the matrix that may act as thermal insulation layers, thus reducing the degraded volatile products' diffusion out of the matrix. In order to achieve this, thermally stable modifiers without the above-mentioned problems should be used. Using such modifiers intercalated or exfoliated dispersions can be prepared which cause higher stabilization than unmodified clays. In the latter case usually microcomposites can be prepared instead of nanocomposites, since it is very difficult for the clay plates to separate and be dispersed as individual plates in the polymer matrix. Furthermore, the exfoliated clay may also promote char formation under degradation, which can act as physical barriers (labyrinth effect) to the volatile products generated during decomposition [42, 43]. Except of the particle dispersion its size plays also important role, as was reveled in some studies. Particles with smaller sizes can be dispersed more homogeneously into polymer matrix and thus the barrier effect is higher.

The stabilization effect of carbon nanotubes (CNTs) is explained by a barrier effect of the nanotubes that usually agglomerate due to van der Waals force and them form aggregates, which hinder the diffusion of the degradation products from the bulk of the polymer onto gas phase [35, 45]. Furthermore, the thermal stabilization effect on carbon nanotubes could be attributed to the increased interfacial interaction between the nanofibres and polymers which lead to an increase of the degradation's activation energy. Besides, there are several preparation methods and types of functionalizations [45]. As an example, one can cite the carboxylated and fluorinated MWCNT systems. In the former, there are higher activation energy value in comparison with neat epoxy resin and the latter. This difference may be attributed to differences in nanotubes dispersion; the fluorinated MWCNT system is more uniformly dispersed in the matrix whereas the more heterogeneously dispersed carboxylated MWCNTs can hinder mobility of the reactive species and disrupt the reaction stoichiometric on the local scale [46]. The improvement thermal stability of such nanocomposites is generally attributed to: (i) the creation of a tortuous path, resulting by the nanoparticles dispersion into the polymer matrix, slowing the diffusion of the produced substances in the material, (ii) the formation of a carbonaceous char insulating the material and slowing the release of the decomposition products [35].

Except of the kind of nanoparticles, its amount is also very crucial for thermal stabilization enhancement. It was reported that at two amounts, usually 2–5 wt%, and depending the type of nanoparticles, the highest thermal stabilization can be achieved. This is due to the fine dispersion of nanoparticles into the polymer matrices and, thus, high surface area of a non-degradable material, as most of the nanoparticles used are, is available to protect the macromolecular chains from thermal decomposition. However, at higher nanoparticles content, due to the formation of aggregates, a deterioration of thermal stability was mentioned in most of the reported nanocomposites in the present review. In this case, it seems that a microcomposite is formed instead of a nanocomposite. Thus, the shielding effect of the nanoparticles is lessened.

In nitrogen, pure epoxy resin thermally degrades through a double step process with maximum rates at 301 and 414 °C. The first step of degradation involves water elimination which results in formation of C–C unsaturations. Carbon-oxygen bonds in beta positions to these unsaturations (allylic bonds) become the thermally weakest bonds in the epoxy network. They break down giving fragmentation of the crosslinked structure that eventually produces fragments small enough to evaporate at these temperatures. Volatilization is, however, limited up to 350 °C because it is contrasted by rearrangements such as cyclization that produces a relatively stable structure which breaks down in the second step of degradation. In this step extensive break down of chemical bonds of the epoxy network takes place including C-phenyl bonds of bisphenol-A, leading to almost complete volatilization. A minor amount of charred residue is formed (ca. 4 %) due to limited recombination to a thermally stable charred material of reactive degrading species during decomposition.

The incorporation of clay in the epoxy system results in increased thermal stability when nanocomposite is cured at 180 °C. As found from TGA analysis the onset of degradation is delayed and the temperature range of degradation has been influenced by the addition of clay. In contrast, however, when the nanocomposite material is post-cured at 220 °C, the addition of nanoclay resulted in a decrease in the thermal stability of the epoxy resin. This can be associated with the dissociated alkyl chains, after being subjected to the high temperature and destabilizing the thermal properties. This finding is critical for maximing the enhancements from nanoclay. All cure cycle temperatures must be carefully selected otherwise detrimental effects may be realized, even when similar levels of exfoliation and mechanical properties are achieved [47]. However, Carrasco and Pages [36] found that the thermal decomposition of cured materials was independent of cure temperature but from the content of clay. With the incorporation of crude clay, the initial temperature degradation and the maximum temperature degradation shifted towards higher temperature. The enhancement on the resin's thermal stability was more significant in the initial stage of decomposition. This behavior may be attributed to the protection of epoxy polymer chains present between hard MMT-clay nanolayers that act as a barrier protecting from volatizing the epoxy polymer matrix. The polymer networks present between the clay layers undergo the restricted segmental motion which reflected in higher thermal stability properties than the urethane modified epoxy (UME) systems.

The effects of OMMT loading on the decomposition behavior were reported from Guo et al. [48]. It was shown that the OOMT (platelets or tactoids) incorporated in the matrix improve the thermal stability of the epoxy resin. When the OMMT loading was lower than 8 phr, the samples exhibited higher thermal stability than those with higher OMMT loading. However Gu and Liang [49] found that onset degradation temperatures of epoxy/clay nanocomposites were lowered, especially in the case when longer alkyl ammonium chains are used as organo-modifiers. In comparison with the neat resin system, an almost identical or slightly decreased onset temperature of degradation for the nanocomposites was observed by Brnardić et al. [50]. The shift was most pronounced in the nanocomposite OMMT 10 wt%.

The addition of unmodified MWCNTs has a negative effect on thermal stability of the epoxy nanocomposites. This is due to the poor affinity between as-received MWCNTs and epoxy resin matrix, which increases vacancies or voids in the nanocomposite. However, the addition of amine-modified MWCNTs enhances the values of the first and second decomposition temperatures, which indicates the thermal stability improvement of the nanocomposite. The reason may be attributed to the fact that amine-modified MWCNTs have a better affinity for the polymer matrix than unmodified MWCNTs. Similar results are reports from Yang and Gu [51] using modified MWCNTs. TGA analysis indicates the thermal stability improvement of the epoxy/MWCNTs nanocomposite. The reason may be due to the fact that triethylenetetramine TETA-grafted MWCNTs possess a good affinity to the epoxy matrix. The incorporation of functionalized CNTs into resin promoted the thermal stability and flame retardancy of the epoxy.

Other nanoparticles like SiO_2, TiO_2, $CaCO_3$ and POSS were reported that can also enhance thermal stability of epoxy nanocomposites. Introduction of nanoscale silica into epoxy resins certainly improved their thermal stability and reduced their weight loss rates, and the improvement can be further enhanced with using phosphorous compound as the curing agent [52]. The activation energies of the thermal degradation reactions of the nanocomposites were calculated from various methods, and the results from the Ozawa method provided reliable data. It was shown that the presence of nanoscale silica leveled up the values of the activation energies of the degradation reactions, and phosphorus group depressed the values of E_a. The presence of POSS changes the thermal degradation of an epoxy resin from first order to a mechanism controlled by diffusion. Thermogravimetric analysis indicates that the POSS-epoxy nanocomposites display high ceramic yields, suggesting improved flame retardancy. The thermal stability of TiO_2/epoxy nanocomposites is depending on the dispersion state of the reinforcement in the matrix and it is correlated with nanoreinforcement loading (0.0015–0.006 % by volume). However, in the case when $CaCO_3$ was added as filler, a marginal decrease in initial decomposition temperature (IDT), temperature of maximum rate of weight loss (T_{max}) and the final decomposition temperature (FDT) attributed to the moisture and coating of $CaCO_3$ filler was observed [32, 53–55].

5 Effect of Reinforcement on Thermal Stability on Thermosetting Nanocomposites

Based on some reported studies, papers different and sometimes contradictory results have been mentioned concerning the effect of the nanoparticles on polymer thermal stability. There are papers suggested that nanoparticles have no obviously effect on thermal stability, some of them suggested a small to substantial enhancement and some others suggested acceleration of thermal decomposition. Thus, the main question remains what is the really effect of nanoparticles on polymer thermal stability? The biggest discrepancy can be found on the effect of clay nanoparticles [22]. There are some research papers suggesting that the addition of clays has different effect on polymers thermal stability as well as on their different thermal degradation stages [34–36, 39, 56]. This is because the clay platelets are very difficult to be dispersed in the polymer matrix and untreated clays microcomposites are formed instead of nanocomposites. Generally, the clay has two opponent effects in the thermal stability of the polymer/clay nanocomposites. The barrier effect which should improve the thermal stability and the promoter effect which would encourage the degradation process and the order is the catalysis effect towards the degradation of the polymer matrix which would decrease the thermal stability.

The drastic improvement in thermal stability would be to the different effect of nanoparticles:

- High surface volume;
- Improving barrier properties due to the clay contribution to tortuosity path and to the reduction of polymer molecular mobility;
- Low permeability and decrease in the rate of evolution of the formed volatile products;
- Formation of high performance carbonaceous silicate char on the nanoparticles surface that insulate the underline material and slows the escape of volatile products generated during the decomposition;
- Absorption of formed gasses into clay plates.

By adding a low fraction to the polymer matrix, the clay layers are well dispersed, the barrier effect is predominant, but with increasing loading, the promoter effect rapidly rises and becomes impressible, so that the thermal stability of the nanocomposites decreases. This accelerating effect is mainly due to:

- The presence of hydroxyl groups on the edges of the clay could be catalyze the polymer decomposition;
- Existence of active catalytic sites influences the polymer degradation;
- By the degradation of the clay a protonated silicate has been formed, which in turn catalyses the degradation of the polymer.

Organoclays have also similar effects on polymers thermal stability and in this case the achieved clay dispersion (intercalated-exfoliated) as well as the used modifier can also alter thermal decomposition of polymers. Cationic compounds used for organomodification of MMT have a negative effect in thermal degradation while the exfoliated structure due to the finer dispersion of the clay nanoparticles can lead to thermal stabilization than intercalated structure.

Except montmorillonite, other clays like layered double hydroxide (LDH) can enhance thermal degradation of polymers. In such nanocomposites there are no reported for accelerating effect of LDH on polymer thermal degradation. This is because LDH can be more easily dispersed in intercalated or exfoliated structures, compared with MMT, into a polymer matrix and thus the stabilization effect is higher [22].

Carbon nanotubes (MWCNTs and SWCNTs) were also in all cases reported that can increase thermal stability of polymers. This was because the CNTs have excellent thermal conductivity. The improving thermal stability of the nanocomposites was also attributed to the formation and stabilization of MWCNTs-bonded macroradicals [45].

Except of the commonly used nanoparticles, other less important, like graphite, carbon nanoparticles, Ag, TiO_2, magnesium hydroxide, $CaCO_3$, ZnO, POSS, boehmite (AlOOH), Al_2O_3, etc., were also found to have substantial stabilization effect on polymer degradation [22].

In a study performed by Bharadwaj et al. [34] the onset of degradation of polymer nanocomposites was slightly but progressively hastened upon addition of clay to the nanocomposites as compared to the neat polymer. The neat polymer is completely decomposed at 400 °C. The nanocomposites degrade at a faster rate in the temperature range 25–400 °C compared to the neat polymer and thereafter the situation reverses. The monotonic increase in the rate of degradation in the nanocomposites may be due to the presence of increasing amount of hydroxyl group in the organic modifier that provides a supply of oxygen.

The main conclusion from all the above is that except of MMT, where contradictory results were reported, in all other nanocomposites the addition of nanoparticles can lead to thermal stabilization effect of polymers during their decomposition. However, in this direction the effect of nanoparticles content is very crucial. In most of the cases the thermal stability enhancement takes place at low loading (4–5 wt%) of nanoparticles while at higher contents thermal stabilization becomes progressively lower. This is because at higher concentrations nanoparticles can form aggregates and thus effective area of nanoparticles in contact with polymer macromolecular is lower. In this case microcomposites may are formed instead of nanocomposites and thus the protective effect of nanoparticles becomes lower [22].

6 Effect of Nano Particle on Thermal Stability on Thermosetting Nanocomposites

Nanoparticles are used as reinforcement and usually increase the thermal stability in comparison to the unfilled thermoset. The clay slows the cure process because influences the nature of the network structure formed because the activated monomer mechanism favors chain transfer processes [57].

According to recent investigation, the thermal degradation of nanocomposites depends on the clay loading, structure and the nature of the ambient gas. A nanocomposite with 2 wt% loading showed one step degradation; whereas 10 wt% clay loading showed two steps and the maxima degradation were at 395 and 397 °C, indicating that degradation started at lower temperature with increasing loading of clay. There are two factors which have opposite influences on the thermal stability of epoxy-clay nanocomposites. The first factor is that the addition of clay to epoxy decreases the curing reactivity of epoxy resin. Lower reactivity of the resin generally results in lower crosslinking density of the cured resin and the longer polymer chains among the cross-linking points. It is known that for a longer polymer chain, the nanocomposites resulted are easier to degrade than the neat epoxy resin. Secondly, silicate layers have good barrier to gases such as oxygen and nitrogen, they can insulate the underlying materials and slow the mass loss rate of decomposition products [49, 58]. Moreover, exfoliated nanocomposites have better barrier properties and thermal stability than intercalated ones. In the case of intercalated nanocomposites (10 wt% clay), the first factor is dominant, whereas for an exfoliated nanocomposites (2 wt% clay), the second factor is dominant.

7 Recent Studies on Thermosetting Nanocomposites

Bafekrpour et al. [59] used carbon nanofibers (CNF)/phenolic nanocomposites coatings applied to phenolic substrate for tailoring thermal behavior under rapid temperature changes. The authors evaluated the effects of thickness and CNF content on the development nanocomposite coating on reducing temperature gradient field for the samples containing 2, 4 and 16 wt% of CNF. In the latter, they used eight layers of same thickness (0.625 mm) with interleaving layers of phenolic resin and nanocomposite coating. The plate consisted in 4 layers of pure phenolic and four layers of CNF. The results showed that the temperature gradient decreased with increasing the number of CNF/phenolic nanocomposite layers located at the bottom of the plate. Also, it was showed that the temperature gradient between the two sides of the plate, when compared to the non-coated phenolic plate with the same total thickness was 18.84 °C lower. The temperature distribution across the thickness of samples became more linear with increasing CNF content in the coating. This is probably due to higher thermal conductivity and a lower heat capacity within the coating, which results in a better heat transfer.

In another study, Bafekrpour and et al. [60] designed and fabricated functionally graded carbon nanofiber-phenolic nanocomposites (FGN) using four different gradient patterns and the same overall carbon nanofiber (CNF) content and geometry. The composition across the thickness of the samples was controlled in eight 600-mm-thick layers with clear interlayer boundaries. They observed that the incorporation of CNF into the phenolic matrix increased the thermal conductivity of the resin. For example, the thermal conductivity of the matrix increased 28.1, 49.5 and 90.4 % with the incorporation of 2, 4 and 16 wt% CNF, at room temperature, respectively. The authors attributed this thermal conductivity enhancement in the phenolic matrix is due to the formation of inter-connected mesh of CNFs.

Badrinarayanan and Kessler [61] explored the potential for development of bulk polymer nanocomposites with tailored thermal expansivity. The authors incorporated zirconium tungstate (ZrW_2O_8) nanoparticles (characterized by a negative coefficient of thermal expansion (CTE) in a low viscosity bisphenol E cyanate ester (BECy) thermosetting resin. In general, the dimensional stability of composites can be enhanced by reducing the differences in the CTE between polymer matrix and fiber. The high CTE of the polymer matrix and the low CTE of the fiber leads to development of residual stresses and matrix microcracking. A potential strategy to diminish these residual stresses involves development of polymer nanocomposites with well dispersed nanoparticles that reduce the extent of disparity in CTE. The authors verified that the incorporation of up to 10 vol% whisker-like nanoparticles results in a 20 % reduction in the CTE of the polymer matrix.

Xing et al. [62] development an UV-curing flame retardant film, which consisted of epoxy acrylate resin (EA) used as an oligomer, tri(acryloyloxyethyl) phosphate (TAEP) and triglycidyl isocyanurate acrylate (TGICA) used as flame retardants (FR). The authors also used alpha-zirconium phosphate α-Zr $(HPO_4)_2H_2O$ (α-ZrP) for reinforced the flame retardancy and improve the thermal properties of the films. They observed that the peak heat release rate (PHRR) and heat release capacity (HRC) of the nanocomposites were significantly reduced. The incorporation of 1, 3 and 5 wt% of α-ZrP reduced the PHRR in 57, 50 and 47 % respectively, meanwhile, the HRC were reduced in 58, 52 and 47 %, respectively. According to the authors, the α-ZrP nanoparticles could further retard the degradation of polymer matrix due to the dispersion of α-ZrP exfoliated or intercalated nanometer sheets. This was due the films formed were more effective protective char layer and reduce the release rate of inflammable gas.

Bora et al. [63] prepared nanocomposites based on graphene oxide (GO) and unsaturated polyester resin (UPR). They verified that the incorporation of 3 wt% of GO improved the major degradation temperature of polyester from 230 to 285 °C. This improvement in thermal stability was attributed to the strong interaction between GO and UPR which restricts the mobility of the polymer matrix segments at the interface. The interaction may be attributed to the formation of hydrogen bonding between oxygen functionality on reinforcement and polymer or probably dipolar interactions between the two components.

Arasa et al. [64] developed thermosetting nanocomposites prepared from diglycidyl ether of bisphenol and c-valerolactone with different organically modified clays that were cationically copolymerized in the presence of scandium, ytterbium or lanthanum triflates as initiators. The authors observed that the incorporation of Cloisite® Na, Cloisite® 30B and Cloisite® A6 improved the temperature of 5 wt% of weight loss in 43, 33 and 41 °C, respectively. The major degradation temperature of neat resin was also improved in 31, 17 and 24 °C, respectively. According to the authors the modified nanoclay enhanced the thermal stability by acting as a superior insulator and mass transport barrier to the volatile products generated during the thermal decomposition.

8 Thermal Application of Thermosetting Nanocomposites

8.1 Ablation Materials

Ablation is the process of removing material from a surface or other erosive process and usually associated with materials for space reentry vehicles and rocket nozzles [65–67]. Ablative materials represent the traditional approach to thermal protection, used for over 40 years in a broad range of applications [68]. The ablative materials are used as thermal protection materials for rocket nozzles, space vehicles and combustion chambers of rocket motors. These materials should withstand very high temperatures in the order of thousands of degrees Celsius, generally between 1000 and 4000 °C, highly thrust and high impact resistance. The final material should be able to form complex shapes and be as light as possible. Currently, the main consumers of ablative materials are military, NASA and commercial space launching company [65].

Most ablative thermal protection system (TPS) materials are reinforced composites where organic resins are used as matrices. When heated, the resin pyrolyzes generating gaseous products, mostly composed by hydrocarbons that permeate the solid diffusing toward the external heated surface and proceed into the boundary layer, where the heat transfer processes take place. The resin pyrolysis also produces a carbonaceous residue, indicated as char. The process is typically endothermic and pyrolysis gases are heated as they percolate toward the surface, thus transferring energy from the solid to the gas [68]. The transformations and energy accommodation mechanisms of an ablative TPS material are schematically illustrated in Fig. 3. The entering of the pyrolysis gases into the boundary layer modifies its properties, generally producing a reduction of convective heating. However, the gases may undergo chemical reactions with the boundary layer gases, thus influencing the net heat transfer to the surface. Furthermore, chemical reactions can result in consumption of the surface material leading to surface recession [67–69].

Two categories of ablative materials are available: melting and non-melting types. In melting type (represented by thermoplastics), the liquid is removed immediately after formation and new surface is exposed, thus the thermal protection

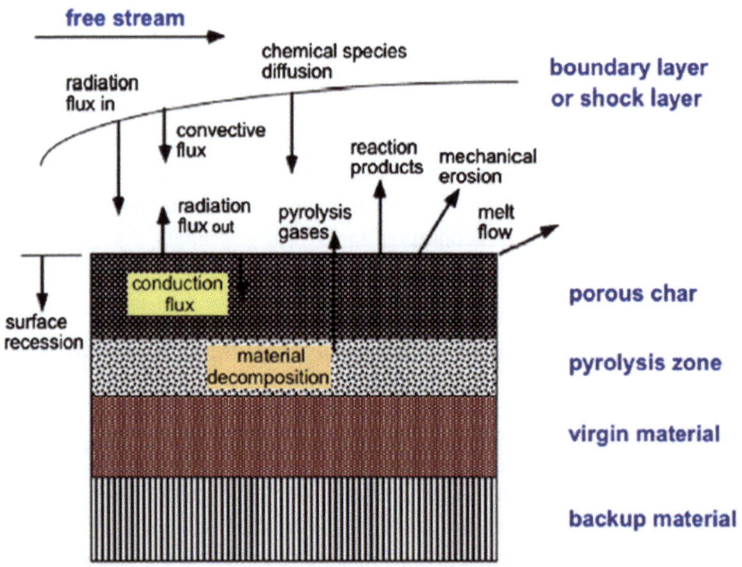

Fig. 3 Energy accommodation mechanisms of ablative thermal protections systems [69]

is not efficient [68]. In non-melting and unreinforced materials, mechanical ablation precedes chemical ablation, and a significant strength reduction is produced as a consequence of the temperature increase. The thermal ablation based on sublimation becomes appreciable only at very high temperatures (usually above 3000 °C). Non-melting ablative materials are made of char-forming thermosetting resins, providing multiple levels of protection, and they are suitable for thermal protection of re-entry vehicles, probes and ballistic missiles that normally are used in high heating rates, however with shorter duration [68, 69]. However, because of their low strength and stiffness, thermosetting resins (carbon/epoxy and carbon/phenolic composites are the most widely used) are usually reinforced with fibers and/or ceramic fillers [68, 70]. However, composite materials produced with nanosized reinforcements have led to a new paradigm for polymeric ablators. Many studies can be found in the literature using nanosized reinforcements for development nanocomposites ablator materials with thermosetting resins.

Koo et al. [71] compared the ablative properties of carbon phenolic composite (MX-4926) to laboratory produced carbon nanofibers (CNFs)/phenolic nanocomposites (CNF-NRAMs). Three CNF loadings, namely 20 %, 24 % and 28 wt% were dispersed in the phenolic matrix without rayon-carbon fiber reinforcement. A subscale rocket motor (heat flux 1140 W/cm^2) was used to study the ablation and insulation characteristics of the ablatives. According to the experimental results, when compared to the MX-4926, all the CNF-NRAMs exhibited lower ablation rates and the lower maximum backside heat-soaked temperature increase. Peak erosion of the higher loaded CNF-NRAM without the rayon fabric was 42 % lower than MX-4926. However, the residual mass of MX-4926 was higher in comparison of CNF-NRAMs. Additionally, the surface temperatures of the CNF-NRAM samples were higher than that of MX-4926.

Liu et al. [72] used POSS nanomodification in phenolic resin with carbon fiber reinforcement. SEM analysis showed the production of best charred surface on burnt samples that enhanced the ablation performance. Novel flake graphite was introduced into barium-phenolic resin by Yu and Wan [73]. Nanocomposites were made by roller-coating technology and its ablation property was tested under long pulse laser radiation. Nanocomposites showed better ablation performance compared to the control system. It was also observed that the size of the graphite flake affected the ablation rate [73]. Srikanth et al. [74] prepared ablative nanocomposites by introducing nanosilica into the phenolic resin with carbon fiber reinforcement. Ablation resistance of nanocomposites increased with the nanosilica content up to 2 wt%. However, beyond this point ablation resistance decreased.

The ablation performance, thermal decomposition and temperature distribution through the thickness of asbestos/phenolic composites modified with layered silicate were compared with traditional asbestos/phenolic composites by Bahramian et al. [75]. Nanofillers were introduced at 3, 4 and 6 wt% content ratio. Samples were tested at the heat flux of 900 W/cm^2. The 6 wt% nanocomposite samples showed the best ablation performance. Natali et al. [70, 76, 77] produced two different mixtures, both consisting of 50 wt% phenolic matrix (PR) and 50 wt% nanofiller. One of them with carbon black (CB) referred to as PR-CB, constituted by 50 %-PR and 50 %-CB. The other with MWCNTs, constituted by 50 %-PR and 50 %-MWCNTs referred to as PR-MWCNT. Their analyses included thermogravimetric analysis, evaluation of the heat capacity, oxy-acetylene test, and the post burning morphology. They observed better performance in nanomodified samples. Koo et al. [78] prepared ablative nanocomposites by using MMT organoclay, POSS and CNF with and without carbon fibers in a phenolic resin. The combination of high loading of MMT organoclay, low loading of POSS, and high loading of CNF in phenolic resin showed better ablation performance than control carbon/phenolic composite.

Recently, Srikanth et al. [79] development thermal protection system (TPS) materials using 3.5, 6.5 and 9.5 wt% of zirconia dispersed in a resol based phenolic resin. The results showed that the presence of zirconia has increased the ablation rate. This is probably due to enhanced conversion of solid char into zirconium carbide (ZrC) and carbon monoxide leading to under protection of the carbon fibers facing the plasma jet used. In another work, Chen et al. [80] improved ablation resistance of carbon-phenolic composites by introducing zirconium diboride (ZrB_2) particles. Results indicated that the employed method of introducing ZrB_2 into carbon-phenolic composites could profoundly improve the ablation performance. The linear ablation rate of ZrB_2 carbon-phenolic composites reduced by 79 % in comparison with carbon-phenolic composites.

8.2 Fire Retardants

The massive use of polymer materials in our everyday life is driven by their remarkable combination of properties, low weight and ease of processing. However, polymers are also known for their relatively high flammability; most often accompanied by the production of corrosive or toxic gases and smoke during combustion [81, 82]. Consequently, improving the fire retardant behavior of polymers is a major challenge for extending their use to most applications. Safety requirements are currently becoming more and more drastic in terms of polymers' reaction to fire and their fire resistance performances. And various flame retardant additives, such as halogenated additives, are being phased out for their proven or suspected adverse effects on the environment [81]. The combined challenge thus consists in developing effective and environmentally friendly flame retardant systems for polymer materials [81–83].

Many polymer composites ignite when exposed to high heat flux, releasing heat that can, in some circumstances, contribute to the growth of the fire. Significant quantities of smoke and toxic fumes may also be released, respectively limiting visibility and posing a health hazard [81, 83]. For these reasons, stringent fire regulations govern the use of composites in aircraft, ships, buildings, land transport, oil and gas facilities, among other applications. These regulations require that the fire reaction properties reach specified levels [81, 83]. The reaction properties that are often used to define the fire hazard include the heat release rate (HRR), time-to-ignition and flame spread rate [81, 82]. These properties are important because they influence the temperature and spread of a fire. Other important properties include smoke density and yield of carbon monoxide gas because these determine survivability [82, 83].

Of the many fire reaction properties, it is generally recognized that HRR is the single most important in controlling fire hazard. HRR is defined as the mass loss rate of the material times its heat of combustion. The amount of heat released from a polymer composite is controlled by the combustion of flammable gas products resulted from the decomposition of the organic components [82]. The resin matrix is generally the main source of flammable volatiles. The reactions of the polymers generally used in composites are endothermic, so these reactions absorb heat. Therefore, the net HRR is a complex property determined by the heat generated by the decomposition reactions of the volatiles released by the decomposing composite less the heat absorbed by the endothermic decomposition reactions of the polymer matrix [82, 83].

Cone calorimetry (CC) is one of the most effective medium-sized polymer fire behavior tests. The principle of cone calorimeter experiments is based on the measurement of the decreasing oxygen concentration in the combustion gases of a sample subjected to a given heat flux, in general from 10 to 100 kW/m^2 [83]. Figure 4 illustrates the experimental set-up of a cone calorimeter. Standardized in the United States (ASTM E 1354), the cone calorimeter test is also the subject of an international standard (ISO 5660).

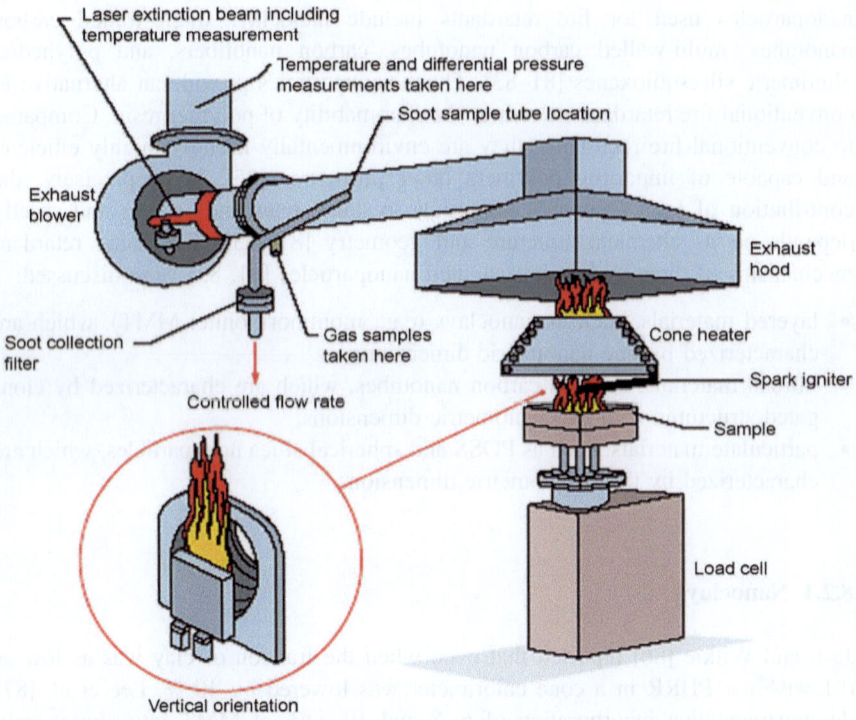

Fig. 4 Schematic representation of the cone calorimeter test [83]

The sample is placed on a load cell in order to evaluate the evolution of mass loss during the experiment [81]. A conical radiant electrical heater uniformly irradiates the sample from above [81, 82]. The combustion is triggered by an electric spark. The combustion gases produced pass through the heating cone and are captured by means of an exhaust duct system with a centrifugal fan and a hood [81]. The gas flow, oxygen, CO and CO_2 concentrations and smoke density are measured in the exhaust duct. The measurements of the gas flow and oxygen concentration are used to calculate the quantity of heat released per unit of time and surface area: HRR is expressed in kW/m^2. The evolution of the HRR over time, in particular the value of its peak/maximum (PHRR or HRR_{max}), is usually taken into account in order to evaluate the fire properties [81–83]. The calculation is based on Huggett's observation that most organic materials release a quantity of heat practically proportional to the quantity of oxygen consumed while burning [81]. Integration of the HRR versus time curve gives the total heat released (THR)—expressed in kJ/m^2 [81, 82]. In addition, the cone calorimeter test also enables characterization of the time to ignition (TTI), time of combustion or extinction (TOF), mass loss during combustion, quantities of CO and CO_2 and total smoke released (TSR) [81, 84].

Nowadays a variety of nanoparticles have been used to fabricate fire retardant nanocomposites with improved mechanical and functional properties. The common nanoparticles used for fire retardants include nanoclay, single-walled carbon nanotubes, multi-walled carbon nanotubes, carbon nanofibers, and polyhedral oligomeric silsesquioxanes [81–83]. These nanoparticles provide an alternative to conventional fire retardants to reduce the flammability of polymer resin. Compared to conventional fire retardants, they are environmentally friendly, highly efficient, and capable of imparting polymers other properties [85]. More precisely, the contribution of each type of nanoparticle to flame retardancy varies and strictly depends on its chemical structure and geometry [81, 85]. The flame retardant mechanisms of three widely investigated nanoparticles [81, 85] were discussed:

- layered materials, such as nanoclays (e.g., montmorillonite: MMT), which are characterized by one nanometric dimension;
- fibrous materials, such as carbon nanotubes, which are characterized by elongated structures with two nanometric dimensions;
- particulate materials, such as POSS and spherical silica nanoparticles, which are characterized by three nanometric dimensions.

8.2.1 Nanoclays

Jash and Wilkie [86] reported that even when the fraction of clay was as low as 0.1 wt% the PHRR in a cone calorimeter was lowered by 40 %. Lee et al. [87] demonstrated that incorporation of 6, 8 and 10 wt% of MMT into epoxy resin increased linearly the char yield from 9.1 to 15.4 % reducing the thermal degradation of the epoxy matrix. Nazaré et al. [88] studied the flammability properties of unsaturated polyester resin with nanoclays using cone calorimetry. The authors verified that the incorporation of 5 wt% of nanoclays reduces the PHRR by 23–27 % and THR values by 4–11 %. While incorporation of condensed-phase flame retardants (such as ammonium polyphosphate, melamine phosphate and alumina trihydrate) reduce the PHRR and THR values of polyester resin, the inclusion of small amounts of nanoclay (5 % w/w) in combination with these char promoting flame retardants causes total reductions of the PHRR of polyester resin in the range 60–70 %. Ammonium polyphosphate, in particular and in combination with polyester-nanoclay hybrids show the best results compared to other flame retardants.

Frache et al. [89] development epoxy resin composites based on a hydrotalcites, $[Mg_{0.67}Al_{0.33}(OH)_2][CO_3]_{0.165}\cdot 0.5H_2O$, containing 5 wt% of hydrotalcites. The sample containing the hydrotalcite showed a decrease of 30 % in the PHRR during cone calorimetry tests. Pereira et al. [90] evaluated the flame retardant behavior of unsaturated polyester (UP) nanocomposites based on layered double hydroxides (LDH). Two different organo-LDH were used (adipate-LDH (A-LDH) and 2-methyl-2-propene-1-sulfonate–LDH (S-LDH)) using 1 and 5 wt% of reinforcement. The fire reaction study, using CC, reveals a reduction in the flammability of UP resin in 46 and 32 % by incorporating 1A-LDH and 5S-LDH, respectively. The incorporation of a relatively low amount of organomodified nanoclay in the polymer matrix creates a protective layer during combustion. The accumulation of the clay on the surface of the matrix acts as a protective barrier that limits heat transfer into the material, volatilization of combustible degradation products and diffusion of oxygen into the material [81].

The main fire retardancy mechanisms in polymer/clay nanocomposites seem to be the formation of a barrier against heat and volatiles by migration of the clay nanolayer toward the material surface, followed by char formation, together with increased melt viscosity for exfoliated nanocomposites [81]. These mechanisms modify the fire properties of the polymer nanocomposite, sometimes improving them and in some other cases worsening them, depending on the type of fire test applied and the nature of the measurements recorded [81, 91]. For example, in CC tests, the incorporation of nanoclays generally retards and decreases the PHRR, but does not reduce the total heat involved and may also decrease the time to ignition [81]. The increased melt viscosity in exfoliated nanocomposites prevents dripping and promotes char formation [81, 91].

8.2.2 Carbon Nanotubes

Rahatekar et al. [92] observed that an addition of only 0.0025 wt% of highly aligned MWCNTs showed a 45 % reduction in the peak mass loss rate during gasification of epoxy/MWCNTs composites when compared with neat epoxy. Reduction in peak of mass/heat release rate is an important factor in controlling flame spread. The authors attribute this effect to highly aligned MWCNTs which can be easily exfoliated using high shear mixing. Zhao and Gou [93] developed multifunctional thermoset composites with polyester resin, glass fiber and CNF sheets. The authors also observed that lower quantity (0.38 wt%) of CNF sheets lowered the HRR.

Lee et al. [87] observed that the addition of 1 wt% of MWCNT into epoxy resin clay nanocomposites increased the char yield from 15.4 to 19.5 % demonstrating the synergistic effects of MMT/MWCNT additives. It is well known that char yield generated on a surface can act as a heat barrier and thermal insulator. Thus, the increased char yield is expected to reduce the degradation of the epoxy polymer. The limit oxygen index (LOI) defined as the minimum fraction of O_2 in a gas mixture of O_2 and N_2 that will support flaming combustion of the samples was also investigated. LOI showed a similar trend with char yield, with additives increasing the oxygen content required for ignition, making it more difficult to ignite the additives/epoxy resin complexes. The layered structured network formed by MMT and MWCNTs likely played an important role as a shield by re-emitting the oxygen radicals from the surface of epoxy to outside, decreasing the rate of polymer degradation.

Wu et al. [94] incorporated SWCNT and MWCNT membranes (buckypaper) and CNF paper onto the surface of epoxy carbon fiber composites. The authors observed that SWCNT buckypaper and CNF paper did not show notable improvement on fire retardancy. However, MWCNT buckypaper acted as an

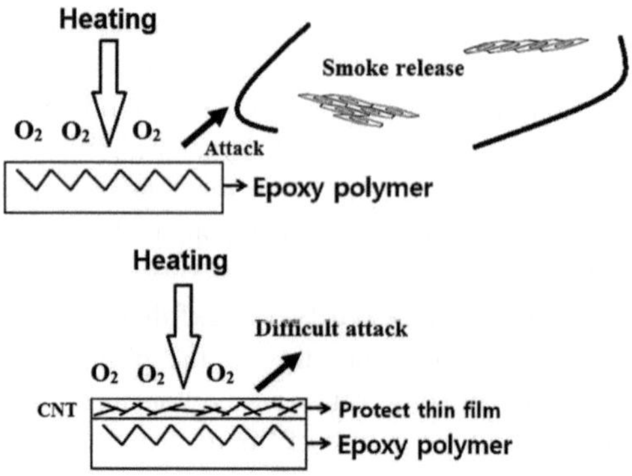

Fig. 5 Suggested mechanism of flame retardants based in epoxy/CNT nanocomposites. Adapted [81, 95]

effective flame retardant shield and reduce the PHRR by more than 60 % and also reduce smoke generation by 50 % during combustion.

According to Im et al. [95] at temperatures above 250 °C the thermal oxidation rate of a polymer depends on oxygen radical initiation through hydrogen abstraction by oxygen in MWCNT and MMT fire retardant nanocomposites. However, the play of oxygen radicals can be hindered by the formation of a thin protective film of MWCNT/carbon char generated by oxidative dehydrogenation on the surface of the epoxy complex, as can be seen in Fig. 5. It is assumed that the free radical scavenger effect of MWCNTs is the main reason for the improvement in the thermal stability of the epoxy resin. Thus, the main role of MWCNTs on the surface of an epoxy complex is the formation of compact charred layers promoted by MWCNTs acting as a heat barrier and as thermal insulation. Even though the charred layer acts as a heat barrier and as thermal insulation, MWCNTs primarily promote the generation of a charred layer through the excellent thermal conductivity of MWCNTs. The thermal energy from the outside can be dispersed well by the presence of MWCNTs, generating the charred layer effectively.

MWCNTs inside the epoxy resin also play an important role in the flame retardant properties. Flame retardation by MWCNTs may hinder the spread of fire to other places due to the formation of gel-type epoxy resin even during combustion [95]. A clay such as MMT has a high specific heat; thus, MMT can act as a heat storage medium [87, 95]. During combustion of the complex, the applied thermal energy can be stored in MMT and hinder the spread of thermal energy to the epoxy resin [87, 95].

8.2.3 Silsesquioxane

Franchini et al. [96] used monofunctional POSS (4 or 9 wt%), containing isobutyl or phenyl bonds, into an epoxy-amine and fire retardant properties were evaluated. The fire behavior is clearly modified by the presence of the POSS clusters for both types of POSS cages, a significant decrease is observed in the PHRR. In the case of

POSS with phenyl groups, the PHRR is decreased by 40 % in this case as compared to 25 % for isobutyl-based POSS network. It was demonstrated that POSS nanoclusters induced an effective fire retardant effect, which was controlled by several factors. The authors showed that POSS bearing phenyl bonds were far more effective than POSS with isobutyl bonds. Also, the presence of a chemical linkage between the phenyl-based POSS clusters and the matrix favored the dispersion of the nanoclusters, resulting in enhanced fire retardancy. The fire retardant effect was only slightly enhanced by increasing the amount of POSS nanoclusters. The authors also verified some interesting information by visual observation of the end of test residues. On cross sections of the residues, it can be seen that the phenyl POSS nanoclusters form during combustion a solid sponge-like structure. This structure is not formed on neat epoxy network or in presence of isobutyl containing POSS clusters. Remarkable reduction of HRR values observed for phenyl POSS hybrid networks is likely to be due to the presence of this char layer formed during combustion. The char layer generated is thought to act as a barrier for both heat flow and mass transport.

Wu et al. [97] synthesized a named functional POSS (NPOSS) with two epoxy ring groups via the reaction between trisilanolisobutyl-POSS and triglycidyl isocyanurate. Then a halogen-free epoxy composite containing silicon/nitrogen was prepared. The results of microscale combustion calorimeter indicate that the presence of 10 wt% of NPOSS in epoxy resin can decrease its PHRR by about 30 %. The authors believed that NPOSS can retards the movement and scission of polymeric chains of epoxy resin and forms a stable charred layer in the condensed phase to prevent the underlying materials from further combustion.

Zhang et al. [98] used cage-type octaphenyl silsesquioxane (OPS) and ladder-type polyphenyl silsesquioxane (PPSQ) as flame-retardants into epoxy resin. For the epoxy/OPS sample, a weak blowing-out effect, by which the sample self-extinguished after around 45 s, was observed. For the epoxy/PPSQ however, no such blowing-out effect was observed. The PHRR of epoxy/OPS is seen to be reduced remarkably from 855 to 626 kW/m^2. However, most unexpectedly, the PHRR of epoxy/PPSQ was measured as 925 kW/m^2, which is higher than that of pure epoxy. The authors observed that OPS seemingly had a better flame retardancy than PPSQ for pure epoxy. Therefore, it seems that the respective structures of OPS and PPSQ play an important role in their flame retardancy and blowing-out effects in epoxy. According to the authors, one possible explanation is in that once PPSQ ladder chain may break down as like a zip is zipped even though the ladder structure is very stable before zipping. The active broken segments of PPSQ accelerate the decomposition of the epoxy matrix and cause high heat release. Relatively, the OPS can produce limited broken segments when one OPS cage breaks [98, 99]. The fire retardancy performances of the corresponding networks appear to be more influenced by the nature of the non-reactive bonds borne by the POSS cages, rather than the presence of a chemical linkage between POSS and matrix. In the case of POSS containing phenyl-groups, the decrease in PHRR observed has been ascribed to the initial tendency of the phenyl-containing POSS to char when decomposing [96, 99].

8.2.4 Metallic Nanoparticles

Rallini et al. [100] modified epoxy matrix with boron carbide nanoparticles at 1 and 5 wt% concentrations and the nanostructured polymer was used as a matrix to impregnate carbon fiber fabrics. The CC tests showed that even after the exposure to a heat-flux of 50 kW/m^2 for 600 s, the composite based on 5 wt% of boron carbide exhibited residual structural integrity as well as significant reduction of the HRR than the control sample. These benefits were related to the chemical reactions of boron carbide when exposed to an oxidizing hyper-thermal environment. B_4C can react with oxygen to produce liquid boron oxide and carbon dioxide. Boron carbide can also react with other combustion products of the polymeric matrix such as carbon monoxide and water vapor to yield boron oxide and amorphous carbon. In this way, the release of gaseous by-products can be reduced and a solid residue can be produced, minimizing the volumetric shrinkage of the charring material. A more resistant residue enabled the reduction of the degradation rate and reduced the heat transmitted to the inner layers of the laminates thus improving both thermal stability and fire resistance. Additionally, the glassy network produced by the liquid boron oxide acted as a protective layer for the carbon fibers, inhibiting their oxidation processes, and also serving as a high temperature adhesive that bound the plies of the composite materials together. In another work, Wang et al. [101] also observed that the incorporation of 10 wt% of boron particles into phenolic resin/clay nanocomposites increases the decomposition temperature at about 42 °C when compared with nanocomposites without boron particles. These results suggested that the incorporation of boron is effective in promoting thermal stability and flame retardancy of thermosetting nanocomposites.

Tibiletti et al. [102] incorporated nano-alumina and submicron alumina trihydrate particles into UP resin. Synergistic effects on thermal stability and HRR were observed for combinations between both submicron filler and nanofiller. The best result for fire behavior was obtained for a global loading of 10 wt% with an equal mass ratio for both types of particles. Combining both particles leads to a 32 % reduction in PHRR. These synergistic effects can be ascribed to physical effects resulting from the arrangement of both kinds of mineral particles of very different median size at the surface of the composite during polymer combustion and ablation. The formation of this mineral barrier promotes catalytic effects ascribed to the huge specific surface area of oxide nanoparticles.

9 Future Trends

The field of nanoscience and nanotechnology is extending the applications of physics, chemistry, biology, engineering and technology into previously unapproached infinitesimal length scales. The polymer nanocomposites have been the exponentially growing field of research for developing materials in last few decades and have been mainly focusing on the structure-property relationships and their development. Since the polymer-nanocomposites have been the staple of modern polymer industry, their durability under various environmental conditions and degradability after their service life are also essential fields of research for the near future [103, 104].

In addition, controlling the dispersion of nanoparticles into polymer matrices is a significant challenge in achieving the dramatic property improvements promised by polymer nanocomposites [103, 105–107]. While nanoparticle dispersion is believed to critically affect properties, it is not apparent that a single state of particle dispersion or organization should optimize any given or all macroscale properties [105, 106]. Thus, optimizing one versus two properties of a composite can require very different morphologies [105]. It suggests that the creation of multifunctional composites requires exquisite control over nanoparticle spatial distribution [105]. Such understanding, which is currently only at a nascent stage, is crucial to the end use of these materials in a diversity of applications [105].

An area where nanocomposites could achieve a dramatic commercial prominence is in advanced composite applications [106]. Modification of the matrix phase with carbon nanotubes at the lower scale of dimensions and carbon nanofibers at a higher dimensional scale would allow for significant increases in mechanical and thermal properties to the overall composite properties [106]. While this would offer some improvement in unidirectional composites, it could be dramatic in the case of cross-ply composites which are the major type of composite structure utilized in advanced composite applications [105, 106]. This concept is under present consideration and could allows a step change in the advanced composite field. The basic concept has already been commercially employed in specialty sports equipment, as tennis rackets and hockey sticks. These improvements are the key to the future aircraft and wind energy turbine applications [106]. This approach is analogous to naturally occurring composite structures where the hierarchical construction method employs several dimensional scales beginning at the nanolevel. The same concept is also relevant to reinforced composites where exfoliated clay could be added to the matrix for unsaturated polyester-fiberglass composites or others fiberglass reinforced matrix polymers [106].

Another field that nanocomposites have the potential to play a significant role in future space systems. Launch vehicles would greatly benefit from appropriately designed nanocomposites that could provide improved barrier properties and gradient morphologies enabling linerless composite cryogenic fuel tanks [103]. Self-rigidizing and self-passivating nanocomposites materials could be used for construction of space vehicle components that are both highly resistant to space-borne particles and resistant to degradation from electromagnetic radiation, while reducing the overall weight of the spacecraft [103]. However, making connections between nanoparticle dispersion and organization with macroscale properties is then a crucial aspect that is only now beginning to be considered [105]. Systematic investigation of the effect of the manner in which the nanocomposites are prepared/assembled on their mechanical response is necessary [103, 105]. Rheology of polymeric chain near solid surfaces should be used to investigate effects the interfacial interaction on the chain dynamics over several length and times scales [105].

According to presented above many products based on thermosetting nanocomposites are already on the market and still others will come. However, relatively little is known about the environmental or industrial health and safety of these products [108–111]. No regulatory requirements beyond those that are required for all chemicals in commerce have been imposed on the manufacture or use of thermosetting nanocomposites [108]. On the other hand, is necessary that dynamic interactions between the scientific and regulatory communities, industry and consumers are crucial to verify if current rules, regulations, and laws are sufficient to adequately assess the risks associated with the full life cycle, including manufacture, use and disposal of thermosetting nanocomposites [108, 109]. These approaches are also necessary for allowing the utilization of the technological, scientific and economical benefits of nanotechnologies [109].

10 Conclusions

The potential of nanocomposite materials in various sectors of research and industry is promising and attracting increasing investment from governments and business in many parts of the world. It is clear that nanocomposite materials have the potential to substantially affect manufacturing processes across a wide range of industries over the medium-to long term.

The new nanocomposite materials can be produced having improved physical, thermal and flammability properties with controlled durability, by having various types of nanoparticles. However, the degradation durability and toxicity of polymer-nanoparticles systems have to be evaluated for each nanoparticle with different polymers matrix under different environmental conditions for realizing the potential of nanomaterials.

To create nanocomposites materials with specific applications the nanoparticles dispersion control into the polymer matrices still remains a critical challenge for researchers. So, the development of nanocomposite materials requires control over nanoparticle distribution in the polymer matrix. Making connections between nanoparticle dispersion, enhanced the macroscale properties and evaluated the end of life of this materials is then a crucial aspects that is only now beginning to be considered by researchers around the world. So, make these connections is essential to better development and application of the nanotechnology in the near future.

References

1. Ornaghi Jr, H.L., Poletto, M., Zattera, A.J., Amico, S.C.: Cellulose. doi:10.1007/s10570-013-0094-1
2. Pistor, V., Ornaghi Jr, H.L., Ferreira, C.A., Zattera, A.J.: J. Appl. Polym. Sci. **125**, E462 (2012)

3. Rossa, L.V., Scienza, L.C., Zattera, A.J.: Polym. Compos. **34**, 450 (2013)
4. Pistor, V., Ornaghi, F.G., Ornaghi Jr, H.L., Zattera, A.J.: Mater. Sci. Eng. A **532**, 339 (2011)
5. Pigatto, C., Almeida Jr, J.H.S., Ornaghi Jr, H.L., Rodríguez, A.L., Mahlmann, C.M., Amico, S.C.: Polym. Compos. **33**, 2262 (2012)
6. Romanzini, D., Lavoratti, A., Ornaghi Jr, H.L., Amico, S.C., Zattera, A.J.: Mater. Des. **47**, 9 (2013)
7. Ornaghi Jr, H.L., Bolner, A.S., Fiorio, R., Zattera, A.J., Amico, S.C.: J. Appl. Polym. Sci. **118**, 887 (2010)
8. Ornaghi Jr, H.L., da Silva, H.S.P., Zattera, A.J., Amico, S.C.: J. Appl. Polym. Sci. **125**, E110 (2012)
9. Piazza, D., Lorandi, N.P., Pasqual, C.I., Scienza, L.C., Zattera, A.J.: Mater. Sci. Eng. A **528**, 6769 (2011)
10. Ornaghi Jr, H.L., Pistor, V., Zattera, A.J., Non-Cryst, J.: Solids **358**, 427 (2012)
11. Poletto, M., Zattera, A.J., Forte, M.M.C., Santana, R.M.C.: Bioresour. Technol. **109**, 148 (2012)
12. Poletto, M., Zattera, A.J., Santana, R.M.C.: Bioresour. Technol. **126**, 7 (2012)
13. Borsoi, C., Scienza, L.C., Zattera, A.J.: J. Appl. Polym. Sci. **128**, 653 (2013)
14. Savage, G.: Eng. Fail. Anal. **17**, 92 (2010)
15. Loutas, T.H., Panopoulou, A., Roulias, D., Kostopoulos, V.: Expert Syst. Appl. **39**, 8412 (2012)
16. Júnior, J.H.S.A., Ornaghi Jr, H.L., Amico, S.C., Amado, F.D.R.: Mater. Des. **42**, 111 (2012)
17. Ornaghi Jr, H.L., da Silva, H.S.P., Zattera, A.J., Amico, S.C.: Mater. Sci. Eng. A **528**, 7285 (2011)
18. Romanzini, D., Ornaghi Jr, H.L., Amico, S.C., Zattera, A.J.: J J. Reinf. Plast. Compos. **31**, 1652 (2012)
19. Romanzini, D., Ornaghi Jr, H.L., Amico, S.C., Zattera, A.J.: Mater. Res. **15**, 415 (2012)
20. Poletto, M., Dettenborn, J., Zeni, M., Zattera, A.J.: Waste Manage. **31**, 779 (2011)
21. Borsoi, C., Berwig, K.H., Scienza, L.C., Zattera, A.J.: Polym. Compos. **34**, 967 (2013)
22. Chrissafis, K., Bikiaris, D.: Thermochim. Acta **523**, 1 (2011)
23. Gangopadhyay, R., De, A.: Chem. Mater. **12**, 608 (2000)
24. Yamamoto, T., Kanda, N., Non-Newton, J.: Fluid **1**, 181 (2012)
25. Ray, S.S., Okamoto, M.: Prog. Polym. Sci. **28**, 1539 (2003)
26. Carrado, K.A.: Appl. Clay Sci. **17**, 1 (2000)
27. Melendez, I.M., Neubauer, E., Angerer, P., Danninger, H., Torralba, J.M.: Compos. Sci. Technol. doi:10.1016/j.compscitech.2011.04.005
28. Lustig, S.R., Boyes, E.D., French, R.H., Gierke, T.D., Harmer, M.A., Hietpas, P.B., Jagota, A., McLean, R.S., Mitcheli, G.P., Onoa, G.B., Sams, K.D.: Nano Lett. **3**, 1007 (2003)
29. Scocchi, G., Posocco, P., Danani, A., Pricl, S., Fermeglia, M.: Fluid Phase Equilib. **261**, 366 (2007)
30. Gilman, J.W.: Appl. Clay Sci. **15**, 31 (1999)
31. Pandey, J.K., Reddy, K.R., Kumar, A.P., Singh, R.P.: Polym. Degrad. Stab. **88**, 234 (2005)
32. Bikiaris, D.: Thermochim. Acta **523**, 25 (2011)
33. Ollier, R., Rodriguez, E., Alvarez, V.: Compos. A **48**, 137 (2013)
34. Bharadwaj, R.K., Mehrabi, A.R., Hamilton, C., Trujillo, C., Murga, M., Fan, R., Chavira, A., Thompson, A.K.: Polymer **43**, 3699 (2002)
35. Becker, O., Varley, R.J., Simon, G.P.: Eur. Polym. J. **40**, 187 (2004)
36. Carrasco, F., Pagès, P.: Polym. Degrad. Stab. **93**, 1000 (2008)
37. Lakshmi, M.S., Narmadha, B., Reddy, B.S.R.: Polym. Degrad. Stab. **93**, 201 (2008)
38. Saitoh, K., Ohashi, K., Oyama, T., Takahashi, A., Kadota, J., Hirano, H.: J. Appl. Polym. Sci. **122**, 666 (2011)
39. Xie, W., Gao, Z., Liu, K., Pan, W.-P., Vaia, R., Hunter, D., Singh, A.: Thermochim. Acta **339**, 367 (2001)
40. Zheng, X., Wilkie, C.A.: Polym. Degrad. Stab. **82**, 441 (2003)
41. Hwu, J.M., Jiang, G.J., Gao, Z.M., Xie, W., Pan, W.-P.: J. Appl. Polym. Sci. **83**, 1702 (2002)

42. Leszczyńska, A., Njuguna, J., Pielichowski, K., Banerjee, J.R.: Thermochim. Acta **453**, 75 (2007)
43. Leszczyńska, A., Njuguna, J., Pielichowski, K., Banerjee, J.R.: Thermochim. Acta **454**, 1 (2007)
44. Xie, W., Gao, Z., Pan, W.-P., Hunter, D., Singh, A., Vaia, R.: Chem. Mater. **13**, 2979 (2001)
45. Sahoo, N.G., Rana, S., Cho, J.W., Li, L., Chan, S.H.: Prog. Polym. Sci. **35**, 837 (2010)
46. Abdalla, M., Dean, D., Robinson, P., Nyairo, E.: Polymer **49**, 3310 (2008)
47. Ingram, S.E., Liggat, J.J., Pethrick, R.A.: Polym. Int. **56**, 1029 (2007)
48. Guo, B., Jia, D., Cai, C.: Eur. Polym. J **40**, 1743 (2004)
49. Gu, A., Liang, G.: Polym. Degrad. Stab. **80**, 383 (2003)
50. Brnardić, I., Macan, J., Ivanković, H., Ivanlović, M.: J. Appl. Polym. Sci. **107**, 1932 (2008)
51. Yang, K., Gu, M.: Polym. Eng. Sci. **49**, 2158 (2009)
52. Liu, Y.-L., Wei, W.-L., Hsu, K.-Y., Ho, W.-H.: Thermochim. Acta **412**, 139 (2004)
53. Montero, B., Ramírez, C., Rico, M., Barral, L., Díez, J., López, J.: Polym. Int. **59**, 112 (2010)
54. Nagendiran, S., Alagar, N., Hamerton, I.: Acta Mater. **58**, 3345 (2010)
55. Chatterjee, A., Islam, M.S.: Mater. Sci. Eng. A **487**, 574 (2008)
56. Hackman, I., Hollaway, L.: Compos. A **37**, 1161 (2006)
57. Arasa, M., Pethrick, R.A., Mantecón, A., Serra, A.: Eur. Polym. J **45**, 1282 (2009)
58. Pinnavaia, J., Beall, G.W.: Polymer Clay Nanocomposites. Wiley, Chichester (2000)
59. Bafekrpor, E., Simon, G.P., Yang, C., Chipara, M., Habsuda, J., Fox, B.: Compos. A **46**, 80 (2013)
60. Bafekrpor, E., Simon, G.P., Yang, C., Chipara, M., Habsuda, J., Fox, B.: Polymer **54**, 3940 (2013)
61. Badrinarayanan, P., Kessler, M.R.: Compos. Sci. Technol. **71**, 1385 (2011)
62. Xing, W., Jie, G., Song, L., Wang, X., Lv, X., Hu, Y.: Mater. Chem. Phys. **125**, 196 (2011)
63. Bora, C., Gogoi, P., Baglari, S., Dolui, S.K.: J. Appl. Polym. Sci. **129**, 3432 (2013)
64. Arasa, N., Pethrick, R.A., Mantecón, A., Serra, A.: Eur. Polym. J **46**, 5 (2013)
65. Tate, J.S., Gaikwad, S., Theodoropoulou, N., Trevino, E., Koo, J.H.: J. Compos. **2013**, ID 403656 (2013)
66. Ogasawara, T., Aoki, T., Hassan, M.S.A., Mizokami, Y., Watanabe, N.: Compos. A **42**, 221 (2011)
67. Bahramian, A.R., Kokabi, M., Famili, M.H.N., Bheshty, M.H.: Polymer **47**, 3661 (2006)
68. Pulci, G., Tirillò, J., Marra, F., Fossati, F., Bartuli, C., Valente, T.: Compos. A **41**, 1483 (2010)
69. Laub, B., Venkatapathy, E.: International Workshop on Planetary Probe Atmospheric Entry and Descent Trajectory Analysis and Science (2003)
70. Natali, M., Monti, M., Puglia, D., Kenny, J.M., Torre, L.: Compos. A **43**, 174 (2012)
71. Koo, J.H., Stretz, H., Weispfenning, J.T., Luo, Z.P., Wootan, W.: AIAA Paper No AIAA-2004-1996 (2004)
72. Liu, Y., Lu, Z., Chen, X., Wang, D., Liu, J., Hu, L.: Proceedings of the 4th IEEE International Conference on Nano/Micro Engineered and Molecular Systems, p. 605 (2009)
73. Yu, Q.-C., Wan, H.: J. Inorg. Mater. **27**, 157 (2012)
74. Srikanth, I., Daniel, A., Kumar, S., Padmavathi, N., Singh, V., Ghosal, P., Kumar, A., Devi, G.R.: Scripta Mater. **63**, 200 (2010)
75. Bahramian, A.R., Kokabi, M., Famili, M.H.N., Bheshty, M.H.: J. Hazard. Mater. **150**, 136 (2008)
76. Natali, M., Monti, M., Kenny, J., Torre, L.: J. Appl. Polym. Sci. **120**, 2632 (2011)
77. Natali, M., Monti, M., Kenny, J., Torre, L.: Compos. A **42**, 1197 (2011)
78. Koo, J.H., Natali, M., Tate, J.S., Allcorn, E.: Int. J. Energ. Mater. Chem. Prop. **12**, 119 (2013)
79. Srikanth, I., Padmavathi, N., Kumar, S., Ghosal, P., Kumar, A., Subrahmanyam, Ch.: Compos. Sci. Technol. **80**, 1 (2013)
80. Chen, Y., Chen, P., Hong, C., Zhang, B., Hui, D.: Compos. B **47**, 320 (2013)

81. Laoutid, F., Bonnaud, L., Alexandre, M., Lopez-Cuesta, J.-M., Dubois, Ph: Mater. Sci. Eng. R **63**, 100 (2009)
82. Mouritz, A.P., Mathys, Z., Gibson, A.G.: Compos. A **37**, 1040 (2006)
83. Pavlidou, S., Papaspyrides, C.D.: Prog. Polym. Sci. **33**, 1119 (2008)
84. Koo, J.H., Venumbaka, S., Cassidy, P.E., Fitch, J.W., Grand, A.F., Bundick, J.: Fire Mater. **24**, 209 (2000)
85. Zhao, Z., Gou, J., Biett, S., Ibeh, C., Hui, D.: Compos. Sci. Technol. **69**, 2081 (2009)
86. Jash, P., Wilkie, C.A.: Polym. Degrad. Stab. **88**, 401 (2005)
87. Lee, S.K., Bai, B.C., Im, J.S., In, S.J., Lee, Y.-S.: J. Ind. Eng. Chem. **16**, 891 (2010)
88. Nazaré, S., Kandola, B.K., Horrocks, A.R.: Polym. Adv. Technol. **17**, 294 (2006)
89. Frache, A., Monticelli, O., Nocchetti, M., Tartaglione, G., Constantino, U.: Polym. Degrad. Stab. **96**, 164 (2011)
90. Pereira, C.M.C., Herrero, M., Labajos, F.M., Marques, A.T., Rives, V.: Polym. Degrad. Stab. **94**, 939 (2009)
91. Geråd, C., Fontaine, G., Bellayer, S., Bourbigot, S.: Polym. Degrad. Stab. **97**, 1366 (2012)
92. Rahatekar, S.S., Zammarano, M., Matko, S., Koziol, K.K., Windle, A.H., Nyden, M., Kashiwagi, T., Gilman, J.G.: Polym. Degrad. Stab. **95**, 870 (2010)
93. Zhao, Z., Gou, J.: Sci. Technol. Adv. Mater. **10**, 1 (2009)
94. Wu, Q., Zhu, W., Zhang, C., Liang, Z., Wang, B.: Carbon **48**, 1799 (2010)
95. Im, J.S., Lee, S.K., In, S.J., Lee, Y.-S.: J. Anal. Appl. Pyrol. **89**, 225 (2010)
96. Franchini, E., Galy, J., Gérad, J.-F., Tabuani, D., Medici, A.: Polym. Degrad. Stab. **94**, 1728 (2009)
97. Wu, K., Song, L., Hu, Y., Lu, H., Kandola, B.K., Kandare, E.: Prog. Org. Coat. **65**, 490 (2009)
98. Zhang, W., Li, X., Jiang, Y., Yang, R.: Polym. Degrad. Stab. **98**, 246 (2013)
99. Zhang, W., Li, X., Jiang, Y., Yang, R.: Polym. Degrad. Stab. **96**, 2167 (2011)
100. Rallini, M., Natali, M., Kenny, J.M., Torre, L.: Polymer **54**, 5154 (2013)
101. Wang, D.-C., Chang, G.-W., Chen, Y.: Polym. Degrad. Stab. **93**, 125 (2008)
102. Tibiletti, L., Longuet, C., Ferry, L., Coutelen, P., Mas, A., Robin, J.-J., Lopez-Cuesta, J.-M.: Polym. Degrad. Stab. **96**, 67 (2011)
103. Kumar, A.P., Depan, D., Tomer, N.S., Singh, R.P.: Prog. Polym. Sci. **34**, 479 (2009)
104. Hojjati, F.H.M., Okamoto, M., Gorga, R.E.: J. Compos. Mater. **40**, 1511 (2006)
105. Jancar, J., Douglas, J.F., Starr, F.W., Kumar, S.K., Cassagnau, P., Lesser, A.J., Sternstein, S.S., Buehler, M.J.: Polymer **51**, 3321 (2010)
106. Paul, D.R., Robeson, L.M.: Polymer **49**, 3187 (2008)
107. Pfaendner, R.: Polym. Degrad. Stab. **95**, 369 (2010)
108. Fairbrother, A., Fairbrother, J.R.: Ecotox. Environ Safe **72**, 1327 (2009)
109. Savolainne, K., Pylkkänen, L., Norppa, H., Falck, G., Lindberg, H., Tuomi, T., Vippola, M., Alenius, H., Hämeri, K., Koivisto, J., Brouwer, D., Mark, D., Bard, D., Berges, M., Jankowska, E., Posniak, M., Farmer, P., Singh, R., Krombach, F., Bihari, P., Kasper, G., Seipenbusch, M.: Saf. Sci. **48**, 957 (2010)
110. Boczkowski, J., Lanone, S.: Adv. Drug Deliver. Rev. **64**, 1694 (2012)
111. Savolainne, K., Alenius, H., Norppa, H., Pylkkänen, L., Tuomi, T., Kasper, G.: Toxicology **269**, 92 (2010)

Effect of Thermo-oxidation on the Mechanical Performance of Polymer Based Composites for High Temperature Applications

Sumana Ghosh

Abstract In the present study the effect of thermo-oxidation on the mechanical properties of polymer based composites has been reported for high temperature applications. The polymer based composites with high thermal stability and future trend towards modification of this type of composites have been discussed here.

Keywords Polymer composites · Thermo-oxidation · Mechanical properties

1 Introduction

Composite materials have ever increasing demand in all branches of engineering. This materials offer new solutions to the complicated engineering problems. Materials are combined in a composite in such a way as to enable us to make better use of their virtues by minimizing their deficiencies. The process optimization is required to reduce the constraints associated with the selection and manufacture of composite materials. The properties of composite materials can be tailored to suit particular design requirements. Complex shapes can be manufactured with the composites easily leading to cheaper and better solutions [1–5].

Few polymers are thermally stable in comparison to the metals and ceramics. Even the most stable polymers like polyimides, or poly(ether ether ketone) (known as PEEK) are degraded by thermal exposure to above about 300 °C. The chemical degradation can not be decreased through the reinforcement of polymer. However, the strength deterioration and creep or visco-elastic deformation can be delayed by the fibre reinforcement. Moreover, polymers have very low mechanical strength and stiffness in bulk form. Except few plastics, the weakest plastics tend to be ductile but the strongest tend to be brittle like metals. Traditionally, polymers are insulators

S. Ghosh (✉)
Bio-ceramics and Coating Division, CSIR–Central Glass and Ceramic Research Institute, Kolkata 700 032, India
e-mail: sumana@cgcri.res.in

P.M. Visakh and Y. Arao (eds.), *Thermal Degradation of Polymer Blends, Composites and Nanocomposites*, Engineering Materials,
DOI 10.1007/978-3-319-03464-5_4

and strength is usually a secondary consideration in their applications. The electrical conductivity of plastics reinforced with carbon fibres is of importance in many aeronautical applications. Most polymers are low-density materials, and the addition of fibres confers no density improvement. Most developments have been made so far in the reinforcing of polymers. It has been observed that there is still scope for improvement [1].

Since polymer composites often possess more desirable mechanical properties than granular or ceramic composites so they are used in various applications including high tensile strength, elastic modulus and therefore flexibility [6]. During the latest decades fiber reinforced polymer composites (FRPC) have been proved to be suitable construction material. Recently, natural fibers have received much more attention as reinforcement in polymer composite for making low cost construction materials all over the world. Natural fibers have served many useful purposes but the demand for utilizing it as reinforcement in polymer matrix is growing in recent years. These natural fibers offer a number of advantages over traditional synthetic fibers because of their superior corrosion resistance, excellent thermo-mechanical properties and high strength to weight ratio, which make them attractive as reinforcement in composite materials [7].

Recent increase in the use of polymer composites has created new challenges to predict their service life for high temperature applications. The present chapter deals with the probable solutions in order to minimize the effect of thermo-oxidation on mechanical properties of polymer based composites for high temperature applications.

2 Polymer Based Composites for High Thermal Stability Applications

Haque et al. [1] investigated the microstructures of carbon fiber bismaleimide composite as well as changes in flexural properties after high temperature exposure for 3000 h to 260 °C in air. The percentage of fiber end area exposed to air in a specimen end section significantly influences the extent of these changes [1]. Lee and Hall [8] estimated the weight change and retention of in-plane shear (±45°) strength of graphite fibre-reinforced cyanate ester resin matrix composites on exposure to high humidity and thermal cycling, respectively. The in-plane shear strength and fatigue lifetime at a given stress amplitude were reduced by a sudden moisture gain associated with extensive matrix/interface cracking. The rate of in-plane shear strength degradation was also measured after thermal cycling to a peak temperature of 150 or 204 °C and static exposure to dry heat [8]. The thermal behavior of liquefied wood polymer composites was studied by Doh et al. [9]. Low-density polyethylene (LDPE), high-density polyethylene (HDPE), and polypropylene (PP) were used with liquefied wood (LW) as polymer matrices. HDPE showed a better thermal stability than PP. Melt index did not show a significant effect on the thermal stability. The melting temperature of virgin LDPE and HDPE decreased with the addition of 10 % LW. The melting temperature of virgin PP also decreased with the addition of LW. Enthalpy of virgin polymers also decreased with the addition of LW. This study proved the necessity of thermal stability for the consolidation process of composite materials [9].

Hanu et al. [10] developed silicone polymer composites filled with mica, glass frit, ferric oxide and/or a combination of these for electrical power cables and other high temperature applications. The thermal stability of the polymer composites was determined by thermogravimetric techniques, thermal conductivity and heat release rate was measured by cone calorimetry. They investigated that fillers e.g. mica and ferric oxide had stabilizing effect on the thermal stability of silicone polymer and they lower the heat release rates during combustion. However, mica was only found to increase time to ignition [10]. Conducting composite systems have been prepared by Elyashevich et al. [11]. Microporous polyethylene films were obtained by melt extrusion with subsequent annealing, uniaxial extension, and thermal fixation. Polyaniline layers were formed by in situ polymerization of aniline onto polyethylene porous support placed into the aqueous reaction mixture. Structural and chemical transformations upon heating of these systems in air without any load application and in vacuum under load have been investigated by thermo-mechanical tests, IR spectrometry and electron microscopy. Composite systems demonstrated a considerably lower shrinkage upon heating than microporous polyethylene substrates. It has been noted that the composites retained mechanical integrity on heating up to temperatures much higher than the polyethylene melting point. It is concluded that thermo-mechanical behavior of the composites is determined by the presence of polyaniline phase on the surface and the bulk of polyethylene support [11].

Single-walled carbon nanotube (SWNT)–poly(vinylidene fluoride) (PVDF) composites were fabricated by dispersion of SWNT in an aqueous surfactant solution, followed by mixing with PVDF powder, filtration and hot pressing. The thermal properties of the composites at various SWNT volume fraction up to 49 % were investigated. The coefficient of thermal expansion (CTE) was decreased with increase of the SWNT content. The thermal conductivity increased with temperature from 25 to 150 °C. The thermal conductivity did not enhance to the level required by heat sink applications. The melting point was not affected significantly by the addition of SWNT. The degree of crystallinity was increased and the decomposition temperature of the matrix was decreased with increasing the amount of SWNT [12]. The thermal stability of wood polymer composites made with extractive-free wood from four different wood species was studied. Hot water (HW) extractives, ethanol/cyclohexane (E/C) extractives and both types of extractives were eliminated. Composites of LLDPE and 10 wt% of wood were prepared using poly vinyl alcohol-co-ethylene (EVOH) as a compatibilizer. The thermal degradation behavior of the composites was characterized with thermogravimetric analysis (TGA). In all the cases, the degradation temperatures increased after removal of the extractives. The removal of E/C extractives was less effective in the thermal stability improvement than the removal of HW extractives. The highest improvement on the thermal stability of WPCs was achieved when both types of extractives (E/C and HW) were removed [13].

The thermal energy storage phase change materials (PCM) based on paraffin/high density polyethylene (HDPE) composites were prepared by using twin-screw extruder technique. The morphology and properties of the PCM composites based on the flame retardant system with expanded graphite (EG) and ammonium polyphosphate (APP) were characterized by Scanning electron microscope (SEM), Differential scanning calorimeter (DSC), Thermogravimetric analyses (TGA) and Cone calorimeter tests. The SEM images showed that paraffin dispersed well in the three-dimensional net structure formed by the HDPE. The EG and APP were well dispersed in the PCM composites. The DSC measurements indicated that the additives of flame retardant had little effect on the peak temperatures of phase change and thermal energy storage property. The TGA results showed that the loadings of the EG and APP increased the temperature of the maximum weight loss and the charred residue of the PCM composites at 650 °C, which contributed to the improved thermal stability properties. Cone calorimeter tests showed that the peak of heat release rate (PHRR) decreased significantly. The homogeneous and compact charred residue structure after combustion contributed to the enhanced thermal stability, improved flammability and increased self-extinguishing properties of the PCM composites [14].

Liu et al. [15] reported the effects of CNTs on the crystallization and melting behavior, thermal stability and fire retardancy properties of polymer composites. Finally, they assessed the challenges for the future and the research that needs to be done to achieve high performance polymer/CNT composites [15]. Su et al. [16] described the thermal degradation/stability of polymer–CNT composites and their development trends. The thermal degradation of polymer–carbon nanotube (CNT) composites plays a crucial role in determining their processing and applications. The mechanisms of thermal degradation/stability improvement by carbon nanotubes are affected by the barrier effect, thermal conductivity of CNT, physical or chemical adsorption, radical scavenging action, and polymer–nanotube interaction. There is a huge scope to further investigate the thermal degradation of polymer–CNTs [16]. Polymer nanocomposites are an important class of polymers that have wide application in a number of different industrial sectors. Nanoscale fillers include layered silicates (such as montmorillonite), nanotubes (mainly carbon nanotubes, CNTs), fullerenes, SiO_2, metal oxides (e.g., TiO_2, Fe_2O_3, Al_2O_3), nanoparticles of metals (e.g., Au, Ag), polyhedral oligomeric silsesquioxane (POSS), semiconductors (e.g., PbS, CdS), carbon black, nanodiamonds, etc. The thermal degradation mechanism of these nanocomposites is related to the kind of used nanoparticles and its amount, the structure of the char formed during polymer degradation, the gas impermeability of inorganic nanoparticles, which inhibit the formation and escape of volatile byproducts during degradation and the interactions between polymer reactive groups and inorganic nanoparticles [17].

Vadukumpully et al. [18] described the fabrication and characterization of ultra-thin composite films of surfactant-wrapped graphene nanoflakes and poly(vinyl chloride). Free-standing composite thin films were prepared by a simple solution blending, drop casting and annealing route. A significant enhancement in the mechanical properties of pure poly(vinyl chloride) films was obtained with 2 wt%

loading of grapheme. 58 % increase in Young's modulus and ~130 % improvement of tensile strength were observed. The enhanced thermal stability was confirmed by thermal analysis of the composite films, which showed an increase in the glass transition temperature of the polymer [18].

The thermal stability of a new family of wood polymer composites (WPC) which use a thermoplastic elastomer matrix (pebax® copolymers) was investigated by Sliwa et al. [19]. These copolymers are poly(ether-b-amide) thermoplastic elastomers which show a significant elongation at break and a melting point below 200 °C and thereby, preventing degradation of wood fibres upon processing. Moreover, these polymers are synthesized from renewable sources and able to interact with wood fibres. Two types of pebax® matrices and two species of wood flour as fillers were used. Composites were made by using a laboratory-size twin-screw extruder to obtain homogeneous composite pellets prior to injection moulding into tensile test samples. The thermal stability of the matrix, wood fibres and composites was investigated using thermogravimetric analysis under air and nitrogen atmosphere. Spectacular improvement of thermal stability of the composites was noted under air atmosphere as opposed to measurement performed under nitrogen. The presence of wood in pebax® hinders the thermo-oxidation in air by the formation of char residue in the earlier stage of degradation. The wood content was also optimized for obtaining the protective synergism [19].

The thermal degradation behavior of polychloroprene rubber (CR) composites based on unmodified and ionic liquid modified multi-walled carbon nanotubes (MWCNTs) was studied by thermogravimetric analysis (TGA) under aerobic and anaerobic (nitrogen) conditions. The CR and its composites showed three stage and four stage degradation in nitrogen and air, respectively. The unmodified CNTs alone did not improve the thermal stability of composites to a great extent. On the contrary, reasonable enhancement was observed in case of modified CNTs/CR composites. This may be attributed to the interfacial interactions of ionic liquid/modified tubes with CR and to the fine dispersion of modified tubes in CR. The degradation products of CR and its composites were analyzed using pyrolysis-gas chromatography-mass spectrometry (Py-GC-MS). The activation energy for thermal decomposition was found to be high for modified CNTs/CR composites. Isothermal degradation of modified CNTs/CR composites at 290 °C for 30 min in nitrogen showed decreased weight loss (14 %) as opposed to CR (31 %) and unmodified CNTs/CR composites (30 %) [20]. The fabrication of flexible epoxy thin film composites was investigated by Boon et al. [21]. The thermal stability of epoxy composites decreased as the NiZn ferrite content in the epoxy was increased due to the catalytic effect of ferrite [21]. The effects of incorporation of clay and carbon nanotube dispersions on the thermal stability and flammability of biodegradable polymers has been reported. This study showed that the homogeneous dispersion of nanoparticles in a biodegradable polymer matrix enhanced the thermal degradation temperature and flammability of environmentally friendly polymer nanocomposites [22].

The effects of octavinyl polyhedral oligomeric silsesquioxane (Vi-POSS), silica (SiO_2), and octaphenyl POSS (Ph-POSS) on the thermal stability of poly(dimethylsiloxane) (PDMS) were studied using thermal gravimetric analysis. Vi-POSS enhanced the thermal stability of PDMS [23]. Composite bipolar plates based on the proper mixing of multi-walled carbon nanotubes (MWNTs), synthetic graphite particles and acrylonitrile–butadiene–styrene (ABS) powder have been produced by hot compression molding [24]. Thermal stability of the composites was examined by thermogravimetric analysis (TGA). The thermal behavior was little affected by the addition of MWNTs [24]. New rigid polypropylene composite foams filled with high amounts of flame-retardant systems based on synthetic hydromagnesite, a basic magnesium carbonate obtained from an industrial by-product were developed. The combination of hydromagnesite with an intumescent additive (ammonium polyphosphate) and layered nanoparticles improved the thermal stability. In particular, the intumescent additive delayed the beginning of the thermal decomposition temperature and the layered nanoparticles split the second step of thermal decomposition in a third peak observed at higher temperatures. Improved flame retardancy was noted in the samples having the intumescent additive [25].

Fire safety properties of polymers were improved by using bismuth subcarbonate ($(BiO)_2CO_3{\cdot}xH_2O$) nanoplate, a bismuth-containing layered nanomaterial. The introduction of $(BiO)_2CO_3{\cdot}xH_2O$ ($\leq$6.2 wt%) into poly(methyl methacrylate) (PMMA) matrix by in situ polymerization method increased the thermal stability, flame retardancy and smoke suppression properties remarkably. Moreover, the onset degradation temperature and mid-point degradation temperature was increased. On the other hand, peak heat release rate, total heat release, toxic volatile organic products and smoke density was decreased. Morphological studies of PMMA/$(BiO)_2CO_3{\cdot}xH_2O$ nanoplate composites showed that $(BiO)_2CO_3{\cdot}xH_2O$ nanoplates were well dispersed in the PMMA matrix. Thermal decomposition behaviors and char analysis demonstrated the catalytic charring effect of $(BiO)_2CO_3{\cdot}xH_2O$ to PMMA matrix. The $(BiO)_2CO_3{\cdot}xH_2O$ nanoplates combines several flame-retardant strategies including the char formation, dilution effect of CO_2 and water, and physical barrier effect, and thus enhanced the thermal stability, flame retardancy and smoke suppression of PMMA/$(BiO)_2CO_3{\cdot}xH_2O$ composites simultaneously [26]. Synthesis of reduced silanized graphene oxide/epoxy-polyurethane (EPUAs/R-Si-GEO) composites with enhanced thermal and mechanical properties was reported by Lin et al. [27]. Graphene oxide (GEO) prepared from natural graphite flakes was modified with methacryloxypropyltrimethoxysilane to prepare silanized GEO (Si-GEO), and was then reduced by $NaHSO_3$ to prepare R-Si-GEO (partially reduced Si-GEO). EPAc/R-Si-GEO (R-Si-GEO/epoxy acrylate copolymers) was synthesized via an in situ polymerization of R-Si-GEO and epoxy acrylic monomers. EPUAs/R-Si-GEO was obtained by curing reaction between EPAc/R-Si-GEO and an isocyanate curing agent. Thermal gravimetric analysis (TGA), tensile strength, elongation at break, and cross-linking density measurements showed that the thermal stability and mechanical properties of EPUAs/R-Si-GEO were greatly enhanced by the addition of R-Si-GEO [27].

Simultaneous influence of polypropylene-graft-maleic anhydride (MAPP) and silane-treated hemp fibers (HF) on morphology, thermal and mechanical properties of high-flow polypropylene (PP) modified with poly[styrene-b-(ethylene-co-butylene)-b-styrene] (SEBS) was studied. Thermal stability of HF was improved after silane treatment and less than 2 % weight loss was observed at 240 °C in composites with 30 wt% HF. Better dispersion of fibers and better efficiency in enhancing static and dynamic mechanical properties of PP, doubling its strength and stiffness were observed in composites with treated fibers compared to untreated ones. High strength and stiffness were showed by PP modified with SEBS and MAPP containing 30 wt% HFs. These composites were studied as an alternative to conventional PP/glass fibers composites for injection molding of small to medium auto parts [28]. Continuous fiber-reinforced polymers are now widely used in many industries and provide properties that are superior to those of traditional ACSR (aluminum conductor steel reinforced) cables [29]. Although composite core cables show good performance in terms of corrosion, the contact of carbon fibers with aluminum promotes galvanic corrosion, which affects the mechanical performance. Three different types of fiber coatings were tested (phenol formaldehyde resin, epoxy-based resin, and epoxy resin with polyester braiding). The use of epoxy resin combined with polyester braiding provided the best resistance to the galvanic corrosion. Investigation of thermal stability revealed that use of phenol formaldehyde resin resulted in a higher glass transition temperature. A post-cure process applied to epoxy-based resin helped to achieve glass transition temperatures up to 200 °C [29]. The nanocomposite materials with polymeric matrices, especially using layered silicates are an alternative to composites with conventional fillers. The investigation regarding the presence of montmorillonite clay in the degradation of polypropylene-montmorillonite nanocomposites (PP/OMMT) showed that the clay contributed to the beginning of exothermic oxidation reactions and to the kinetics decrease of volatile release and its formation [30].

3 Effect of Thermal Oxidation on the Mechanical Performance of Polymer Based Composites

Thermo-oxidative performance of metal-coated polymers and composites has been described by some researchers [31]. They presented the design, fabrication and oxidation resistance performance of metal barrier coatings on high temperature polymers and their composites. The metal thin-films were selected for their ductility, low oxygen solubility and diffusivity. The coatings were fabricated with thermal and e-beam evaporation processes. The bond between the metal topcoats and the polymeric substrates was enhanced with surface roughening and other metallic interlayers. The retardation in oxidation in coated specimens was characterized with isothermal aging studies. Experimental results showed that the oxidation resistance was strongly related to coating durability as the durable coatings substantially decelerated the oxidative degradation [31].

Effects of thermal treatment on the mechanical properties of poly(*p*-phenylene benzobisoxazole) fiber reinforced phenolic resin composite materials was studied by Bian et al. [32]. They fabricated the unidirectional PBO fiber reinforced phenolic resin composite material laminates and exposed in a muffle furnace of 300, 550, 700, and 800 °C for 5 min, respectively in order to study the effects of thermal treatment on mechanical properties of the composites. After thermal treatments at 300, 550 and 700 °C for 5 min, the flexural strength was reduced by 17, 37 and 80 %, respectively, the flexural modulus was decreased by 5, 14 and 48 %, respectively, and the interlaminar shear strength was lowered by 12, 48 and 80 %, respectively. The phenolic resin began to pyrolyze and shrink resulting in the irreversible damage of the composites after thermal treatment at 300 °C. After thermal treatment at 550 °C, the phenolic resin pyrolyzed mostly. But the PBO fiber had no obvious pyrolyze while the interface had severely broken. As the phenolic resin formed amorphous carbonaceous and PBO fiber pyrolyzed mostly so the mechanical properties dropped dramatically after thermal treatment at 700 °C. After heating at 800 °C for 5 min, the fiber was nearly totally pyrolyzed and kept fibrous carbonaceous and the specimen became very brittle [32].

The effect of thermo-oxidation on the local mechanical behavior of two polymer materials was investigated through the employment of instrumented Ultra-Micro Indentation (UMI) and Confocal Interferometric Microscopy (CIM). The UMI curves gave good indication of the local short-time behavior whereas the CIM observation provided some information about the long-time behavior. Significant modification of the local mechanical behavior for both studied epoxy resins with oxidation level was noted [33]. The effects of Cloisite 20A (C20A) nano-clay compounding on the thermo-oxidative degradation and thermal oxidation induced residual stresses were studied for a thermoset polymer. A variant of bismaleimide (BMI) was used having high temperature airframe applications. Isothermal aging experiments were conducted at 250 °C on nano-clay modified BMI with 3 wt% C20A and baseline BMI in different oxygen environments (0, 20 and 60 % O_2) for more than 1000 h for the examination of the influence of nano-clay compounding on the thermo-oxidative behavior of BMI. It was observed that nano-clay enhanced the thermal stability of thermoset polymer BMI by reducing oxygen diffusion and oxidative degradation during isothermal aging. A significant ($\sim$25 %) reduction in the residual radial and hoop stresses for the 3 wt% C20A modified BMI was observed because the dispersed nano-clay improved the resin shrinkage [34]. The thermo-oxidation of carbon-fiber-reinforced polymers (CFRPs) exposed to high temperatures up to 150 °C and high oxygen pressures up to 5 bars was investigated [35]. Unidirectional IM7/977-2 composite specimens were aged at 150 °C under atmospheric air and under oxygen pressure (1.7 bars and 5 bars). Degradation phenomena was investigated through periodic tests after different aging times. The thermo-oxidation-induced degradation strongly depends on aging time, distance between fibers and partial oxygen pressure [35].

The chemical structure of ambient cured DGEBA/TEPA and thermal–mechanical properties during thermo-oxidative aging were investigated. It was found to be difficult for this ambient cured DGEBA/TEPA system to cure completely even after a long period. Dynamic Mechanical Thermal Analysis (DMTA) curves demonstrated a post-cure reaction occurred in the early period of aging and then an additional transition peak emerged at a lower temperature, whose intensity increased with the aging time. The additional peak in DMTA curves were related to the lower molecule weight compounds. X-ray Photoelectron Spectroscopy (XPS) results further demonstrated that additional peak was formed due to the degradation of main chains near the sample surface and the effect of oxidation increased with decreasing depth [36]. Vu et al. reported [37] the effects of thermo-oxidation on matrix cracking in cross-ply $[0/90]_S$ composite laminates. IM7/977-2 carbon/epoxy samples were firstly aged at 150 °C under 1.7 bars of oxygen for 24, 48 and 96 h, respectively. Quasi-static tensile tests were then carried out on un-aged and aged samples. The number of matrix cracks was counted during the tensile tests to establish the evolution of the crack density with respect to the applied stress. A numerical model was utilized to evaluate the critical energy release rate of un-aged and aged laminates. A reduction of the critical energy release rate of the aged samples was measured as compared to the un-aged samples [37].

Fibre-reinforced polymers are subjected to the various ageing factors e.g. temperature, pressure, oxygen or moisture during their service life, which generally result in a decrease of their properties. Ammar-Khodja et al. [38] studied the combined actions of two aging parameters such as temperature and oxygen. Thermo-oxidation was analysed by ageing in air plain and multi-hole laminates for 9000 h. Since perforated panels had a greater surface exposed to the oxidative environment the degradation due to oxygen was increased. Thermo-oxidative ageing led to oxidation and cracking of the laminates surfaces, weight loss and decrease of both compression and SBS failure strengths. The glass transition temperature remained constant. Degradation was more prominent for the multi-hole panels. Greater decrease of the physical and mechanical properties for these laminates was observed [38]. High performance multifunctional composites can be developed through the multi-scale hybridization of carbon nanotubes (CNTs) with micro-particles in polymers. Li et al. [39] worked in this direction. In their work, hybrid fillers comprised of CNTs directly grown on alumina micro-spheres by chemical vapor deposition were incorporated into epoxy matrix that was then reinforced with woven glass fibers. The hierarchical composites with 0.5 wt% hybrid loading was observed to exhibit an improvement of 11 and 19 % in interlaminar shear strength and flexural modulus, respectively. Further, the glass transition temperature was increased by 15 °C and the storage modulus at 50 °C was increased by 20 %. These are mainly attributed to the improvements of matrix properties due to good dispersion of hybrids and their resistance towards the formation and development of matrix cracks. This study revealed the potential of multi-scale carbon hybrids in improving the mechanical and thermo-mechanical properties of the fiber-reinforced composites [39].

Polymeric matrix composites should be properly designed to resist degradation due to physical aging, chemical changes and thermo-oxidation for long term durability requirements at the elevated temperatures [40]. The interaction between oxidation and damage during high temperature aging of polymeric matrix composites was described by some researchers. They used carbon fiber-reinforced PMR-15 composites. The substantial oxidation growth was observed in the areas where discrete cracking was present. Damage evolutions, the interaction of damage and oxygen diffusivity are the critical factors, which must be considered for reduction of oxidation growth in composite materials. A model-based analysis of the oxidation in composites was presented to determine the relative effects of the matrix oxidation, role of fiber and interface effects and that of the damage growth [40]. Gigliotti et al. [40] carried out the modelling, simulation and experimental characterization of local shrinkage strains and stresses induced by thermo-oxidation phenomena in the IM7/977-2 carbon/epoxy composite material at elevated temperatures. A good agreement was found between the simulated and measured profiles. Based upon the analysis of the thermo-oxidation induced stress field it was understood that the environment has an influence on the onset of damage in composite materials at elevated temperature [41]. The thermo-oxidative degradation of polylactide (PLA) films was studied by Rasselet et al. [42] between 70 and 150 °C. They have noted that the oxidative degradation of PLA led to a random chain scission responsible for a reduction of the molar mass. A correlation between molar mass and strain at break during oxidation was established. PLA displays a brittle behavior when molar mass falls below 40 kg mol^{-1}, which was in agreement with relationships linking the critical value for embrittlement with the molar mass [42].

Researchers are trying to improve the physical properties of the CNT/epoxy composites. Although carbon nanotubes (CNTs) have superior properties the interfacial bonding between the CNTs and the polymer matrix is weak. Kim et al. [43] treated CNTs by an acidic solution to remove impurities and subsequently by amine treatment or plasma oxidation for the improvement of interfacial bonding and dispersion of nanotubes in the epoxy matrix. The mechanical properties of the modified CNT/epoxy composites and rheological properties of nanotube containing epoxy resin were improved. This may be ascribed to the modification of CNTs, which improved dispersion and interaction between the CNT and the epoxy resin [43]. Latex mixing and co-coagulation approach followed by static hot-press and twin roll mixing process was utilized to produce natural rubber (NR)-reduced graphene oxide (rGO) composites. In this process, fine control of filler dispersion was obtained and the composites exhibited a three-dimensional rGO network or alternatively a homogeneous dispersion of single rGO platelets. The composites with rGO segregated network had good resistance to oxygen and water vapor permeation. As compared to pristine rubber and composites the mechanical properties were improved with the homogeneous dispersion of single rGO platelets. The experimental results confirmed that the morphology of filler has a prominent influence on the properties of natural rubber composites [44].

It is already established that thermal-oxidative degradation of the matrix can strongly affect the apparent interfacial shear strength (IFSS) in glass fibre-polypropylene (GF–PP). Yang et al. [45] took different approaches to further investigate this phenomenon. Strong dimensional dependence was noted in case of property deterioration caused by the degradation. The degraded and non-degraded samples were characterized in terms of the thermal mechanical properties and crystallinity in order to determine the effect of polymer degradation on the measured IFSS. Comparison of the degraded and non-degraded PP microbond samples for IFSS clearly demonstrated the effect of thermal-oxidative degradation on adhesion [45].

Dominkovics et al. [46] carried out several series of experiments to study the effect of components on the stability of PP/layered silicate nanocomposites. The amount of organophilic montmorillonite (OMMT) varied between 0 and 6 while that of maleated polypropylene (MAPP) was between 0 and 50 vol%. The composites were prepared in an internal mixer at 190 °C and mixing speed and time were varied to investigate the effect of processing conditions on the stability. They showed that both OMMT and MAPP accelerated degradation during processing and deteriorated the properties of PP composites. Residual stability decreased drastically with increasing amounts of both components while chain scission led to the decrease of viscosity and inferior strength and deformability. The most probable reason for decreased stability was the reaction of the components with the stabilizers. Processing conditions has an influence on degradation considerably. Increasing shear rate and longer residence times resulted in pronounced degradation. It can be said that the basic stabilization of commercial grade polypropylenes is insufficient to protect the polymer against degradation. Additional stabilization processing is required to obtain products with superior quality [46].

Bullions et al. [47] had reported the results of an investigation of thermal-oxidative aging effects on the properties of a carbon fiber-reinforced high-performance phenylethynyl-terminated poly(etherimide). Composite panels and neat resin samples were aged at 204 °C in four different oxygen partial pressure environments such as 0, 2.84, 20.2, and 40.4 kPa with the aging time of 1750, 3500, and 5000 h. It was observed that degradation was much less and minimal mass loss had taken place from both composite and neat-resin specimens after the thermal-oxidative aging. However, the transverse flexural strength decreased with increasing oxygen partial pressure and aging time. The retention of transverse flexural strength was stated as a function of oxygen partial pressure and aging time; retention was proportional to (aging time)$^{0.29}$. Profiles of Vickers Hardness across the cross sections of aged neat resin samples showed severe degradation with increased oxygen partial pressure. It was noticed that 5000 h of aging led to increment of the glass transition temperature, which decreased with increasing oxygen partial pressure [47].

4 Future Trends

The future trend is towards the modification of polymer composites as well as application of suitable coating on polymer matrix composites. Combined sol–gel/sealing treatment process was utilized by Huang et al. [48] to fabricate thermal barrier coating on the polyimide matrix composites. Delamination and spallation of the ZrO_2/phosphates dual coating were not observed even after thermal cycling at 400 °C for 135 cycles and the average bonding strength was 5.83 MPa. Thermal oxidation of the substrate coated with the ZrO_2/phosphates dual coating was minimized at about 900 °C [48]. Graphene is of considerable interest because it can produce a dramatic improvement in properties at very low filler content. Recent advances in the modification of graphene and the fabrication of graphene-based polymer nanocomposites have been reported by the researchers [49]. The utilization of modified graphene/graphene oxide in the fabrication of nanocomposites with different polymer matrixes have been demonstrated. Different organic polymers have been used to fabricate graphene filled polymer nanocomposites using various methods. The percolation threshold can be achieved at very lower filler loading in the case of modified graphene-based polymer nanocomposites [49].

Polybenzoxazine nanocomposites highly filled with nano-SiO_2 particles were investigated for their mechanical and thermal properties as a function of filler loading. The nanocomposites were prepared by high shear mixing followed by compression molding. Relatively high micro-hardness of the PBA-a/nano-SiO_2 composites up to about 600 MPa was achieved. Finally, the significant increment in the glass transition temperature (T_g) of the PBA-a/nano-SiO_2 composites was also observed with the ΔT_g up to 16 °C at the nano-SiO_2 loading of 30 wt%. The resulting PBA-a/nano-SiO_2 composite can be used as coating material in electronic packaging or other related applications [50]. Aluminosilicates are important materials for applications related to adsorbents, water softeners, catalysis and mechanical and thermal reinforcement because of their high surface area, excellent thermal and hydrothermal stability, high shape-selectivity and superior ion-exchange capability. Recently, their use as polymer fillers has allowed their application to innovative areas such as medical and biological fields as well as in sensors, filtration membranes, energy storage and novel catalysis routes. Further, wide versatility and tailoring possibilities of both filler and matrix indicates this area to be promising technologies of the near future [51].

Single-polymer composites (SPCs) represent an emerging family within the polymeric composite materials. SPCs are classified in respect to their composition (one- and two-constituents), and preforms (non-consolidated and consolidated). SPCs are composed of amorphous or semicrystalline matrices and semicrystalline reinforcements. Methods to widen the temperature difference between the matrix- and reinforcement-giving materials of the same polymer (one-constituent) or same polymer type (two-constituent approach) have been investigated. The development of SPCs is increased by the need of engineering parts in different applications having low density and recyclability. Recently, development of SPCs is supported by novel preform preparation, consolidation and production possibilities [52]. Research activities related to functionally graded materials (FGMs) are rapidly increasing in recent years. The properties of carbon nanotubes such as high strength, high stiffness, high aspect ratio and low density have made them an excellent reinforcement for composite materials. Functionally graded carbon nanotube reinforced composite (FG-CNTRC) materials are being focused now to find out its prospect for future use [53].

Potential of nanocellulose has been established as the next generation renewable reinforcement for the production of renewable high performance biocomposites by the researchers. The tensile modulus and strength of most cellulose nanocomposites has been reported linearly with the tensile modulus and strength of the cellulose nanopaper structures. Uniform dispersion of individual cellulose nanofibres in the polymer matrix may improve the composite properties [54]. In last few decades research regarding the polymer–nanoparticles/nanocomposites is exponentially growing. Since modern polymer industry is very much dependent on the polymer–nanocomposites so their durability under various environmental conditions and degradability after their service life are essential fields of research. Clay minerals and carbon nanotubes are more often used for improving physical, mechanical and thermal properties of the polymers. The nanoparticulates have been incorporated into polymer as nano-additives for degradation and stabilization of the polymers. In this respect, the degradation and durability of polymers has been examined by Kumar et al. [55] in the presence of nanoparticles/nanocomposites under different environmental conditions.

There is a growing need for new and improved composite materials technologies due to increasing integration of microelectronics. A major challenge is to combine low thermal resistance joints with sufficient thermomechanical decoupling and reliability. Zandén et al. [56] fabricated and characterized a new type of solder matrix nano polymer composite (SMNPC) aiming to address these challenges. The SMNPC was fabricated into preforms through liquid-phase infiltration of a Sn–Ag–Cu matrix into a silver nanoparticle coated electrospun polyimide fibre mesh. The composite showed high heat transfer capability, close to that of a direct soldered interface, lower elastic modulus compared to pure Sn–Ag–Cu alloy, and reliable thermo-mechanical performance during thermal cycling. The experimental results indicated that the developed SMNPC can be a useful composite alternative compared to conventional solders and polymer matrix materials for thermal management applications [56]. Interest in greener materials has re-boosted the utilisation of natural fibres as reinforcement for renewable polymers. However, such bio-based composites often fail to deliver good mechanical performance. Hierarchical nanocomposites can be formed by using three strategies such as surface microfibrillation of (ligno) cellulosic fibres; dispersing microfibrillated cellulose within the matrix of conventional fibre reinforced composites; and attaching nano-sized bacterial cellulose onto natural fibres. The property performance gap can be minimized between renewable and petroleum-derived materials by creating hierarchical structures within composites for commercial use [57].

Yuan et al. [58] utilized a novel approach to fabricate covalently functionalized graphene oxide (fGO)/polypropylene (PP) nanocomposites. Graphene oxide was modified with *p*-phenylenediamine and cyanuric chloride and then grafted with maleic anhydride grafted polypropylene (MAPP). It was observed that the fGO had good dispersion with exfoliated and intercalated nanostructure and strong interfacial adhesion in PP. Significant increment of thermal stability of the nanocomposites was obtained at low fGO loading. The addition of only 0.5 wt% fGO increased the storage modulus and heat deflection temperature (HDT) of PP by 15.4 % and 11 °C, respectively. However, the elongation at break of nano-composites was decreased and the tensile strength remained unchanged with increasing loading of fGO. The decrease in ductility was attributed to the reduction of mobility of lamellae by fGO. The orientation of polymer chains and lamellae with respect to the direction of action of the force during deformation were inhibited in the presence of fGO [58]. Cellulose nanocrystals (CNCs) are reinforcing fillers of considerable interest for polymers due to their high modulus and potential for sustainable production. Xu et al. [59] prepared CNC-based composites with a waterborne epoxy resin matrix and characterized. The glass transition temperature (T_g) and modulus for the composites increased with increasing CNC content. The tensile strength increased from 40 to 60 MPa, suggesting good adhesion between epoxy and CNC surfaces exposed to the matrix. Further, additional water content resulting from CNC addition was not observed. Based on the experimental results it can be said that CNCs can improve thermo-mechanical performance of waterborne epoxy polymers. Thus, these polymer composites are promising as reinforcing fillers in structural materials and coatings [59].

5 Conclusion

In summary, it can be concluded that despite substantial research in the field of polymer based composites there is huge scope for more research for further enhancement of their thermo-mechanical performance.

Acknowledgments The authors are grateful to Mr. K. Dasgupta, Director, CSIR–Central Glass and Ceramic Research Institute (CSIR–CGCRI), Kolkata–700 032, India, for his kind permission to publish this work.

References

1. Haque, M.H., Upadhyaya, P., Roy, S., Ware, T., Voit, W., Lu, H.: The changes in flexural properties and microstructures of carbon fiber bismaleimide composite after exposure to a high temperature. Compos. Struct. **108**, 57–64 (2014)
2. La Mantia, F.P., Morreale, M.: Green composites: a brief review. Compos. A **42**, 579–588 (2011)
3. Salavatian, M., Smith, L.: An improved analytical model for shear modulus of fiber reinforced laminates with damage. Compos. Sci. Technol. **105**, 9–14 (2014)
4. Yu, T., Jiang, N., Li, Y.: Functionalized multi-walled carbon nanotube for improving the flame retardancy of ramie/poly(lactic acid) composite. Compos. Sci. Technol. **104**, 26–33 (2014)
5. Srikanth, I., Padmavathi, N., Kumar, S., Ghosal, P., Kumar, A., Subrahmanyam, Ch.: Mechanical, thermal and ablative properties of zirconia, CNT modified carbon/phenolic composites. Compos. Sci. Technol. 1–7 (2013)
6. Bell, J.M., Goh, R.G.S, Waclawik, E.R., Giulianini, M., Motta, N.: Polymer-carbon nanotube composites: basic science and applications. In: Cairney, J.M., Ringer, S.P., Wuhrer, R. (eds.) Materials Forum, vol. 32, pp. 144–152 (2008)
7. Harle, S.M.: The performance of natural fiber reinforced polymer composites: review. Int. J. Civ. Eng. Res. **5**, 285–288 (2014)
8. Lee, B.L., Holl, M.W.: Effects of moisture and thermal cycling on in-plane shear properties of graphite fibre-reinforced cyanate ester resin composites. Compos. A: Appl. Sci. Manuf. **27**, 1015–1022 (1996)
9. Doh, G.-H., Lee, S.-Y., Kang, I.-A., Kong, Y.-T.: Thermal behavior of liquefied wood polymer composites (LWPC). Compos. Struct. **68**, 103–108 (2005)

10. Hanu, L.G., Simon, G.P., Cheng, Y.-B.: Thermal stability and flammability of silicone polymer composites. Polym. Degrad. Stab. **91**, 1373–1379 (2006)
11. Elyashevich, G.K., Sidorovich, A.V., Smirnov, M.A., Kuryndin, I.S., Bobrova, N.V., Trchová, M., Stejskal, J.: Thermal and structural stability of composite systems based on polyaniline deposited on porous polyethylene films. Polym. Degrad. Stab. **91**, 2786–2792 (2006)
12. Xu, Y., Ray, G., Abdel-Magid, B.: Thermal behavior of single-walled carbon nanotube polymer–matrix composites. Compos. A: Appl. Sci. Manuf. **37**, 114–121 (2006)
13. Shebani, A.N., van Reenen, A.J., Meincken, M.: The effect of wood extractives on the thermal stability of different wood-LLDPE composites. Thermochim. Acta **481**, 52–56 (2009)
14. Cai, Y., Wei, Q., Huang, F., Lin, S., Chen, F., Gao, W.: Thermal stability, latent heat and flame retardant properties of the thermal energy storage phase change materials based on paraffin/high density polyethylene composites. Renew. Energy **34**, 2117–2123 (2009)
15. Liu, T.X., Huang, S.: Morphology and thermal behavior of polymer/carbon nanotube composites. Physical Properties and Applications of Polymer Nanocomposites, pp. 529–562. Woodhead Publishing, Cambridge (2010)
16. Su, S.P., Xu, Y.H., China, P.R., Wilkie, C.A.: Thermal degradation of polymer–carbon nanotube composites. Polymer–Carbon Nanotube Composites, pp. 482–510 (2011)
17. Chrissafis, D.B.: Can nanoparticles really enhance thermal stability of polymers? Part I: an overview on thermal decomposition of addition polymers. Thermochim. Acta **523**, 1–24 (2011)
18. Vadukumpully, S., Paul, J., Mahanta, N., Valiyaveettil, S.: Flexible conductive graphene/poly (vinyl chloride) composite thin films with high mechanical strength and thermal stability. Carbon **49**, 198–205 (2011)
19. Sliwa, F., Bounia, N.E., Marin, G., Charrier, F., Malet, F.: A new generation of wood polymer composite with improved thermal stability. Polym. Degrad. Stab. **97**, 496–503 (2012)
20. Subramaniam, K., Das, A., Häußler, L., Harnisch, C., Stöckelhuber, K.W., Heinrich, G.: Enhanced thermal stability of polychloroprene rubber composites with ionic liquid modified MWCNTs. Polym. Degrad. Stab. **97**, 776–785 (2012)
21. Boon, M.S., Serena Saw, W.P., Mariatti, M.: Magnetic, dielectric and thermal stability of Ni–Zn ferrite-epoxy composite thin films for electronic applications. J. Magn. Magn. Mater. **324**, 755–760 (2012)
22. Ray, S.S.: Thermal stability and flammability of environmentally friendly polymer nanocomposites using biodegradable polymer matrices and clay/carbon nanotube (CNT) reinforcements. Environmentally Friendly Polymer Nanocomposites, pp. 295–327. Woodhead Publishing, Cambridge (2013)
23. Yang, D., Zhang, W., Yao, R., Jiang, B.: Thermal stability enhancement mechanism of poly (dimethylsiloxane) composite by incorporating octavinyl polyhedral oligomeric silsesquioxanes. Polym. Degrad. Stab. **98**, 109–114 (2013)
24. de Oliveira, M.C.L., Ett, G., Antunes, R.A.: Corrosion and thermal stability of multi-walled carbon nanotube-graphite-acrylonitrile-butadiene-styrene composite bipolar plates for polymer electrolyte membrane fuel cells. J. Power Sources **221**, 345–355 (2013)
25. Realinho, V., Haurie, L., Antunes, M., Velasco, J.I.: Thermal stability and fire behaviour of flame retardant high density rigid foams based on hydromagnesite-filled polypropylene composites. Compos. B Eng. **58**, 553–558 (2014)
26. Jiang, S., Gui, Z., Shi, Y., Zhou, K., Yuan, B., Bao, C., Lo, S., Hu, Y.: Bismuth subcarbonate nanoplates for thermal stability, fire retardancy and smoke suppression applications in polymers: a new strategy. Polym. Degrad. Stab. **107**, 1–9 (2014)
27. Lin, J., Zhang, P., Zheng, C., Wu, X., Mao, T., Zhu, M., Wang, H., Feng, D., Qian, S., Cai, X.: Reduced silanized graphene oxide/epoxy-polyurethane composites with enhanced thermal and mechanical properties. Appl. Surf. Sci. **316**, 114–123 (2014)
28. Panaitescu, D.M., Vuluga, Z., Ghiurea, M., Iorga, M., Nicolae, C., Gabor, R.: Influence of compatibilizing system on morphology, thermal and mechanical properties of high flow polypropylene reinforced with short hemp fibers. Compos. B Eng. **69**, 286–295 (2015)

29. Santos, T.F.A., Vasconcelos, G.C., de Souza, W.A., Costa, M.L., Botelho, E.C.: Suitability of carbon fiber-reinforced polymers as power cable cores: galvanic corrosion and thermal stability evaluation. Mater. Des. **65**, 780–788 (2015)
30. Fitaroni, L.B., de Lima, J.A., Cruz, S.A., Waldman, W.R.: Thermal stability of polypropylene–montmorillonite clay nanocomposites: limitation of the thermogravimetric analysis. Polym. Degrad. Stab. **111**, 102–108 (2015)
31. An, N., Tandon, G.P., Pochiraju, K.V.: Thermo-oxidative performance of metal-coated polymers and composites. Surf. Coat. Technol. **232**, 166–172 (2013)
32. Bian, L., Xiao, J., Zeng, J., Xing, S., Yin, C., Jia, A.: Effects of thermal treatment on the mechanical properties of poly(p-phenylene benzobisoxazole) fiber reinforced phenolic resin composite materials. Mater. Des. 230–235. Elsevier, Amsterdam (2014)
33. Minervino, M., Gigliotti, M., Lafarie-Frenot, M.C., Grandidier, J.C.: The effect of thermo-oxidation on the mechanical behaviour of polymer epoxy materials. Polym. Testing **32**, 1020–1028 (2013)
34. Upadhyaya, P., Roy, S., Haque, M.H., Lu, H.: Influence of nano-clay compounding on thermo-oxidative stability and mechanical properties of a thermoset polymer system. Compos. Sci. Technol. **84**, 8–14 (2013)
35. Vu, D.Q., Gigliotti, M., Lafarie-Frenot, M.C.: Experimental characterization of thermo-oxidation-induced shrinkage and damage in polymer–matrix composite. Compos. A: Appl. Sci. Manuf. **43**, 577–586 (2012)
36. Li, K., Wang, K., Zhan, M., Xu, W.: The change of thermal–mechanical properties and chemical structure of ambient cured DGEBA/TEPA under accelerated thermo-oxidative aging. Polym. Degrad. Stab. **98**, 2340–2346 (2013)
37. Vu, D.-Q., Gigliotti, M., Lafarie-Frenot, M.C.: The effect of thermo-oxidation on matrix cracking of cross-ply $[0/90]_S$ composite laminates. Compos. A: Appl. Sci. Manuf. **44**, 114–121 (2013)
38. Ammar-Khodja, I., Picard, C., Fois, M., Marais, C., Netchitaïlo, P.: Preliminary results on thermo-oxidative ageing of multi-hole carbon/epoxy composites. Compos. Sci. Technol. **69**, 1427–1431 (2009)
39. Li, W., Dichiara, A., Zha, J., Su, Z., Bai, J.: On improvement of mechanical and thermo-mechanical properties of glass fabric/epoxy composites by incorporating CNT–Al_2O_3 hybrids. Compos. Sci. Technol. **103**, 36–43 (2014)
40. Pochiraju, K., Tandon, G.P.: Interaction of oxidation and damage in high temperature polymeric matrix composites. Compos. A: Appl. Sci. Manuf. **40**, 1931–1940 (2009)
41. Gigliotti, M., Olivier, L., Vu, D.Q., Grandidier, J.-C., Lafarie-Frenot, M.C.: Local shrinkage and stress induced by thermo-oxidation in composite materials at high temperatures. J. Mech. Phys. Solids **59**, 696–712 (2011)
42. Rasselet, D., Ruellan, A., Guinault, A., Miquelard-Garnier, G., Sollogoub, C., Fayolle, B.: Oxidative degradation of polylactide (PLA) and its effects on physical and mechanical properties. Eur. Polymer J. **50**, 109–116 (2014)
43. Kim, J.A., Seong, D.G., Kang, T.J., Youn, J.R.: Effects of surface modification on rheological and mechanical properties of CNT/epoxy composites. Carbon **44**, 1898–1905 (2006)
44. Yan, N., Buonocore, G., Lavorgna, M., Kaciulis, S., Balijepalli, S.K., Zhan, Y., Xia, H., Ambrosio, L.: The role of reduced graphene oxide on chemical, mechanical and barrier properties of natural rubber composites. Compos. Sci. Technol. **102**, 74–81 (2014)
45. Yang, L., Thomason, J.L., Zhu, W.: The influence of thermo-oxidative degradation on the measured interface strength of glass fibre-polypropylene. Compos. A: Appl. Sci. Manuf. **42**, 1293–1300 (2011)
46. Dominkovics, Z., Hári, J., Fekete, E., Pukánszky, B.: Thermo-oxidative stability of polypropylene/layered silicate nanocomposites. Polym. Degrad. Stab. **96**, 581–587 (2011)
47. Bullions, T.A., McGrath, J.E., Loos, A.C.: Thermal-oxidative aging effects on the properties of a carbon fiber-reinforced phenylethynyl-terminated poly(etherimide). Compos. Sci. Technol. **63**, 1737–1748 (2003)

48. Huang, W., Zou, B., Zhao, Y., Meng, X., Wang, C., Cao, X., Wang, Z.: Fabrication of novel thermal barrier coating on polymer composites via the combined sol–gel/sealing treatment process. Appl. Surf. Sci. **258**, 9058–9066 (2012)
49. Kuilla, T., Bhadra, S., Yao, D., Kim, N.H., Bose, S., Lee, J.H.: Recent advances in graphene based polymer composites. Prog. Polym. Sci. **35**, 1350–1375 (2010)
50. Dueramae, I., Jubsilp, C., Takeichi, T., Rimdusit, S.: High thermal and mechanical properties enhancement obtained in highly filled polybenzoxazine nanocomposites with fumed silica. Compos. B Eng. **56**, 197–206 (2014)
51. Lopes, A.C., Martins, P., Lanceros-Mendez, S.: Aluminosilicate and aluminosilicate based polymer composites: present status, applications and future trends. Prog. Surf. Sci. **89**, 239–277 (2014)
52. Karger-Kocsis, J., Bárány, T.: Single-polymer composites (SPCs): status and future trends. Compos. Sci. Technol. **92**, 77–94 (2014)
53. Liew, K.M., Lei, Z.X., Zhang, L.W.: Mechanical analysis of functionally graded carbon nanotube reinforced composites: a review. Compos. Struct. **120**, 90–97 (2015)
54. Lee, K.-Y., Aitomäki, Y., Berglund, L.A., Oksman, K., Bismarck, A.: On the use of nanocellulose as reinforcement in polymer matrix composites. Compos. Sci. Technol. **105**, 15–27 (2014)
55. Kumar, A.P., Depan, D., Tomer, N.S., Singh, R.P.: Nanoscale particles for polymer degradation and stabilization—trends and future perspectives. Prog. Polym. Sci. **34**, 479–515 (2009)
56. Zandén, C., Luo, X., Ye, L., Liu, J.: A new solder matrix nano polymer composite for thermal management applications. Compos. Sci. Technol. **94**, 54–61 (2014)
57. Lee, K.-Y., Bismarck, A.: Creating hierarchical structures in cellulosic fibre reinforced polymer composites for advanced performance. Natural Fibre Composites, pp. 84–102. Woodhead Publishing, Cambridge (2014)
58. Yuan, B., Bao, C., Song, L., Hong, N., Liew, K.M., Hu, Y.: Preparation of functionalized graphene oxide/polypropylene nanocomposite with significantly improved thermal stability and studies on the crystallization behavior and mechanical properties. Chem. Eng. J. **237**, 411–420 (2014)
59. Xu, S., Girouard, N., Schueneman, G., Shofner, M. L., Meredith, J.C.: Mechanical and thermal properties of waterborne epoxy composites containing cellulose nanocrystals. Polymer **54,** 6589–6598 (2013)

Analysis for Thermal Degradation of a Polymer by Factor Analysis

Akifumi Uda

Abstract This chapter introduces an application of multivariate curve resolution (MCR) technique based on a factor analysis. Not only series of IR spectra but also two-dimensional data (series of nuclear magnetic resonance (NMR), mass spectrometry (MS), and X-ray diffraction (XRD)) can deal with same manner (further more two-dimensional data generated by hyphenated techniques such as gas chromatography/mass spectrometry (GC/MS) and liquid chromatography/ultraviolet (LC/UV) analysis, which combine two functions based on different principles, namely, chromatography, which has a separating function, and spectrometry, which provides information related to molecular structure). By using MCR techniques appropriately, the mixture data is resolved into some essential elements (chemical components, transient states and phases). The results can reveal a true chemical characteristic in your study.

1 Introduction

Instrumental analysis includes various spectroscopic and structural measurement methods such as infrared (IR) spectroscopy, Raman spectroscopy, nuclear magnetic resonance (NMR), mass spectrometry (MS), and X-ray diffraction (XRD). However, to fully utilize the enormous amounts of data provided by these sensitive, electronics-based measurement instruments, improved technologies for data analysis is required. In this chapter, we propose the application of multivariate curve resolution (MCR), a factor analysis technique, to IR spectral data observed during the thermal degradation of a polymer. Conventional analytical methods involve examining spectral peaks at specific temperatures during heating and tracking the targeted wavenumber

A. Uda (✉)
Advanced Technology Development Laboratory, Mizushima R&D Center,
Mitsubishi Chemical Corporation, Okayama, Japan
e-mail: 1304604@cc.m-kagaku.co.jp

P.M. Visakh and Y. Arao (eds.), *Thermal Degradation of Polymer Blends, Composites and Nanocomposites*, Engineering Materials,
DOI 10.1007/978-3-319-03464-5_5

peaks individually for temperature-dependent changes. In contrast, our approach reveals more substantive and comprehensive information by batch processing the multivariate spectra.

2 Spectral Data and the Methods of Analysis

In conventional analysis, one instance of analysis produces one data item, such as single-value data (scalars) obtained by titration or weighing, or qualitative data (category variables) obtained from a color reaction. Owing to advancements in analysis instrumentation, vector data that indicate a series of profiles can be obtained from one instance of analysis in chromatography or spectrometry. Furthermore, matrix data are the result of processing in cases where hyphenated analysis instruments are used or time-resolved spectral analysis is performed. Deconvolution is the splitting of a single chromatograph (one-dimensional data) by fitting multiple Gaussian functions, and the MCR method introduced here is a technique for applying resolution to the analysis of two-dimensional matrix data.

2.1 The Series of Spectra

The horizontal axis of a spectrum generally represents a specific range of frequencies or wavelengths. It is a collection of numeric absorbance data corresponding to that axis. The data can be saved in a spreadsheet as one row or column of spectrum absorbance data. A spectrum is a collection of N numeric values (variables) from a1 to aN and can be thought of as an N-dimensional vector as expressed by Eq. (1).

$$\mathbf{s}_j = (a_1,\ a_2,\ \ldots,\ a_N) \tag{1}$$

Measurement for M samples under varied conditions results in an M × N matrix structure in which N spectra are expressed in M rows.

$$\begin{pmatrix} a_{11} & a_{12} & \cdots & a_{1N} \\ a_{21} & a_{22} & & a_{2N} \\ \vdots & & \ddots & \vdots \\ a_{M1} & a_{M2} & \cdots & a_{MN} \end{pmatrix} \tag{2}$$

Figure 1 shows a series of time-resolved IR spectra observed in the course of heating; the data structure is in the form of Eq. (2). This data structure is the same as in the case of XRD observed while providing perturbation that changes the crystal layer spacing (Fig. 2).

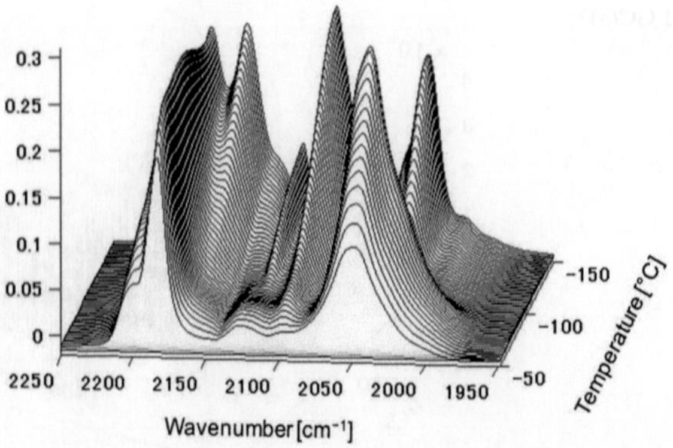

Fig. 1 Observed IR spectra for temperature perturbation

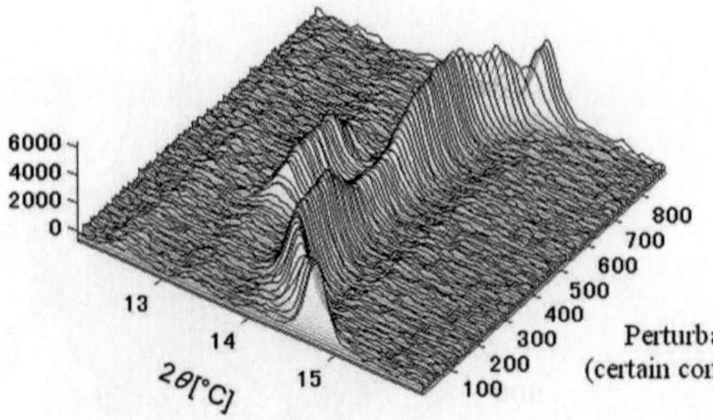

Fig. 2 XRD measured while providing perturbation that changes the crystal lattice layer spacing

Two-dimensional data can also be generated by hyphenated techniques such as gas chromatography/mass spectrometry (GC/MS) and liquid chromatography/ultraviolet (LC/UV) analysis, which combine two functions based on different principles, namely, chromatography, which has a separating function, and spectrometry, which provides information related to molecular structure. The results obtained by performing mass spectrometry while resolving a mixed gas through GC gave data in the form of a matrix structure, as shown in Fig. 3. The LC/UV data shown in Fig. 4 are similarly observed by UV-Vis spectrophotometry while resolving a mixed liquid through LC.

Researchers in the field of analysis generally refer to this as three-dimensional data (my company is no exception), but more accurately, it is a three-dimensional representation of two-dimensional data. Because three-dimensional data result in a block data structure (suggestive of the Rubik's cube), it is difficult to represent these data in a single diagram, and thus, data compilation techniques such as slicing the data parallel to each facet are necessary.

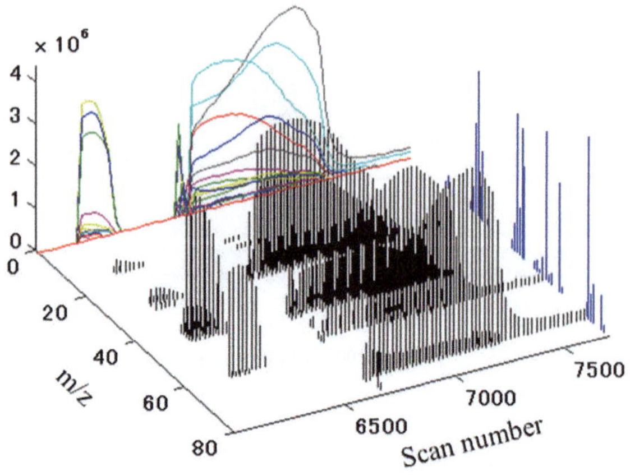

Fig. 3 Observed GC/MS spectra

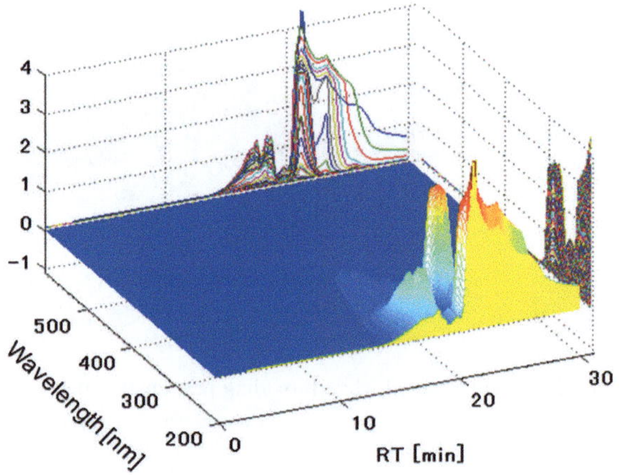

Fig. 4 Observed LC/UV spectra

2.2 Various Algorithm for MCR (Multivariate Curve Resolution)

Curve resolution techniques include soft modeling and hard modeling. Hard modeling is an approach formularized to the problem of fitting data to a hypothesized function system described by parameters. To clarify the difference from the hard modeling approach, curve resolution techniques employing soft modeling, sometimes called self-modeling curve resolution (SMCR), are used.

The first study on curve resolution, carried out by Kaiser [1] in 1958, proposed the varimax method, wherein factor rotation was used in factor analysis. Studies by Lawton and Sylvestre of Kodak clearly picked up on curve resolution technology as a means of reaction analysis in chemistry (1971, 1974) [2]. The idea of employing rotating matrices was first used in iterative target transformation factor analysis

(ITTFA, 1984, 1985) [3, 4] proposed by Gemperline of East Carolina University and Vandeginste et al. of Unilever Corporation. Rotating matrices were also used in the subsequently developed evolving factor analysis (EFA) and simple-to-use interactive mixture analysis (SIMPLISMA) [5] methods, as well as the initial heuristic evolving latent projections (HELP) [6] technique. ITTFA is a method of finding the true solution by using varimax to hypothesize a target that serves as the initial value and then repeating target transformation factor analysis (TTFA) using rotating matrices. However, an analysis technique that proactively uses eigenvalues of submatrices was first employed in EFA by Maeder in 1987. Window factor analysis (WFA), which was developed for data analysis of LC photo diode arrays by Malinowski [7] in 1992, is thought to share a common concept in that it uses eigenvalues of submatrices. SIMPLISMA, presented by Windig of Kodak in 1991, uses the concept of peak purity as the key to extracting components, which was a novel extension to the concepts used in the analysis methods proposed up to that point. HELP, which was proposed in 1991 by Kvalheim of the University of Bergen and Liang et al. of Hunan University, is a unique method that uses spectra from a selected region (bands of only a single component are observed) that is effective for analyzing hyphenated analysis data. The orthogonal projection approach (OPA) proposed by Massart [8] of Bruges University (Belgium) in 1996, which uses the concept of spectral dissimilarity in peak extraction, is also an algorithm that has garnered attention.

The structures of these algorithms can be seen as combinations of three approaches: (1) use of rotating matrices, (2) extraction of spectral purity, and (3) use of eigenvalue information of submatrices. The algorithms that have been proposed so far are listed in chronological order in Table 1.

Table 1 Various algorithms for self-modeling curve resolution

Period	Developer	Algorithm
1958	Kaiser	VARIMAX
1964	Vandeginste	ITTFA
1985	Gemperline	ITTFA
1987	Maeder	EFA
1991	Windig	SIMPLISMA (commercial)
1992	Malinowski	WFA
1992	Kvalheim	HELP (Xtricator:commercial)
1993	Kvalheim and Massart	ETA
1993	Hopke	ALS
1993	Vanslyke	EPCIA
1993	Excoffier	MCA
1995	Kalivas	SVEP
1995	ASI corp.	ConcIRT (commercial)
1996	Massart	OPA

2.3 The Matrix Formula of MCR

As an example, an ordinal overview of the MCR algorithm using matrices is presented. If we suppose the series of spectra **X** measured in Fig. 5 (m = 154 samples × n = 3319 wavenumber) consists of some elements (chemical components), and a linear space (a hypothesis), **X** represents the direct sum of the coprime subspaces.

In this instance, the coprime subspace equates to pure chemical components. In cases where the spectral matrix **X** includes a latent number of components k, it can be expressed as the direct sum of k subspaces $\mathbf{X}_i$ ($i = 1, 2, \ldots, k$) in Eq. (3).

$$\mathbf{X} = \mathbf{X}_1 + \mathbf{X}_2 + \cdots + \mathbf{X}_k \tag{3}$$

In Eq. (3) each element (coprime submatrix or chemical component) $\mathbf{X}_i$ can be resolved in two ways: score, representing the thermal axis; and loading, a spectral wavenumber. These aspects are shown in Fig. 5 as a mutual multiplication from $c_i \times s_i^T$ to $c_k \times s_k^T$ (where $k = 4$). Matrix **C** consists of an (m samples) × (k components) matrix, where the $c_1 - c_k$ vectors have k number column size ($m \times 1$); in Fig. 5, m = 154. Matrix **S** is comprised of a (k components) × (n band numbers) matrix, and the $s_1^T - s_k^T$ vectors possess a row size of number n ($1 \times n$) corresponding to band number (wavenumber, wavelength, etc., n = 3319 in Fig. 5). Matrix **E** is the residual error that could not be fit to the observed spectra. Considering k components, the summation from $\mathbf{X}_1(c_1 \times s_1^T)$ to $\mathbf{X}_k(c_k \times s_k^T)$ attempts to reconstruct the original spectra (shown in matrix $\mathbf{X}_{\mathrm{rec}}$); matrix **E** corresponds to the

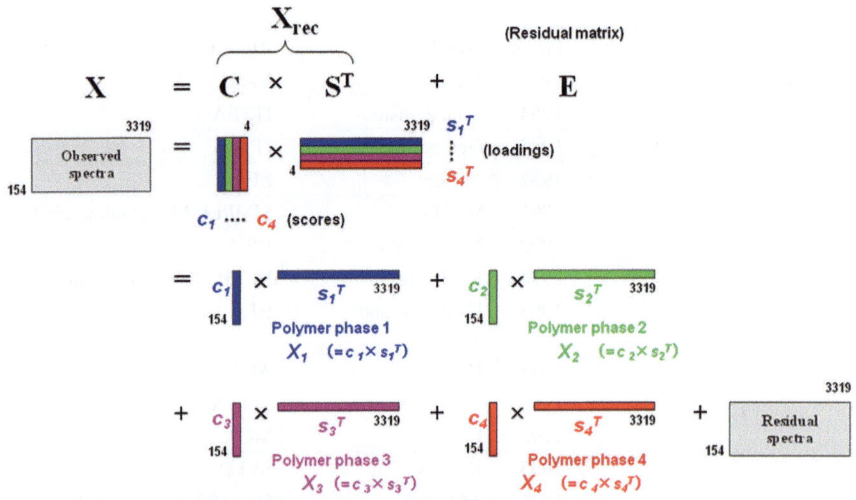

Fig. 5 The diagrammatic of resolved column vectors (*scores*) and low vectors (*loadings*)

difference between the observed spectra **X** and the reconstructed spectra $\mathbf{X}_{\mathrm{rec}}$. The numerical summation in the case of k components is shown in Eq. (4):

$$\begin{aligned}\mathbf{X} = \mathbf{C}\mathbf{S}^T = \sum_{i=1}^{k} c_i s_i^T = c_1 s_1^T + c_2 s_2^T + \cdots + c_k s_k^T + \mathbf{E} \\ = \mathbf{X}_1 + \mathbf{X}_2 + \cdots + \mathbf{X}_k + \mathbf{E} = \mathbf{X}_{\mathrm{rec}} + \mathbf{E}\end{aligned} \tag{4}$$

3 Application of Infrared Spectra to the Thermal Degradation of Poly(Vinyl Alcohol) (PVA)

3.1 Present Research on Thermal Degradation of PVA

To illustrate the approach described in Sect. 2.3, data from IR spectra measured in the thermal degradation of PVA was applied to the model. PVA is one of the most environmentally benign polymers due to its biodegradability and thermodegradability [9, 10], existing as a semicrystalline polymer with two different states: crystalline and amorphous. The crystalline phase changes its specific volume at both the glass transition temperature T_g and the melting point T_m, but this semicrystalline polymer does not possess a clearly defined melting point [9, 11]. The results of applying various analytical techniques (Fourier transform infrared spectroscopy (FTIR), thermogravimetry (TG), mass spectrometry (MS), and differential scanning calorimetry (DSC) [11–13]) to such samples has determined that the thermal degradation of PVA in both the molten and solid state produces volatile saturated and unsaturated aldehydes and ketones at random [11]. Perturbed-correlation moving-window two-dimensional (PCMW2D) correlation spectroscopy [13, 14] can be applied to analyze the degradation behavior in three regions: C–H deformation (1200–1500 cm^{-1}), C=O stretching (1500–1900 cm^{-1}), and O–H stretching (3500–3700 cm^{-1}), but the properties of each region are discussed separately. Although the temperature-dependent spectra contain detailed molecular information about the thermal perturbation, extracting useful information from the observed spectra is problematic due to the mixture of components and the overlapping of their spectra. To circumvent such problems, the MCR technique is applied to two-way data such as the IR spectra in this case.

3.2 Measurement of IR Spectra

The PVA in Fig. 6a used in this experiment was purchased from Nacalai Tesque, Inc. and is atactic with a degree of polymerization of 2.0×10^3 and a degree of saponification of 99–100 mol%. As the polymer backbone of PVA is formed by the polymerization of vinyl acetate (Fig. 6b) before saponification, the tacticity of PVA

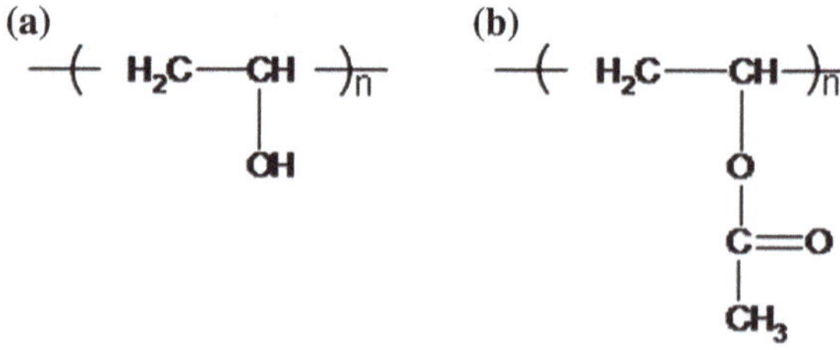

Fig. 6 Chemical structures of **a** poly(vinyl alcohol) and **b** poly vinyl acetate (*Source* [17])

becomes atactic (including a small amount of syndiotacticity) via steric repulsions between acetyl groups. The film sample was annealed in a vacuum oven at 120 °C for 6 h. The melting temperature (T_m) for the PVA sample measured by DSC was 205 °C.

IR absorption spectra of the film sample were measured using a Thermo Electron Nexus 870 FTIR spectrometer. A total of 128 scans were measured to obtain each spectrum at a resolution of 2 cm^{-1}. Spectra were measured under a nitrogen purge (oxygen less than 1 %) from 800 to 4000 cm^{-1} across a temperature range from 140 to 280 °C in 1 °C increments, with a heating rate of 0.25 °C min^{-1}.

3.3 Observed IR Spectra and the Ordinal Approach

The temperature-dependent spectra measured from 140 to 280 °C (in 1 °C increments) are illustrated in Fig. 7a, b. Figure 7a is a contour map looking down from above, while Fig. 7b is the ordinal expression. Each colored line in Fig. 7a indicates a noteworthy vibrational band.

Fig. 7 Temperature-dependent IR spectra of a PVA film

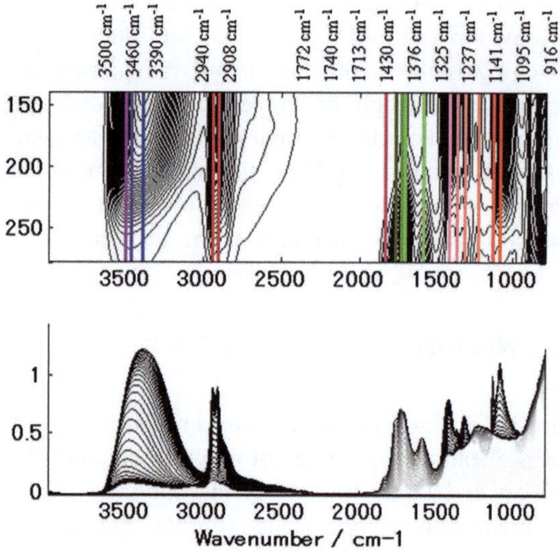

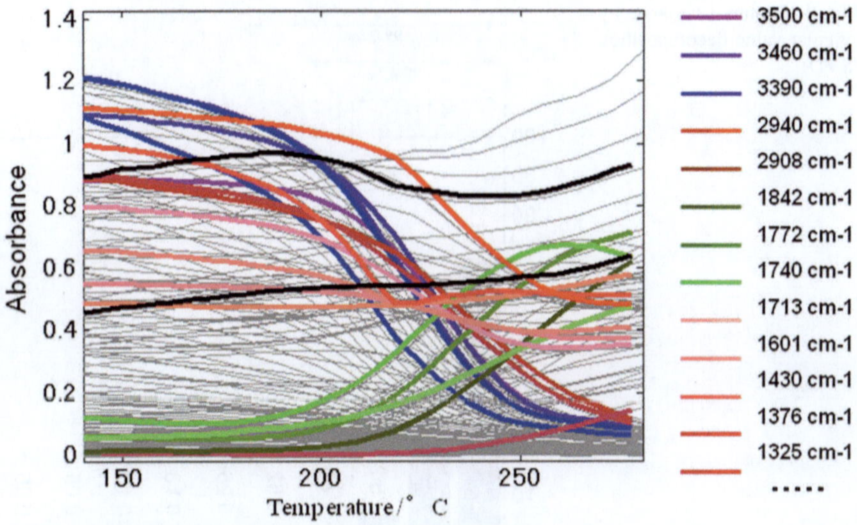

Fig. 8 Concentration profiles at characteristic wavenumber

Measured concentration profiles at characteristic wavenumbers (showing the profiles of IR spectra as a function of temperature) are shown in Fig. 8; the ordinal approach examines these profiles depending upon temperature. However, as various peak changes are influenced by each other, the inherent polymer phase changes are buried in a mixture of profiles; the MCR technique was applied to these IR spectra in order to extract the concealed phases of the components.

3.4 MCR of the Temperature-Dependent IR Spectra of a PVA Film

To estimate the number of components, singular value decomposition (SVD) was applied to the IR data **X** measured during the PVA degradation process. Using singular values from 1 to 3, 96.7 % of all variance was captured; extending the range to 1–4 allowed the capture of 98.4 % of the variance (Fig. 9).

However, while using singular values from 1 to 5 captured 99.0 %, the fifth spectral pattern became noisy, demonstrating that the fifth singular value began to capture the noise level, suggesting that four components are adequate for the IR data matrix **X**. The OPA [8] was then applied to the IR data; the OPA algorithm extracts key spectral patterns by using the dissimilarity criterion, where the dissimilarity for spectra was determined by the determinant matrix [15]. Using the dissimilarity criterion, four dissimilar spectra extracted by the OPA algorithm were utilized in the investigation [16].

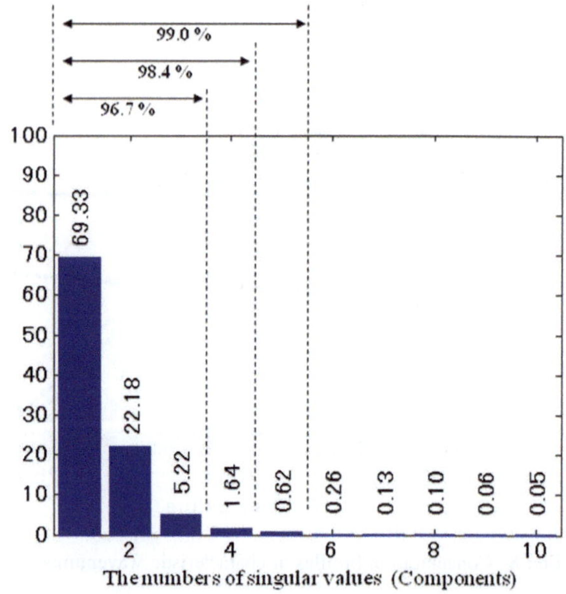

Fig. 9 Captured variance by singular value decomposition (SVD)

The spectra resolved by the OPA approach are shown in Fig. 10; four concentration profiles c_1-c_4 are shown in Fig. 10a, while Fig. 10b presents four spectral patterns, $s_1^T-s_4^T$, from the temperature-dependent spectra of the PVA film [17].

Each pairing of concentration profile and spectral pattern ($c_1-s_1^T$, $c_2-s_2^T$, $c_3-s_3^T$, and $c_4-s_4^T$, respectively) corresponds to the polymer phases 1, 2, 3, and 4, respectively which were alluded to briefly in Fig. 5. The concentration matrix C includes the column vectors c_1-c_4, and spectral matrix S includes the row vectors. The product of c_1 $(m \times 1)$ and s_1^T $(1 \times n)$ makes an $(m \times n)$ matrix $\mathbf{X}_1$ $(=c_1 \times s_1^T)$ for the first chemical component. Using the same procedure, the remaining components

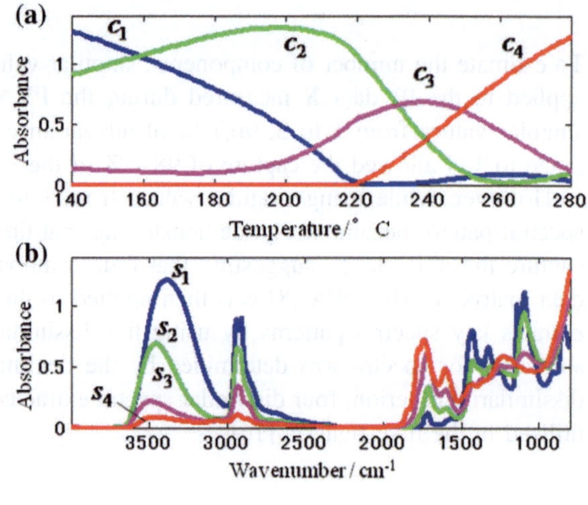

Fig. 10 **a** Concentration profiles and **b** spectra patterns obtained by MCR from the temperature-dependent IR spectra of the PVA film (Reprinted from [17])

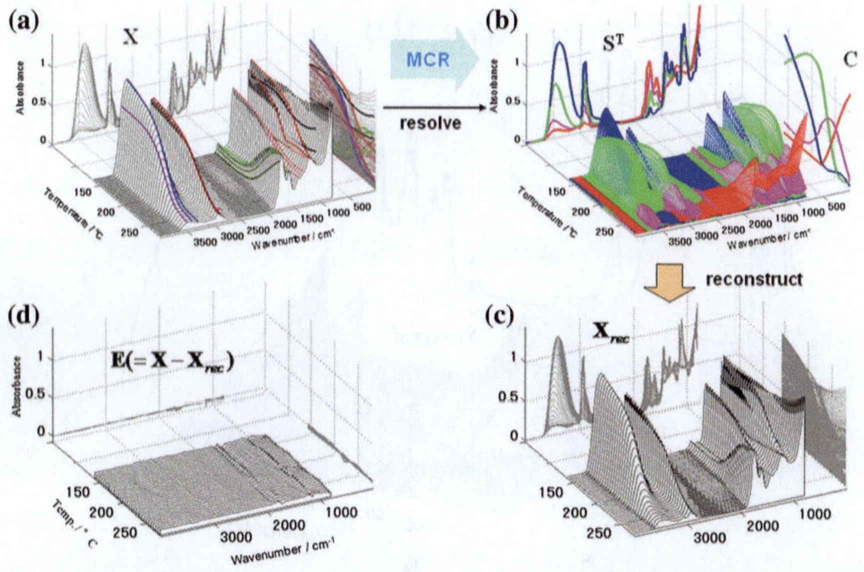

Fig. 11 **a** Observed spectra, **b** resolved spectra, **c** reconstructed spectra and **d** residues (A part reprinted from [17])

were calculated from $\mathbf{X}_2$ $(=c_2 \times s_2^T)$ as the second component to $\mathbf{X}_4$ $(=c_4 \times s_4^T)$ as the fourth. Via the summation from $\mathbf{X}_1$ to $\mathbf{X}_4$, the original matrix $\mathbf{X}_{\mathrm{rec}}$ $(= \mathbf{X}_1 + \cdots + \mathbf{X}_4)$ was reproduced. The process for resolving and reproducing spectra is shown moving from Fig. 11a–c [17]. The difference between the observed spectra **X** (Fig. 11a) and the reconstructed spectra $\mathbf{X}_{\mathrm{rec}}$ (Fig. 11c) is called the residual spectrum (matrix) **E** $(=\mathbf{X} - \mathbf{X}_{\mathrm{rec}})$, which is shown in Fig. 11d. The maximum error was within ±8.8 %, which is an allowable level for this qualitative study. As observed in Fig. 11d [17], the reproducibility of the original IR spectra is adequate.

3.4.1 Resolved Spectrum $\mathbf{s}_1$: Polymer Phase 1, Solid State (Crystalline)

The spectrum s_1 of the first chemical component, the concentration profile c_1 resolved by MCR, and the 3D map of the generated matrix $\mathbf{X}_1$ $(=c_1 \times s_1^T)$ are shown in Fig. 12; the 3D map shows the decline of s_1 absorbance. The smallest peak at 1713 cm^{-1} is due to the C=O stretching vibration; it is clear that PVA itself does not contain a C=O group. Because the production of a 100 mol% degree of saponification is technically difficult [13], usually the small C=O residual in polyvinyl acetate (PVAc, Fig. 6b) from the saponification reaction includes the C=O band. However, it is not clear whether the small amount of residual PVAc groups show absorbance values close to 0.1 or not. Since the measurement of IR spectra began at 140 °C (at which point crystalline material starts to melt), there exists a possibility that some part of the PVA formed C=O moieties via the

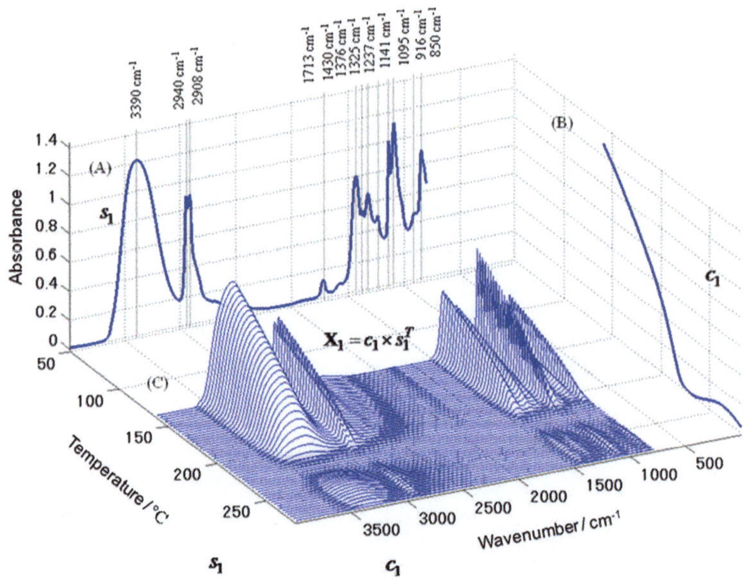

Fig. 12 (*A*) Spectrum, (*B*) concentration and (*C*) spectral changes in polymer phase 1

elimination of adjacent hydroxyl units. Two sharp peaks at 1141 and 1095 cm^{-1} in the fingerprint region are associated with C–O stretching modes from the crystalline and amorphous phases, respectively [18]. The absorbance ratio D_{1141}/D_{1095} is well known as an index for the degree of crystallinity [14, 19, 20]. As the ratio increases, the density and the degree of crystallinity also increase. Furthermore, the ratio of two peaks at 916 and 850 cm^{-1} (D_{916}/D_{850}) is used as an indicator for the amount of syndiotacticity in the polymer structure [21]. To summarize the assignment of spectrum s_1: given (i) the broad O–H peak at ca. 3390 cm^{-1}, (ii) no newly formed C=O peaks, and (iii) the crystallinity and syndiotacticity, component 1 exhibits the features of the crystalline structure of PVA as the polymer phase.

3.4.2 Resolved Spectrum s_2: Polymer Phase 2, Rubber State (Amorphous)

The spectra for the second component are presented in Fig. 13; the spectrum s_2, the concentration profile c_2 and the 3D map of the generated matrix $\mathbf{X}_2$ $(=c_2 \times s_2^T)$ are presented analogously to Fig. 12. Comparing spectrum s_2 with spectrum s_1 (shown in Fig. 12), it can be seen that the O–H stretching band is shifted by 70 cm^{-1} to a higher wavenumber and occurs at 3460 cm^{-1}. Hydroxyl groups yield broad peaks when they engage in strong hydrogen bonding, due to variations in strength of the hydrogen bonds. As the hydrogen bonds weaken via the thermal degradation of the polymer backbone, the O–H stretching peak becomes sharp when released from

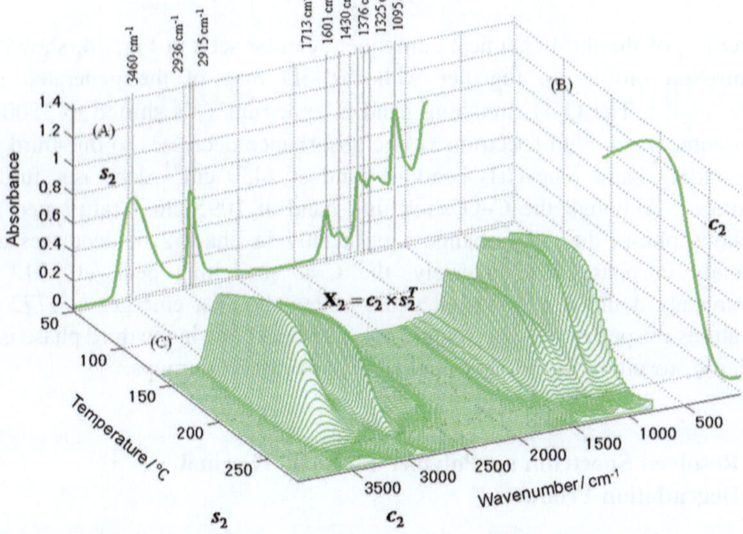

Fig. 13 (*A*) Spectrum, (*B*) concentration and (*C*) spectral changes in polymer phase 2

the effect of hydrogen bonding, shifting to higher wavenumber. However, in this instance, since the thermal degradation occurs simultaneously, the O–H band shifts to a higher wavenumber and attenuates before sharpening. Furthermore, the absorbance of s_2 becomes about half of s_1. The pair of peaks at 2940 and 2908 cm^{-1} in spectrum s_1 are due to antisymmetric and symmetric CH_2 stretching vibrations, respectively; in the case of spectrum s_2, the two bands are slightly shifted to around 2936 and 2915 cm^{-1}, implying an environmental change of the adjacent hydrogens attached to the polymer backbone. Both the O–H bending band at 1325 cm^{-1} and the CH_2 bending band at 1430 cm^{-1} also decrease. The 1713 cm^{-1} peak (which weakly appears in s_1) increases in s_2, due in part to residual polyvinyl acetate, and also to a carbonyl group newly produced by the thermal degradation of the polymer. From the combination of the C=O band at 1713 cm^{-1} and a shoulder in the 2900–2800 cm^{-1} region, we suggest the presence of acetyl group (CH_3CO^-). An emerging peak at 1601 cm^{-1} is associated with thermal degradation products containing a conjugated C=C bond [11, 13]. It can be seen that the C–O stretching band at around 1141 cm^{-1} arising from the crystalline phase disappears in spectrum s_2, while the adjacent band at around 1095 cm^{-1} band changes slightly, due to amorphous phase C–O stretching. This behavior suggests that most of the crystalline phase is transformed into an amorphous phase. The changes in spectrum s_2 imply that the polymer backbone undergoes a remarkable structural change while simultaneously relaxing the restricted crystalline state. The production of carbonyl groups indicates the onset of the thermal degradation of PVA.

3.4.3 Resolved Spectrum s_3: Polymer Phase 3, Molten State (Fluid)

Spectrum s_3 of the third chemical component can be seen in Fig. 14, showing the concentration profile c_3, together with the 3D map of the generated matrix $\mathbf{X}_3$ $(=c_3 \times s_3^T)$. The O–H stretching band in spectrum s_3 is shifted to 3500 cm^{-1} and, in comparison with spectrum s_1, the absorbance decreases to one-third of the corresponding band. The CH_2 bending band at 1430 cm^{-1} decreases further in spectrum s_3. Although the C–O stretching band at 1095 cm^{-1} (attributed to the amorphous phase) does not change significantly in phase 2, it decreases almost completely in phase 3. Conversely, the C=O stretching peak at 1713 cm^{-1} approximately doubles in intensity, while a new shoulder emerges at 1772 cm^{-1}. The features of spectrum s_3 suggest the polymer backbone in the third phase is close to collapse, resulting in the rapid and formation of C=O groups.

3.4.4 Resolved Spectrum s_4: Polymer Phase 4, Thermal Degradation Products

Spectrum s_4 of the fourth chemical component is shown in Fig. 15, presenting the concentration profile c_4 and the 3D map of the generated matrix $\mathbf{X}_4$ $(=c_4 \times s_4^T)$ in the same form as the three previous phases. In the fourth phase, the O–H stretching band at 3500 cm^{-1} becomes very weak and the CH_2 bending band at 1430 cm^{-1} decreases significantly. Conversely, the peak at 1772 cm^{-1} approximately doubles in intensity compared with the corresponding peak in spectrum s_3, while an intense

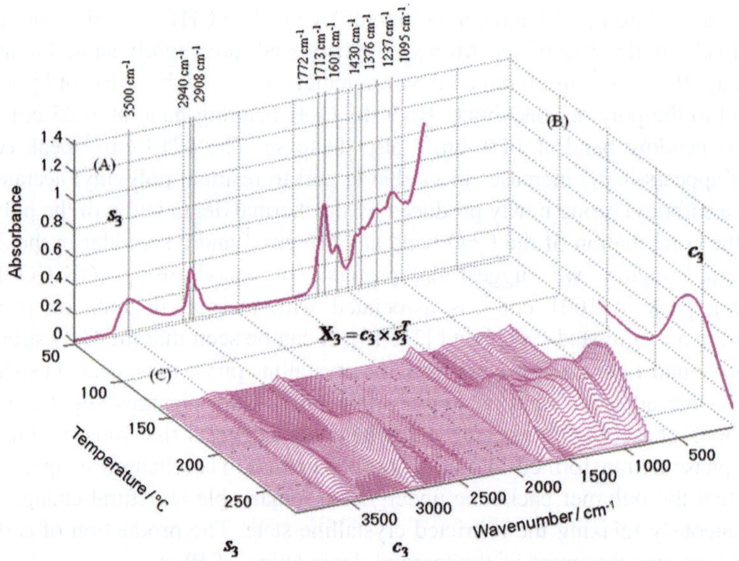

Fig. 14 (*A*) Spectrum, (*B*) concentration and (*C*) spectral changes in polymer phase 3

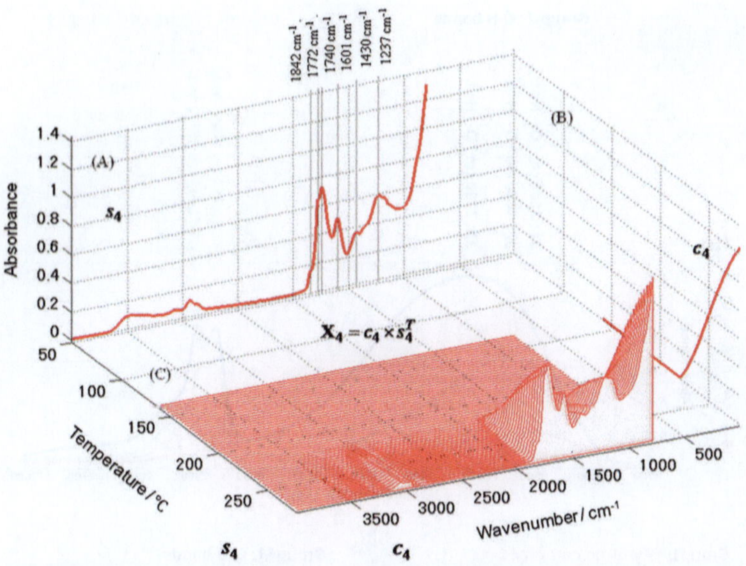

Fig. 15 (*A*) Spectrum, (*B*) concentration and (*C*) spectral changes in polymer phase 4

new peak at 1740 cm^{-1} appears. These peaks imply the formation of a large number compounds containing C=O groups, such as saturated aldehydes and ketones, aromatic aldehydes and ketones [22, 23], and unsaturated aldehydes and ketones by a combination of cross-linking and Diels-Alder cycloaddition reactions [11–13]. The band observed in spectra s_2 and s_3 at 1601 cm^{-1} increases by about three times in spectrum s_4.

3.4.5 Overview of Decomposition from Phase 1 to Phase 4

Expanded spectra for s_1-s_4 are shown for the 800–1900 and 2600–3700 cm^{-1} regions in Fig. 16a, b [17].

The vibrational modes based on the same atomic group are labeled with the same character; for example, the OH bands are labeled J1, J2, J3, and J4. Similarly, the CH bands are labeled with K1, K2, K3, K4, and K5; pyrolytic carbonyl C=O bands with L1, L2, and L3; and C–O bands with M1, M2, M3, and M4 to group characteristic peaks together.

Although the bands labeled J1–J3 are assigned to OH stretching modes, the bands become very broad due to the intermolecular interactions of OH groups. The three peaks, s_1: J3 (3390 cm^{-1}); s_2: J2 (3460 cm^{-1}); and s_3: J1 (3500 cm^{-1}) gradually decrease in intensity and shift to a higher wavenumber, disappearing entirely in phase 4. In this group J (Fig. 16a), it is noteworthy that peak sharpening is not observed, while the shift to higher wavenumber due to a decrease in the intermolecular hydrogen bonding of OH groups with temperature is observed. This

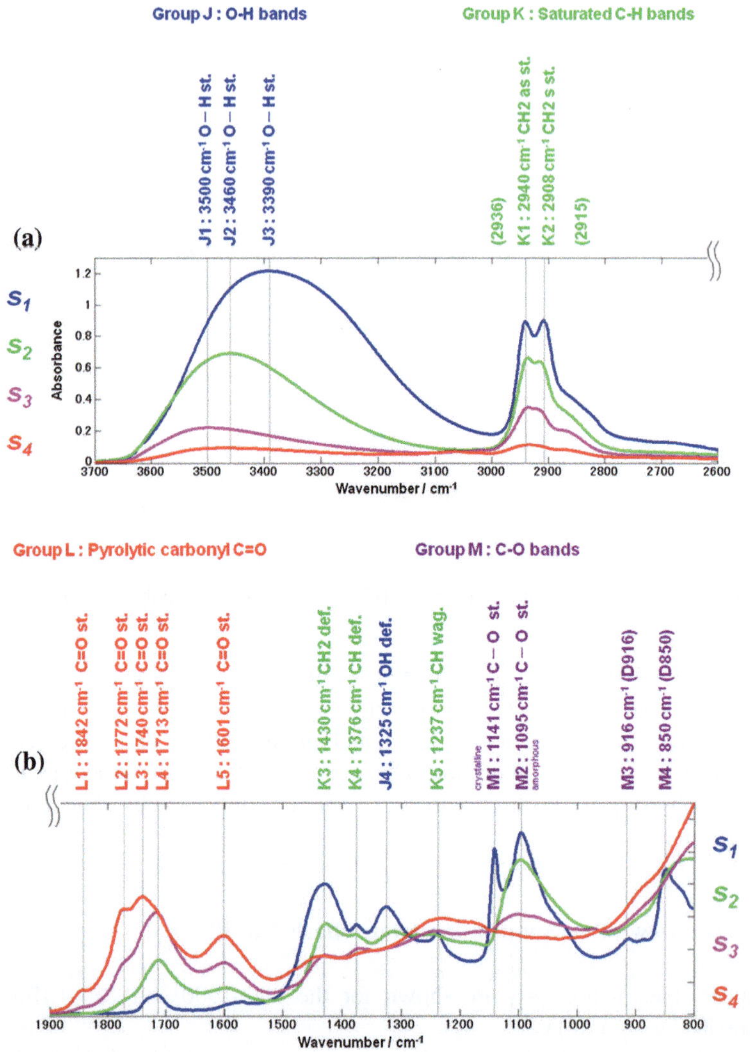

Fig. 16 Enlarged IR spectra from S1 to S4, **a** 3700–2600 and **b** 1900–1800 cm^{-1} regions (A part modified from [17])

behavior suggests that the OH elimination from the polymer backbone occurs before releasing from the H-bonding interaction. Peak J4 (1325 cm^{-1}), assigned to the OH bending mode, also decreases, which further corroborates OH elimination.

Group K arises due to saturated hydrocarbon CH units. Bands K1 (2940 cm^{-1}) and K2 (2908 cm^{-1}) are derived from an adjacent hydrogen CH_2 stretching mode. Band K3 (1430 cm^{-1}) corresponds to the CH_2 bending mode, while band K4 (1376 cm^{-1}) is assigned to the CH bending mode. These modes K1–K4 possess vibrations characteristic of hydrogen bonded directly to the polymer backbone.

Band K5 (1237 cm^{-1}, due to CH "wagging" vibrations) increases slightly, the change capturing the vibration of side groups related to hydrocarbon degradation. From the changes in the CH bands, it is clear that the PVA polymer backbone degrades due to the elimination of hydroxyl side groups.

Group L is due to carbonyl species, while also capturing some degradation products. Peak L3 (1713 cm^{-1}) in spectrum s_2 approximately doubles in intensity in spectrum s_3. Small shoulder peaks L1 (1772 cm^{-1}) and L2 (1740 cm^{-1}) are simultaneously observed to appear in spectrum s_3. These slight shoulders (L1 and L2) in spectrum s_3 have greatly developed in spectrum s_4, and peak L3 (1740 cm^{-1}) reaches a maximum. Group L captures the transition to different thermal degradation products by increasing decomposition temperature.

Group M shows the degree of crystallinity. Band M1 (1141 cm^{-1}) is derived from the crystalline phase, while band M2 (1095 cm^{-1}) originates from the amorphous phase; both peaks are observed in spectrum s_1. Initially, band M1 (1141 cm^{-1}) decreases in spectrum s_2, while band M2 (1095 cm^{-1}) continuously decreases into spectrum s_3. This sequence implies that the polymer backbone is destroyed by thermal degradation after the crystalline phase transitions to the amorphous phase. Saturated and unsaturated aldehydes, ketones, and the formation of vinyl esters rearranged in the end group are recognized as thermal degradation products [11–13]. The sequences are summarized graphically in Figs. 17 and 18. Plausible candidates for the structure of degradation products are shown in Fig. 19, however the assignment of peaks has not yet been resolved.

	atm gr.	assign	w.n.	Polymer Phase 1	Polymer Phase 2	Polymer Phase 3	Polymer Phase 4
No.1	J1	O-H st.	3500			↓	
No.2	J2	O-H st.	3460		↓		
No.3	J3	O-H st.	3390	↓			
No.4	K1	CH2 as st.	2940	↓	↓	↓	
No.5	K2	CH2 s.st	2908	↓	↓	↓	
No.6	L1	C=O st.	1842			↑	↑
No.7	L2	C=O st.	1772			↑	↑
No.8	L3	C=O st.	1740		↑	↑	↑
No.9	L4	C=O st.	1713		↑	↑	↑
No.10	L5	C=O st.	1601		↑	↑	↑
No.11	K3	CH2 def.	1430	↓	↓		
No.12	K4	CH def.	1376	↓	↓		
No.13	J4	OH def.	1325	↓	↓		
No.14	K5	CH wag.	1237				↑
No.15	M1	C-O crys.	1141	↓			
No.16	M2	C-O amor.	1095	↓	↓	↓	↓
No.17	M3	D916	916		↑	↑	↑
No.18	M4	D850	850				

Polymer Phase 1 (crystal structure)
#1 O-H broad peak around 3390 cm^{-1}
#2 No newly formed C=O peaks
#3 Crystallinity and syndiotacticity

Polymer Phase 2 (thermal degradation starts)
#1 O-H stretching band is shifted by 70 cm^{-1}
#2 O-H CH_2 bending band 1430 cm^{-1} decrease
#3 C-O stretching band 1141 cm^{-1} disappears
crystalline

Polymer Phase 3 (polymer backbone collapse)
#1 O-H stretching band is shifted to 3500 cm^{-1}
#2 C-O stretching band 1095 cm^{-1} decreases
amorphous
#3 C=O stretching 1713 cm^{-1} about two times

Polymer Phase 4 (crystal structure)
#1 O-H stretching band becomes very weak
#2 C=O stretching 1772 cm^{-1} about two times
#3 C=O stretching 1601 cm^{-1} about three times

Fig. 17 Spectral assignments corresponding to each phase

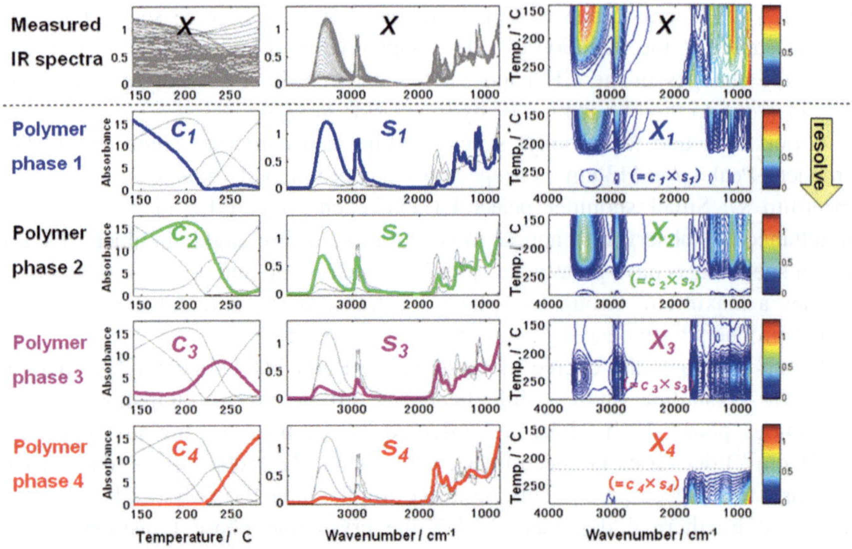

Fig. 18 Measured spectra and resolved spectra by MCR

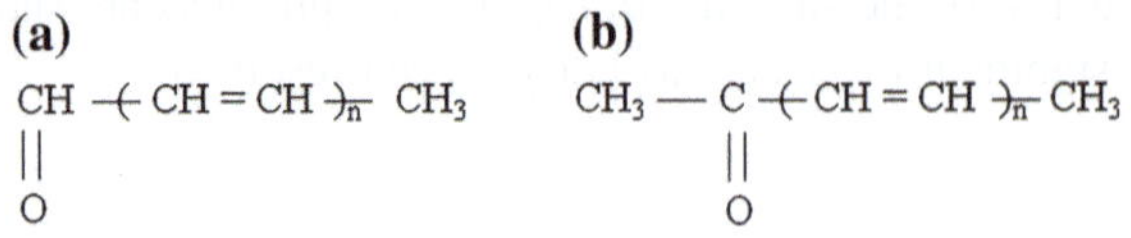

where n = 0, 1, 2, 3, etc

Fig. 19 Thermal decomposition products of PVA **a** aldehydes **b** methyl ketones

3.5 An Idea of Phase Diagrams for the Thermal Degradation Process

Figure 20a shows the ordinal relationship between the specific volume and the crystalline melting point, crystalline phase, and amorphous phase of the polymer. Figure 20b [17] plots the crystallinity estimated every 5 °C by IR spectra [13]. From 195 °C, the crystallinity decreases rapidly, indicating the onset of melting; the figure shows that the bulk of the material was transformed into the amorphous phase at ca. 220 °C. Figure 20c [17] is the same concentration matrix C (shown in Fig. 10a) resolved by MCR from 140 to 280 °C. The c_1 profile in Fig. 20c resembles the decreasing crystallinity profile in Fig. 20b, and the peak temperature of c_2 fits the point where the crystallinity rapidly decreases at ca. 195 °C. The profile c_3 increases above 195 °C (the point the crystallinity begins to fall) before decreasing after passing a peak at 240 °C. Although profile c_4 increases at ca. 220 °C, the temperature is consistent with the ending of the transition to the

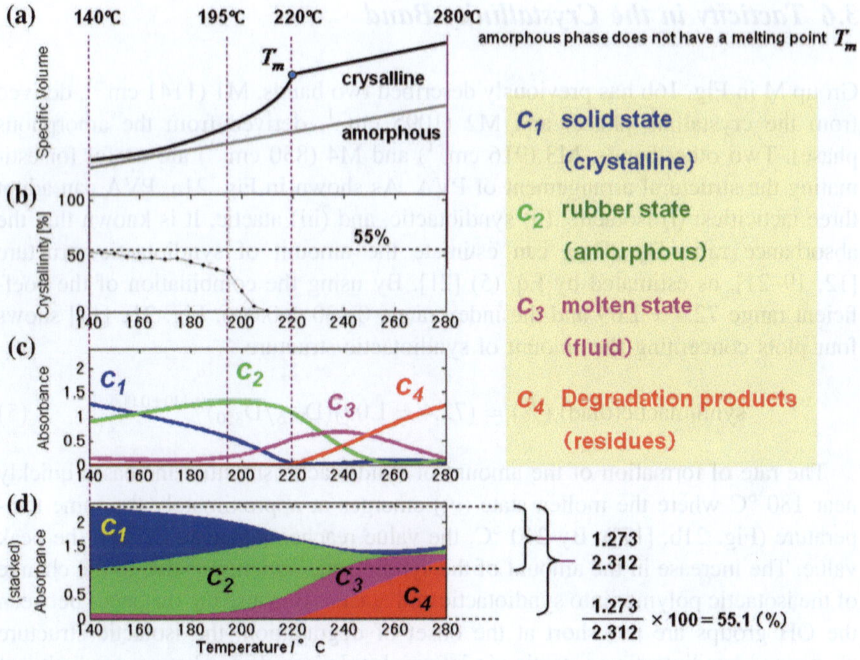

Fig. 20 **a** Specific volume, **b** crystallinity, **c** MCR profile, and **d** stacked profiles (Reprinted from [17])

amorphous state (shown in c_1). From these results, the temperature of 220 °C implies the melting point, T_m (Fig. 20a).

Figure 20d [17] presents the stacking of profiles c_1–c_4 shown in Fig. 20c. The stacking ratio of c_1 at 140 °C is about 55 % (=1.273/2.312 × 100), consistent with the crystallinity in Fig. 20b implying that the c_1 profile represents the polymer prior to the decomposition of the solid state (crystalline phase). As the thermal motion increases with temperature, the crystalline phase is changed into the amorphous phase of the rubber state, caused by the release of intramolecular interactions. Shortly afterward, the amorphous phase becomes a fluid state, before the gradual onset of thermal degradation. Comparing the similar or dissimilar peaks in spectra s_2 and s_3 with the corresponding profiles c_2 and c_3, phases 2 and 3 are suggested to be in rubber (amorphous) and molten states, respectively. It is clear that phase 4 is consistent with polymer thermal degradation residues. A semicrystalline polymer adopts a mixed phase structure containing both crystalline and amorphous phases. When each phase possesses a different degradation time by melting, they evolve into states of coexistence. In an ordinal approach, only examining a mixture of peaks is possible. When MCR is utilized, the mixture of peaks is resolved into the inherent components, and the four essential outlines emerge. The difference in the stacked amounts between the beginning and end is due to the formation of volatile degradation products at the expense of the semicrystalline starting material.

3.6 Tacticity in the Crystallinity Band

Group M in Fig. 16b has previously described two bands, M1 (1141 cm^{-1}, derived from the crystalline phase) and M2 (1095 cm^{-1}, derived from the amorphous phase). Two other bands, M3 (916 cm^{-1}) and M4 (850 cm^{-1}) are useful for estimating the structural arrangement of PVA. As shown in Fig. 21a, PVA can adopt three tacticities: (i) isotactic, (ii) syndiotactic, and (iii) atactic. It is known that the absorbance ratio D_{916}/D_{850} can estimate the amount of syndiotactic structure [12, 19–21], as estimated by Eq. (5) [21]. By using the combination of the coefficient range 72.4 ± 1.09 and the index range 0.430 ± 0.006, Fig. 21c [17] shows four plots concerning the amount of syndiotactic structure.

$$\text{syndiotactic(diad)}\ (\%) = (72.4 \pm 1.09)(D_{916}/D_{850})^{0.430\pm0.006} \tag{5}$$

The rate of formation of the amount of syndiotactic structure increases quickly near 180 °C where the molten state (c_3) emerges at approximately the same temperature (Fig. 21b, [17]). By 240 °C, the value reaches a plateau beyond the peak value. The increase in the amount of the syndiotactic structure is due to the change of the isotactic polymer into syndiotactic and atactic. Because the distances between the OH groups are too short at the onset of degradation, the isotactic structure changes to syndiotactic or atactic via intramolecular repulsions between substituent groups. As a result of accelerated thermal degradation above 240 °C, the remaining

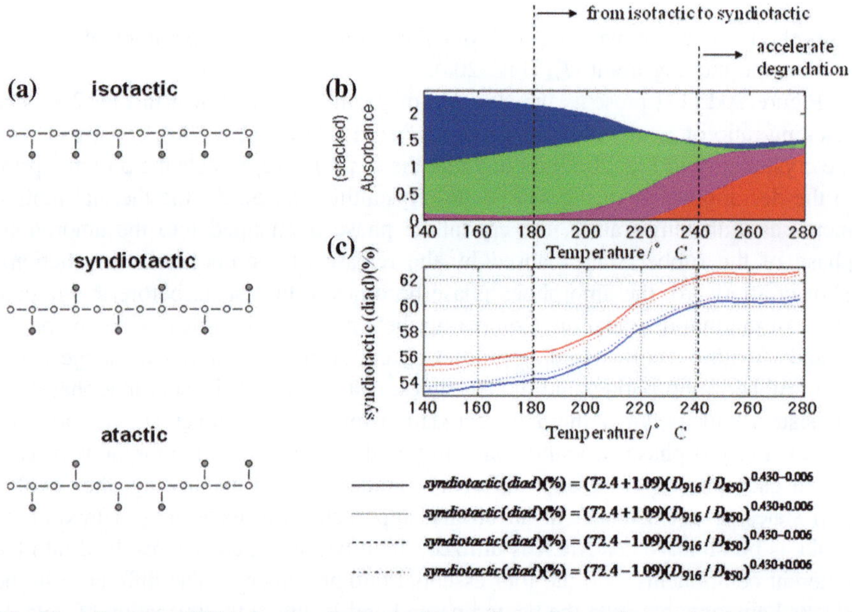

Fig. 21 **a** Tacticity of PVA, **b** concentration profiles on four polymer phase and **c** change range of the amount of syndiotactic structure (Reprinted from [17])

M4 band (850 cm^{-1}) disappears in the last stages of decomposition, whereupon the absorbance ratio D_{916}/D_{850} cannot capture the properties of the crystalline state.

4 Summary

Starting from an explanation of multidimensional data structures, an example for solving IR spectra obtained during the study of thermal polymer degradation (using PVA as a specific example) was introduced. By using MCR techniques appropriately, the data was resolved into four essential polymer phases. The four phases of the components were ascribed to (1) the solid state (crystalline), (2) the rubber state (amorphous), (3) the molten state (fluid), and (4) thermal degradation products (residues). Stacking profiles c_1–c_4 from the phase diagram of the thermal degradation process allowed a visualization of the process. At a temperature roughly around the melt temperature of this polymer, each phase exists in a transitional mixture; the MCR technique is able to reveal truly transient characteristics.

References

1. Kaiser, H.F.: Psychometrika **23** (1958)
2. Sylvestre, E.A., Lawton, W.H., Maggio, M.S.: Technometrics **16**(3) (1974)
3. Gemperline, P.J.: J. Chem. Inf. Comput. Sci. **24** (1984)
4. Vandeginste B.G.M. et al.: J. Chemon. **1** (1987)
5. Windig, W., Guilement, J.: Anal. Chem. **63** (1991)
6. Kvalheim, O.M., Liang, Y.: Anal. Chem. **64**, 936 (1992)
7. Malinowski, E.R.: Factor Analysis in Chemistry, 3rd edn. Wiley, New York (2002)
8. Sanchez, F.C., Toft, J., van den Bogaert, B., Massart, D.L.: Anal. Chem. Chem. **68**, 79–85 (1996)
9. Finch, C.A.: Polyvinyl Alcohol: Developments. Wiley, Chichester, West Sussex (1992)
10. Chiellini, E., Corti, A., D'Antone, S., Solaro, R.: Prog. Polym. Sci. **28**, 963 (2003)
11. Holland, B.J., Hay, J.N.: Polymer **42**, 6775 (2001)
12. Kenney, J.F., Willcockson, G.W.: J. Polym. Sci. **4**, 679 (1966)
13. Morita, S., Kitagawa, K., Noda, I., Ozaki, Y.: J. Mol. Struct. **30**, 883 (2008)
14. Morita, S., Shinzawa, H., Noda, I., Ozaki, Y.: Appl. Spectrosc. **60**, 398 (2006)
15. Sanchez, F.C., Hancewicz, T., Vandeginste, B.G.M., Massart, D.L.: Anal. Chem. Chem. **69**, 1477 (1997)
16. Vandeginste, B.G.M., Massart, D.L., Buydens, L.M.C., De Jong, S., Lewi, P.J., Smeyers-Verbeke, J.: Handbook of Chemometrics and Qualimetrics: Part B. Elsevier, Amsterdam (1998)
17. Uda, A., Morita, S., Ozaki, Y.: Polymer **54**, 2130–2137 (2013)
18. Sakurada, I., Nukushina, Y., Sone, Y.: J. Polym. Sci. **12**, 506 (1955)
19. Kenney, J.F., Holland, V.F.: J. Polym. Sci. **4**, 699 (1966)
20. Fujii, K., Mochizuki, T., Imoto, S., Ukita, M., Matsumoto, M.: J. Polym. Sci. Part A: Polym. Chem. **2**, 2327 (1964)

21. Murahashi, S., Nozakura, S., Washimi, M., Yuuki, H., Hata, S.: J. Polym. Sci. Part B: Polym. Phys. **4**, 65 (1966)
22. Gilbert, J.B., Kipling, J.J.: Fuel **12**, 249–260 (1962)
23. Tsuchiya, Y., Sumi, K.: J. Polym. Sci. Part A: Plym. Chem. **7** (1969)

Radiation Effects on Polymer-Based Systems

Traian Zaharescu

Abstract The applications of the radiation processing involve the deep modifications in the irradiated materials starting with molecular scissions. The specific answers illustrate the progress of chemical changes onto the foreseen goals, which are related to the exposure parameters: dose, dose rate, environment. The type of bonds influences the radiation resistance by their energies and the reactivities of intermediates are the main criterion on which the selection of technological purpose is based. The initial structure of radiation processed material, the radiation stability of main component, the proposed formulation or the manufacture conditions are responsible for the amplitude of conversion or for the process yields. The essential aspects of radiochemical behavior of processed polymers are related to the followed mechanisms, which have particular features provided by material functionality. The main processes involving radiation effects: crosslinking and grafting as well as degradation are analyzed on different polymer structures, on various irradiation parameters, on the material availabilities to high modification levels under high energy radiation exposure. The radiation processing of polymers by their exposure to different types of irradiation sources is presented for the illustration of general possibilities offered by industrial applications. This review is relevant for the extension of applications which can be adapted to several conditions. The presented examples are start points for further studies in which raw materials can be changed as well as blend formulations. The open directions are available based on the provided information.

1 Introduction

High energy radiation induces the "clean" modifications and efficient modification source of many applications. Along the time, several books and reviews presenting the effects induced during high energy irradiation were issued [1–11]. The energy

T. Zaharescu (✉)
Department of Radiation Processing, National Institute or Electrical Engineering
(INCDIE ICPE CA), 313 Splaiul Unirii, P. O. Box 149, 030138 Bucharest, Romania
e-mail: traian.zaharescu@icpe-ca.ro; traian_zaharescu@yahoo.com

P.M. Visakh and Y. Arao (eds.), *Thermal Degradation of Polymer Blends, Composites and Nanocomposites*, Engineering Materials,
DOI 10.1007/978-3-319-03464-5_6

transferred from incident radiation onto macromolecules allows appearing a high concentration of radical on the track. The radiochemical yield is a measure of stability; its value describes the number of events occurred as the result of the absorption of 100 eV energy. In the radiation field the polymer structures and reactivity are the feature connected to initial molecular configuration. New formed radicals are involved in further complex processes, which define the final radiolysis products.

The resistance of polymers against the action of ionizing radiation places them in a stability sequence, which depicts the material capacity to be modified. The successful irradiation processes present several advantages relative to thermal and chemical processes: high productivity without overloading reagents, reduced energy consumption, nonsignificant environmental pollution, sudden break of process under strict control of product quality, the lack of wastes. The concept of "high energy radiation" is based on the large amount of energy that is deposited on macromolecules by incidental rays (X, γ, accelerated electrons). This transferred energy exceeds many times the values of ionizing potentials and bond energies (10–15 eV). This received energy will be localized and the scission of involving bond occurs. Because during the ionization or excitation these processes take place on radiation track, several spurs appear forming cylindrical columns (Fig. 1).

The analysis of the advantages and disadvantages of EB accelerators and different γ-irradiators reveals various aspects that are taken into consideration:

- the penetrations of these two types of radiation are different and the LET is related to the mass of rays (Fig. 2).
- the electron accelerators adjust irradiation beams at desired intensity and energy, while gamma irradiators allows the exposure at certain energy, higher incidental energy is provided by ^{60}Co,
- the nuclide sources with which irradiators are provided are consumed according to the half life of source isotope,
- the construction of gamma irradiators presumes large protection walls, that means higher investment,
- the electron accelerators assures convenient high dose rate shortening processing times,
- the maintenance of accelerators is cheaper than the price of source replacement,

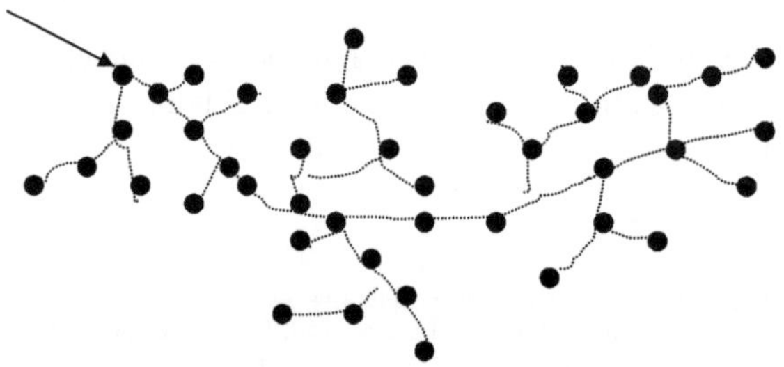

Fig. 1 The distribution of primary events along the radiation track

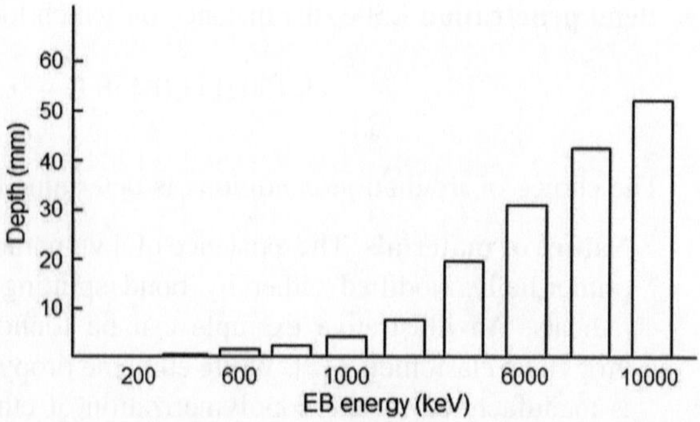

Fig. 2 Penetration of accelerated electrons versus energy

- the rate of conveyer movement for exposure in accelerators can be automatically changed modifying the dose range as required.

Radiation processing is largely applied for various purposes, when the purpose of further operation requires an advanced wear in the limited conditions. The large areas of activities are fitted to radiation processing, which long term usage is combined with high performances of material. The specific dose values characterize the limitation of overdose that induces degradation. The type of processing is associated to the certain dose range, whose frame must be respected (Table 1).

There are some empirical relationships with which penetration ranges can be calculated [12]:

- **optimal penetration** L (the material thickness whole outlet dose equals inlet dose):

$$L\ (\text{optimal}) = 0.404\ E - 0.161$$

- **demi penetration** L 50 (the distance on which the final dose is half of inlet dose):

$$L\ (50) = 0.435\ E - 0.152$$

Table 1 Some principal application areas associated with characteristic dose ranges [10]

Type of operation	Dose range (kGy)
Disinfection	0.25–1
Food preservation	1–25
Sterilization	20–30
Curing of coatings	20–50
Polymerization	25–50
Hydrogel preparation	20–40
Grafting	20–50
Crosslinking	50–150
Degradation	500–1500

- **demi penetration** L50e (the distance on which local dose is half of inlet dose):

$$L\ (50e) = 0.458\ E - 0.152$$

The choice of irradiation conditions is determined by many factors:

(1) **Nature of materials**. The existence of a vulnerable sites on backbone, they are primordially modified either by bond splitting or by removing as primary radicals. An illustrative example can be found in the domain of ethylene-propylene elastomers [13]. While ethylene propylene copolymer (EPR), which is manufactured by the copolymerization of ethylene with propylene, it presents a saturated molecular structure. In contrast, ethylene-propylene terpolymer (EPDM) is produced by copolymerization of ethylene with propylene in the presence of a third diene component (for example ethylidene 2-norbornene), which brings a certain unsaturation level.

(2) **Chemistry of environment**. The presence of oxidant atmosphere (usually air) determines the start of oxidative degradation from the start of irradiation. An inert atmosphere like vacuum or nitrogen does not provide any source of secondary reactions with radicals that appear during radiolysis. The diffusion of oxygen into irradiation material influences the distribution of oxidation products that takes a parabolic shape with the maximum amount at the both external sides and minimum is placed on the symmetry axe [14]. The penetration of molecular oxygen allows the reactions of free radicals with it and the peroxyl radical are formed. They are the initiators of further oxidation, which advances as a chain process.
The availability of molecules to scission and, consequently, for the competitive oxidation in air imposes the decision concerning the selection of irradiation conditions, which will lead to different final state of processed material. Moreover, if the irradiation is performed in aqueous surrounding, the intermediates resulted from water radiolysis (e^-_{aq}, H, HO, HO_2, HO^-, H_2O_2 [15]) specifically react with polymer substrate and the result significantly differs from the product obtained during the irradiation at similar dose condition, but in air.

(3) **procedure of synthesis**. The density of energy transfer is an important factor, which determines the direction of modifications. If the applied dose rate is low, the competition between oxidation and crosslinking is favorable to degradation, because the local concentration of free radicals is low and the amount of oxygen diffuses into polymer is sufficient high for reaction with radicals placed at long distance to each other. In these cases, it is preferable the polymer processing by electron beam irradiation in stead of γ-irradiation al low dose rates. The saturated elastomers like ethylene-propylene copolymer (EPR) are crosslinkable during high dose rate exposure, but on the low dose rate range they are oxidized [16].

The formation of radical intermediates influences not only the chemical state of irradiated polymers, but it brings also about physical characteristics: solubility, fusion enthalpy, heat capacity, crystallinity, mechanical features, electrical properties: permittivity, dielectric loss, breakdown tension, thermal conductivity. The new formed structures may be unlike in comparison with initial material, because intermolecular bridges, new organic functions appeared on macromolecular backbones, general molecular order or free volumes attain other values.

The formation of radical intermediates influences not only the chemical state of irradiated polymers, but it brings also about physical characteristics: solubility, fusion enthalpy, heat capacity, crystallinity, mechanical features, electrical properties: permittivity, dielectric loss, breakdown tension, thermal conductivity. The new formed structures may be unlike in comparison with initial material, because intermolecular bridges, new organic functions appeared on macromolecular backbones, general molecular order or free volumes attain other values.

(4) **dose conditions**. The choice of dose values must be the result of material stability. According with the formulation, with the state of stabilization, with the chemistry of irradiation atmosphere, with the sample composition, with the type of pretreatments, with the dose rate, the modifications occurred in polymer substrate attains certain level, which is the consequence of simultaneous actions of experimental factors. Figure 3 proves the differences that exist between the effect of dose rate. The increase of crosslinking starts from a certain dose named gel dose. It keeps characteristic values. For example, LDPE presents start gelation under irradiation at about 10–15 kGy, while ethylene-propylene elastomers have measurable gel fraction from 5–8 kGy. The presence of additives or crosslinkers modifies essentially the progress in the radiolytical modification of polymers.

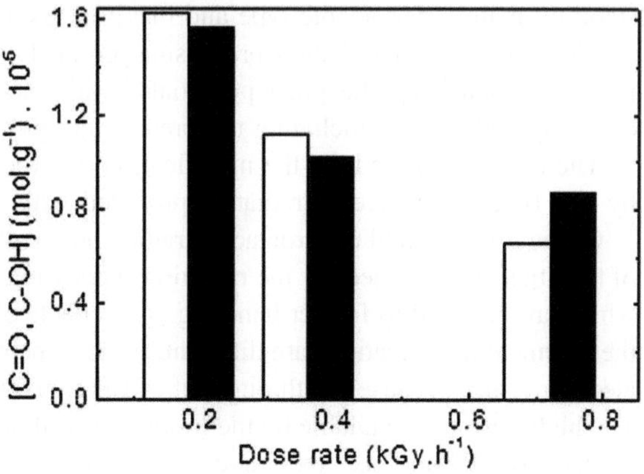

Fig. 3 Accumulation of the main oxygenated functions during g-irradiation of EPR. Total dose: 50 kGy

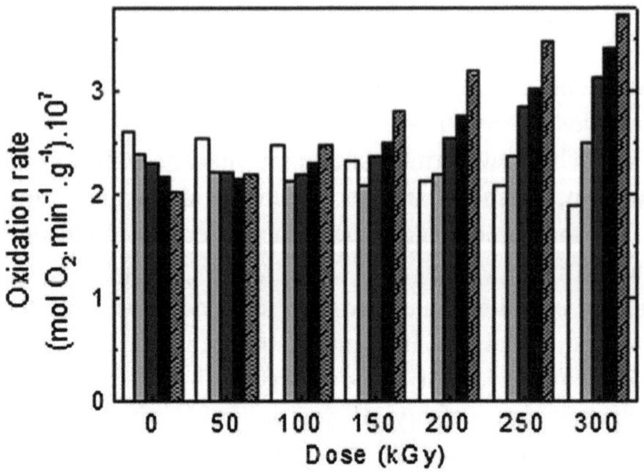

Fig. 4 Modification in oxidation rates of polymer systems (EPDM:PP), (*white*) EPDM, (*light grey*) EPDM:PP = 80:20, (*dark grey*) EPDM:PP = 60:40, (*black*) EPDM:PP = 40:60, (*hachured*) EPDM:PP = 20:80

(5) **sample composition**. Two blending components presenting different radiation stability plays opposite roles during the radiolysis of their mixtures. The component with a lower stability becomes a source of radicals for the other couple partner. On the low total dose range, a crosslinking process is possible, because the concentration of oxygen is not sufficient for feeding oxidative degradation. But, the progress in irradiation makes possible a significant oxidation and the ageing is accelerated, when the concentration of low stability component is significant. Figure 4 exemplifies the contribution of polypropylene to the evolution of chemical states of EPDM/PP blends [17].

The peculiar factor that determines the behavior of polymer under irradiation is radiochemical yield. It depicts the number of events that occur as the result of the absorption of 100 eV, G. The greater the values of G, the less stable the polymer [12]. The consequences of irradiation conditions related to the tremendous radiochemical yields place various macromolecular materials in different groups of stability: mainly degradable type and manly crosslinkable type [12, 18].

The directions of radiation processing are analysed by various reports [19–25] for the emphasizing the great potential of this technique in the development of economy and science including the present and promising applications.

The energy required for the modification of exposed materials can be provided by two types of sources: irradiators provided with radioisotopes, usually ^{60}Co or ^{137}Cs, on one side and electron accelerators, on the other side. These two categories of facilities are designed for the radiation processing of great amounts of materials, which are destined to further handling. The effects of dose rate on the efficiency in the chemical modifications are different, if the same material is exposed to γ-rays of electron beam, because the throughput of EB exposure produces high productivity. The high dose rate available by the irradiation with accelerators, about 100 kGy s^{-1} reduces drastically the oxidative degradation that it would occurred during γ-irradiation, even the dose rate is at an acceptable level (10 kGy h^{-1}) [26]. Though electron accelerators suppose a large investment for the start, the further utilization satisfies the technological requirements.

The great troublesome problem in the electron beam (EB) irradiation is the depth of penetration [27], which depends not only by the incident particle energy, but also on the material density. This problem can be solved by double face irradiation, when exposure dose can be considered uniform along profundity [28]. The main problem in the selection of the type of accelerator is the nominal power, which must be rigorously correlated with its applications [29]. The factors that influence the choice of accelerators are output power, energy of EB, room temperature for accomplishment of process, number of passes and rate of conveyer [30]. A variety of industrial electron accelerators can now provide electron energies from 0.3 MeV to more than 10 MeV with average beam power capabilities up to 300 kW.

The monitoring of dose can be performed with different techniques: chemical systems (Fricke and ceric solutions [31]), radiochromic compounds (dyes [32, 33]), polymeric tapes (polyethylene [34], poly(vinyl chloride) [35, 36], poly(methyl methacrylate) [37–40], epoxy resin [41]), radiation thermoluminescence phosphors [42, 43]. Several requirements are imposed for a proper dosimeter: similarity with processed material in respect with linear energy transfer, reproducibility, sensitivity, lack of the influence of humidity, stability after irradiation, easy to calibrate, appropriate dose range and dose rate, linearity and independency on the type of radiation, the response being constant in time, lack of post-irradiation modification.

Each dosimeter has a certain dose range, on which the specificity recommends it as proper system for accurate dose measurement [44]. Due to their easy availability, handling and processing, polymers may be preferred [45]. For comparison, some practical aspects of X-rays and gamma rays, such as product penetration, dose uniformity, utilization efficiency and processing capacity were reported [46].

Industrial scale of radiation processing is accomplished also on the sterilization of polymeric medical wear [47], food packaged in polymer bags [48] or electrical insulation of wires and cables [49, 50]. The radiation technologies are based on the possibility of attaining improved properties and extension of service life is attained.

2 Improvement in Thermal Properties of Polymeric Materials/Composites

2.1 *Crosslinking*

Starting from the crosslinking of polyethylene [51] the improvement in the functional characteristics of polymers by crosslinking keeps various aspects. The general overview on the involvement of free radicals that are formed during radiolysis concerns the reactions of intermediates to each other or with oxygen [52]. The accumulation of cured fraction or oxygen-containing products depends on the reactivity of radicals. The accumulation of insoluble fraction, which consists of three dimensions network built up by new intermolecular bridges, turns processed materials onto desired behavior exemplified by the mechanical properties, namely tensile strength, scratch resistance, the performances at higher temperatures, the chemical strength by the diminution of rate diffusion of organic solvents, the reduction in the amount of penetrated gases, memory shape retention, electrical properties.

The basic equation which describes the decreasing in soluble fraction of irradiated polymers was reported by Charlesby and Pinner as linear function of $S + S^{1/2}$ on reciprocal value of dose [52]:

$$S + \sqrt{S} = \frac{P_0}{Q_0} + \frac{1}{Q_0 . P_{n,0} . D}$$

where S is the soluble fraction that can be calculate for the abstraction of gelated mass from total sample weight, P_0 and Q_0 represent the radiochemical yields of scission and crosslinking, respectively, $P_{n,0}$ is the initial polymerization degree and D is irradiation dose expressed in kGy. The ratio between radiochemical yield of scission and radiochemical yield of crosslinking illustrates the availability of polymers for crosslinking or degradation [52].

The content of gel fraction will influence material properties. The theoretical approach starts with the the evaluation of radiochemical crosslinking yield:

$$G_n(\text{crosslinked units}) = 0.96 \times 10^6\, q_0/w$$

where G_n(crosslinked units) represents the number of crosslinked units per 100 eV absorbed energy, q_0 is susceptibility of polymeric system to be crosslinked and w is the average gravimetric weight of polymer. It is defined as the proportion of crosslinked monomer for dose unit. Because the bridging takes place involving two units, the radiochemical crosslinking yield is:

$$G(\text{crosslinking}) = 0.48 \times 10^6\, q_0/w$$

Consequently, the dependency of molecular weight on absorbed dose is:

$$M_c = 0.48 \times 10^6/G.D$$

The last relationship can be used for the calculation of radiochemical yield of crosslinking based on the measurement of average weight molecular weight.

The most commercialized polymer is polyethylene, which present a very good stability under irradiation. Because there are manufactured different sorts of polyethylene, they present unlike radiochemical strength.

The processability of polyolefins can be improved by the irradiation in the presence of crosslinker, for example TAIC [53]. Many other polymers can form efficiently increased amounts of insoluble fraction in the presence of various functional crosslinkers: ethylene–vinyl alcohol copolymer with TAIC, DEGDA, NPGDA and TMPTA [54] at maximum 65 % gel fraction at 200 kGy, poly(ε-caprolactone) and poly(butylene succinate-co-adipate) with TMAIC [55] forming insoluble content around 60–80 % at 100 kGy, polyamide 6 with TAC [56] generating a crosslinked phase of 95 % at 80 kGy, poly(L-lactic acid) with TAIC, TMAIC, TMPTA, TMPTMA, HDDA, EG [57] at different gelation degrees between 25 and 85 % obtained at 100 kGy. The addition of TAIC and SiO_2 in the formulation of poly(L-lactic acid) allows to attain a gel fraction of 100 % at 40 kGy [58].

Low density polyethylene is the most frequently used because of its low branching degree. The radiation processing increases the stability by the continuous increasing of the crosslink density from zero in pristine material to 0.204, 1.022 and 4.807 mol dm^{-3} at 100, 200 and 400 kGy, respectively [59]. Of a great importance for the effect of irradiation is the orientation of molecules that determines the differences in the repartition of crystalline phase in polyethylenes. The gel fraction in LDPE, LLDPE and HDPE depends not only on the received dose, but also on the drawing ratios [60]. The differences in the accumulation of insoluble fraction are explained by the unlike values of crystallinities.

The irradiation environment plays an important role in the evolution of polymer stability. While unsaturated hydrocarbon like acetylene [61] or divinyl benzene [62] is present in the material surrounding and provides radicals for the formation of intermolecular bridges, oxidative atmosphere, oxygen or air, promotes oxidation as the result of diffusion inside the polymer matrix. The distribution profile for carbonyl products that generated during irradiation takes a parabolic form [63]. The source of radicals may be one of the components of blends, which presents a lower stability. This case can be illustrated by various blends, EPDM/PP [64], EPDN-NR [65]. These polymer mixture show the maximum level of crosslinking at about 120–150 kGy.

The blends consisting of SBR and EPDM in different proportions are compatibilised by crosslinking under γ-irradiation [66]. The viscosimetric measurements and the swelling investigations gave revealed the appearance of a new phase by the connections of molecular chains with chemical bridges, much stronger that physical interactions. Even though structural differences exist, the crosslinked fraction consists of different morphologic components, graft copolymer, block copolymer and spatial copolymer. The increase in the concentration of EPDM from 0 to 100 %, the molecular masses and crosslinking densities become 13,564 and 449, and 3.69 and 111, respectively, obtained for irradiated mixtures at total dose of 400 kGy.

The compatibilisation of different polymers by irradiation can be achieved by EB irradiation starting from latex components, for example, natural rubber and acrylonitrile-butadiene rubber in the presence of EPTA [67]. The variation of the normalized absorbances at 1625 cm^{-1} (vinylene C=C) and 1450 cm^{-1} ($-CH_2-$) scissor against the absorbance at 2223 cm^{-1} (–CN stretch) with the radiation dose (0–500 kGy) has demonstrated that at 100 kGy the process reaches the steady state. In the Charlesby-Pinner representations straight lines could be drawn at each of three concentrations (0, 2 and 4 phr) of crosslinking promoter.

The effect of TMPTA on the crosslinking of PET and HDPE by γ-irradiation was described by the accumulation of gel fraction in their blends that keeps the intermediate maximum value, 10 % obtained at 30 kGy [68]. Even the total dose increases at 70 kGy, the amount of insoluble fraction remained constant for the studied formulation, PET/HDPE = 80/20. The best mechanical properties were measured at this limit dose, 30 kGy. It was supposed that further irradiation of crosslinked PET/HDPE systems the degradation by scission of new formed bonds occurred. This behaviour can be found in the cases of components with different structures and polarities.

The radiation technology applied for the modification of crystalline or semi-crystalline polymers allows manufacturing crosslink shape-memory products [69]. The exposure to high energy radiation promotes crosslinking of cylindrical tubes at certain dimensions. After heating the stain appears. This radiation processing can be applied successfully to natural rubber [70, 71], polyethylenes [72–74], ethylene vinyl acetate copolymer [75], poly(ε-caprolactone) [76]. The efficient crosslinking of this kind of smart materials is attained, if the formulation contains a sensitizer [77] or a multifunctional additive [78]. The applied dose in the systems consisting of modifying polymer and additive is significantly lower than the nonmodified compound, because the higher gel content is obtained much easier and the number of crosslinks is higher. Consequently, the charge with which material presses inner body is corresponding greater. The applications of SMPs cover different areas, for example: medical treatment (tight immobilisation of tissues or bones, sterilisation), electrical engineering (joint cables for continuous connection, electro-active sensors, encapsulation of electronic parts), mechanical engineering (temperature markers, assembling different equipment parts, fabrication of geometric structures), handling and preservation of food, smart textiles [79, 80]. The shrinkability, the accumulation of gel fraction, the convenient mechanical properties, the chemical stability against oxidation are some features that must be shown by an appropriate SMP [81].

In the medical praxis, UHMWPE have been used successfully as one half of the bearing couple (against metallic alloys or ceramics) in total hip and total knee joint replacements crosslinked under irradiation [82]. Even γ-irradiated UHMWPE at doses higher than 100 kGy presents the incipient fracture development [83], the alkyl macroradicals are involved in crosslinking and in a smaller proportion they promote oxidation [84, 85]. The prosthesis manufactures by the irradiation of UHMWPE have long durability, because the application of radiation treatment induces an increased crystallinity and promotes sterilization in the whole volume of material.

The functional properties of polymers can be ameliorated by the irradiation of their nanocomposites [86, 87]. The addition of MWCNT to low density polyethylene increases the radiation resistance in comparison to the pure LDPE, which was dependent on the MWCNT content [88]. WCNT nanocomposites were gamma irradiated at 90 kGy to improve the interaction between MWCNTs and the polymer matrix [89]. The irradiation produced a 38 % decrease in the toughness of neat UHMWPE. The incorporation of MWCNTs did not significantly affect the melting point of the neat UHMWPE but decreased the degree of crystallinity of the raw UHMWPE, which was related to a reduction in the UHMWPE lamellar density.

The tensile tests showed a 38 % increase in the Young's modulus in the reinforced nanocomposites and a small decrease in toughness (5 %). The addition of carbon nanotubes in polypropylene brings about conductivities of the order of 10^{-2} S/cm, but its scavenger effect reduces the number of radicals generated by irradiation, lessening the strain hardening effect [90]. The presence of CB and silica in the formulation of NBR/CSM blends at maximum concentration of 30 phr changed satisfactorily polymer properties: 152 % increase in tensile strength, 116 %, in elongation at break and 142 % modulus at 100 % elongation according to synergistic effect between the fillers [91]. Ethylene-vinyl acetate copolymer (EVA) flame retarded by a combination of intumescent flame retardants (IFR) and organically modified montmorillonite (OMMT) crosslinked by EB irradiation shows a synergistic effect of IFR and OMMT on the flame retardancy. With the addition of 1 wt% OMMT and 24 wt% IFR, the LOI value of EVA/IFR/OMMT nanocomposite increased from 30.5 to 33.5 % [92]. The reinforcement of polypropylene with 5 % POSS after g-irradiation at 50 kGy showed improved thermal stability due to the formation of crosslinked network between PP and POSS by radiation [93]. The mixing carboxyl-terminated butadiene-acrylonitrile (CTBN) with nano-clay to improve the toughness and mechanical strength of bisphenol A type epoxy followed by further γ-irradiation at high doses (500, 1000 and 1500 kGy) has pointed out the contribution of nanofiller to the stabilization by intercalation of macromolecules into clay layers [94].

Nanocomposites of two different kinds of rubber (acrylonitrile-butadiene rubber NBR and styrene butadiene rubber SBR)/organo-montmorillonite nanocomposites modified by hexadecyltrimethyl ammonium bromide have a remarkable resistance when they are radiochemically processes in the presence of TMPTMA [95]. Similar results were reported on epoxy resin modified with titania presents a very good resistance to the thermal oxidation after receiving 50 kGy γ-dose [96]. The oxidation in neat material starts quickly and progresses with a great oxidation rate. By the addition of titania nanoparticles, the host resin presents a decreased oxidation rates at filler concentrations up to 10 %. A very smooth oxidation takes place in irradiated epoxy resin/titania hybrids because inorganic phase acts as efficient adsorbent agent relative to the radicals formed during radiolysis.

One largely commercialized elastomer, EPDM, can attain improved tensile strength, when it is irradiated as hybrid composites with nanoclay particles [97]. The reinforcement of filler in the case of ATH induced a de-cohesion inside polymers with direct consequences on [98]. Moreover, at room temperature, i.e. below the melting temperature, all the consequences of ageing by gamma irradiation are strongly attenuated by the presence of a semi-crystalline microstructure, the morphology of which is not too strongly modified by irradiation. The content of silica nanoparticles modifies the kinetics of the degradation of EPDM substrate through the complex modification of the filler–filler and filler–matrix interactions involved in the mechanical properties of the filler network.

The differences that exist between polyolefins, namely LDPE, PP and EPDM in the presence of various concentrations of silica demonstrate the structural contributions of branching and unsaturation to the durability of this kind of nanocomposites [99]. An opposite effect for the thermal stability of irradiated PP was observed in the presence of titania, which promotes oxidation initiated by radiation processing [100].

Polylactide (PLA) and PLA-Cloisite 30B (C30B) nanocomposites under g-irradiation present significant modifications with dose. The thermal stability is higher for the nanocomposite, even the decrease of oxidation temperature is noticed for the both formulations [101].

The progress in the oxidation of LDPE modified with maleic anhydride and alumina proceeds somewhat similar at various doses, because oxygen diffusion is hindered by filler nanoparticles [102]. The noticeable difference between pristine and modified LDPE consists of the presence of maleic anhydride, which interacts with molecular chains due to the electronegativity of oxygen atoms. The same radiation dose affects differently the dielectric behavior of the nanocomposites depending on the filler content. The dose of 50 kGy applied on LDPE-g-AM filled with 5 wt% nano-Al_2O_3 leads to a relative permittivity smaller than unfilled LDPE. γ-Radiation can lead to a decrease in the dielectric losses of LDPE Al_2O_3 nanocomposites for properly chosen combination dose–filler content.

The γ-irradiation of natural rubber (Malaysian type) in the presence of nanotubes does not modified the accumulation of gel content, which reaches an upper limit (98 %) at about 25 kGy, but mechanical properties are influenced by the nanofiller loading [103]. The uniform crosslinking in homogenous distribution of CNT confirms the high quality of radiation processed NR.

The radiation exposure makes possible the crosslinking of poly(ε-caprolactone)/polyhedral silsesquioxane nanocomposites by electron beam irradiation [104]. The addition of POSS nanoparticles gathers the PCL in a hard fraction, which explains the highest values in tensile strength at 100 kGy and the sharp decrease in the elongation at break after receiving 50 kGy, equivalent with the gel dose. In comparison to the virgin PCL with a tensile strength of about 20 MPa, the tensile strength of the crosslinked PCL/POSS nanocomposites increased to 25.8 MPa with an increasing POSS content and absorbed dose to 100 kGy, whereas their elongation-at-break was considerably reduced. As the result of intermolecular strength the swelling of composites goes down as the content of POSS increases.

For the manufacture of long term products, such as automobile and aerospace components, sporting goods, etc., it is recommended the grafting of nanoparticles on polymer backbones. This goal is accomplished in the case of PP [105] and PU [106] by the exposure of hybrid materials to the action of γ-radiation.

Vulcanization of rubber blend of acrylonitrile butadiene rubber (NBR) and ethylene-propylene diene monomer (EPDM) rubber (50/50) loaded with increasing contents, up to 100 phr, of reinforcing filler, namely, high abrasion furnace (HAF) carbon black and further subjected to gamma radiation doses up to 250 kGy was performed [107]. Mechanical properties, namely, tensile strength, tensile modulus at 100 % elongation (M100), and hardness have been followed up as a function of irradiation dose and degree of loading with filler.

The radiation treatment of polymer materials, even they are as receives or modified with nanofillers converts them into engineering products with high economical values [29, 108].

The necessity of new form of polymers, which would be behaved friendly, was solved by hydrogels, three-dimensional networks of covalently bound hydrophilic polymer chains. Typical simple materials applied for general-purpose hydrogels are ethylene oxide, vinyl alcohol, vinylpyrrolidone, hydroxyethyl methacrylat, cross-linked natural polymers, mainly polysaccharides. The propagation of crosslinking takes place through the activities of different radicals. The accumulation of gel during irradiation follows the Rosiak, Olejniczak and Charlesby, similar with the law Charlesby-Pinner [109]. By the application of Charlesby-Pinner equation on the results obtained for the radiation processing of PVA/PVP blends it has demonstrated that the gel fraction of blend is placed between the similar values of components. Gel doses were measured for the polymers and they are 11 kGy for PVA, 3.7 kGy for PVP and 4.6 kGy for a mixture of PVA and PVP with a mole fraction of PVP of 0.19 [110]. The addition of a crosslinking agent, for example acrylic acid, into the PAA/PVP irradiated blends stimulates gelation, even though values of their gel doses increase smoothly [111].

The preparation of hydrogels can be achieved by various procedures. The general manner through which hydrogels are prepared is polymerization. The double bonds represent the start points of radical polymerization and crosslinking. One example is *N*-isopropylacrylamide, which generates hydrogels in aqueous solutions [112]. During irradiation the water radiolysis intermediates (hydrated electrons, hydroxyl radicals and hydrogen atoms) react with monomer producing α-carboxyalkyl radicals. They become the initiators of polymerization.

During radiolytic exposure a complex chain of reactions occurs. The primary intermediates appear either due to the scission of molecules by energy depositing on components, or by transfer of radicals [113]. The attack of hydroxyl radicals followed by the abstraction of proton becomes a source of crosslinking precursors. The most vulnerable position is tertiary carbon, which generates the highest proportion of radicals. They act as precursors of final products according with Fig. 5.

PVME microgel particles can be prepared by irradiation of a phase separated diluted aqueous PVME solution above their lower critical solution temperature

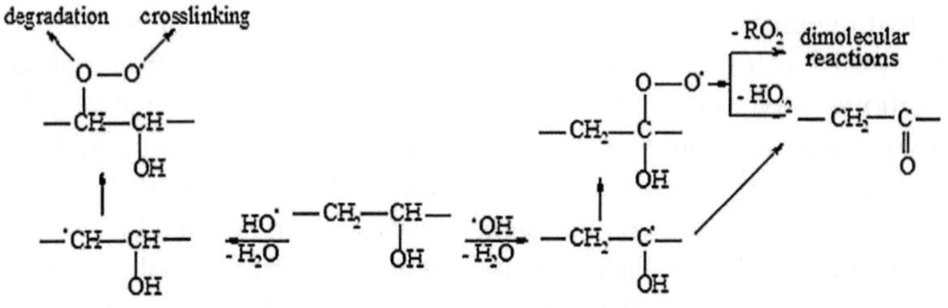

Fig. 5 Transfer reactions caused by radiolytic hydroxyl radicals which are the start of hydrogel formation

(37 °C) following the mechanism described in Fig. 5. The macroradicals can appear by the scission of C–H bonds of carbon atoms composing vinyl methyl ether monomer. When system receives 10.5 kGy the average weight molecular weight increases from 57.10^3 g mol^{-1} to about 10^6 g mol^{-1} [114].

The applications of hydrogels in the production of medical items, resulting materials must have several features, which recommends them: non-toxicity, functionability, sterilizability, biocompatibility [115]. These characteristics are requires for wound dressings, drug delivery systems, transdermal systems, injectable polymers, implants, dental and ophthalmic applications, stimuli-responsive systems, hydrogel hybrid-type organs.

Radiation processed PVA–polysaccharides hydrogels can be prepared at maximum dose of 30 kGy [116]. Using concentration of polysaccharides as low as 0.5–2 % resulted in increase of tensile strength from 45 to 411 g cm^{-2}, elongation from 30 to 410 % and water uptake from 25 to 157 % with respect to PVA gel without polysaccharides. Formulations containing 7–9 % PVA, 0.5–1.5 % carrageenan and 0.5–1 % agar gave highly effective usable hydrogel dressings.

A series of antibacterial hydrogels were fabricated from an aqueous solution of $AgNO_3$, gelatine and carboxymethyl chitosan (CM-chitosan) by radiation-induced reduction and crosslinking at ambient temperature [117]. For the total dose of 30 kGy, the gel fraction was situated around 75 % invariant relative to the concentration of $AgNO_3$, and diameter of particles was between 2 and 5 nm. The swelling ratio is higher (90–100 %) at the concentrations of silver nitrate exceeding 1 %.

Hydrogel wound dressings have been prepared using the gamma rays irradiation technique. The dressings are composed of poly(vinyl pyrrolidone) (PVP), poly (ethylene glycol) (PEG) and agar [118, 119]. At 50 kGy the stress at break increases with 40 %, while swelling degree is enhanced significantly with the concentration of PEG and with the time of immersion. Interpenetrating polymer networks (IPNs) can be synthesized by irradiating acrylonitrile solutions of poly(*N*-vinyl-2-pyrrolidone) with Co-γ rays [120]. Thermal gravimetric analysis of resulting IPNs has proved the enhancement in the stability of materials, which can be related to the intermolecular bonds formed during irradiation. Polyvinyl alcohol and polyvinyl pyrrolidone (PVA–PVP) blended hydrogel for wound dressing has been prepared by using gamma rays irradiation. WVTR values of PVA–PVP blended hydrogel are around 80–200 g m^{-2} h^{-1}, which place these radiation processed systems qualitatively above commercial products [121].

The hydrogelmatrix can be obtained by γ-irradiation, which induces crosslinking simultaneously with the in situ reduction of Ag^+ ions initiated by the products of water radiolysis (e^-_{aq}, OH^-, $H^\cdot$, H^+, H_2, H_2O_2). For the radiochemical gelation of vinyl pyrrolidone two radical entities are involved in different proportion (Fig. 6).

The stress/strain representations are linear PVP and PVP/Ag^+ hydrogels. Obtained values of the effective crosslink density in the range of 52.8–54.0 mol m^{-3} and those corresponding to the molar mass between crosslinks in the range of 15.5–15.9 kg mol^{-1} for pure PVP hydrogel prove the development of good hydrogel

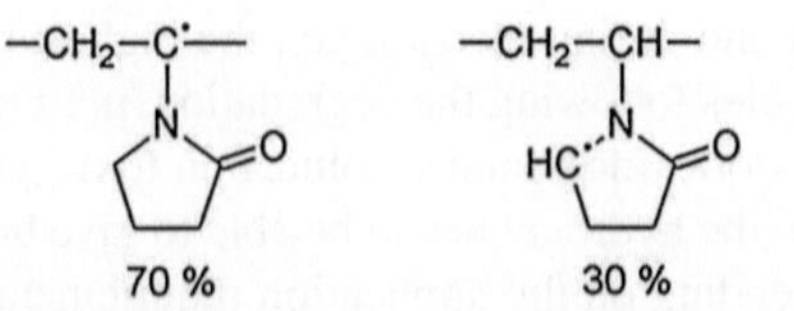

Fig. 6 Radical intermediates promoting crosslinking in irradiated vinyl pyrrolidone

material. Similar results were reported for the hydrogel containing simultaneous silver and gold ions for the increasing in the antibacterial activity [122].

The polyphenol trans-resveratrol is a natural phytoalexin, which is found in red wine and in a wide variety of plant species. Resveratrol displays a wide array of biological activities, such as modulation of lipid metabolism, anti-inflammatory and antioxidant activities. The results of gel fraction and swelling degree were approximately 90 and 1600 %, respectively for applied dose not exceeding 1 kGy [123]. The swelling degree attains maximum value after about 6.5 h. These information concerning the resveratrol stability suggest that it has showed no structural decomposition by the primary and secondary radicals of water radiolysis. The polymeric matrix composed by PVP, PEG and agar showed suitable physical and chemical characteristics to resveratrol immobilization to compose a hydrogel dressing for dermatological use.

Protein-crosslinking whether done by enzymatic or chemically induced pathways increases the overall stability of proteins. Papain, a proteolytic enzyme (EC 3.4.22.2) of biotechnological and biomedical relevance was selected for the delivery of papain under γ-irradiation as globular crosslinked protein [124, 125]. The size of papain particles enlarges due to the radiation processing from 172 nm in the case of nonirradiated ethanol solution to 910 nm after receiving 40 kGy. The use of ethanol combines the protein aggregation, essential for the involved process and radioprotection due to its intrinsic scavenger property, capturing specific radicals generated by water radiolysis, known to cause a deep impact on protein integrity.

The applications of hydrogel in the retention of metal ions were intensively studied [126]. pH-sensitive hydrogels based on poly(N-vinyl-2-pyrrolidon) (PVP), acrylic acid (AAc) and styrene (Sty) were prepared by gamma irradiation [127]. PVP/(AAc-co-Sty) hydrogels were subjected to radiation modification to use them as adsorbent materials for removal of heavy metal ions from aqueous solution. Effect of functionalization of hydrogels by sulfonation (Sf), partial hydrolysis with alkaline solution (NaOH) and treated with the two processes (NaOH/Sf) on metal ion uptake was evaluated, and it results in appreciable uptake of Co^{2+}, Cu^{2+} and Fe^{3+} ions from aqueous solution.

In the same category of reports the drug delivery hydrogels can be considered [128–133]. They must show some characteristics, which get them proper for foreseen goals [134, 135]: the lack of homopolymer, the highest absorption capacity (maximum equilibrium swelling) in saline, desired rate of absorption (preferred particle size and porosity) depending on the application requirement, the highest absorbency under load and reasonable remote rate the highest durability and stability in the swelling environment and during the storage., the highest biodegradability without formation of toxic species following the degradation, pH-neutrality after swelling in water, colorlessness, odorlessness, and absolute non-toxic, photo stability, re-wetting capability (if required) the hydrogel has to be able to give back the imbibed solution or to maintain it; depending on the application requirement (e.g., in agricultural or hygienic applications), low production price. The adequate usage of these hydrogels in the post-operatory application is the adhesive features for easy withdrawing from surgery intervention site [136].

The recycling of polymers receives a large attention because it solves two problems, one is the economy of raw materials, and the other is the decrease of environmental pollution. A review on the contribution of radiation processing to the reclaiming polymer wastes were published previously [137]. The effects of an electron radiation dose (up to 300 kGy) and compatibilizer on the Charpy impact strength (σ_c) and tensile-impact strength (σ_t) of composites made of the following recycled polymers: low-density polyethylene (LDPE), high-density polyethylene (HDPE), polypropylene (PP), polystyrene (PS), and poly(ethylene terephthalate) (PET) indicate the processing potential of radiation treatment applied to polymers [138]. Styrene–ethylene/butylene–styrene elastomer grafted with maleic anhydride (SEBS-*g*-MA) and trimethylol propane trimethacrylate (TMPTA) were used as the compatibilizers, added at 10 and 1 wt%, respectively. It was found that, under the influence of SEBS-*g*-MA, the values of σ_c and σ_t for the studied composites increased by over three and almost five times, respectively. The electron radiation and TMPTA had no noticeable effect on these quantities, which might be due to the protective properties of aromatic rings included in the macromolecules of PS and PET. It was demonstrated that the inclusion of SEBS-*g*-MA enabled achievement of satisfactory impact strength of composites made of recycled polyolefins, PS and PET, while it was not necessary to use the electron radiation and TMPTA.

Basically, the recovery of polymer wastes is based on the compatibilization, when radiation acts as the radical formatting agent [64]. The radiation compatibilization of high-density polyethylene (HDPE)/ethylene–vinyl acetate (EVA) copolymer blends can be achieved, because the vinyl acetate content of EVA was beneficial to radiation crosslinking [139]. The gel content has a maximum value at 30 kGy for all relative concentrations of components. The compatibility checked by TGA investigations as well as mechanical properties showed that the two plastics coexist homogeneously due to the intermolecular bridges appeared during irradiation. At higher doses like 500 kGy, degradation of blends takes place.

Recycled and pristine low-density polyethylene samples were crosslinked by ^{60}Co gamma rays in the presence of two commercial sensitizers: trimethylolpropanetriacrylate (TMPTA) and trimethylolpropane trimethacrylate (TMPTMA), and another laboratory-synthesized sensitizer, hexakisalylaminocyclotriphosphazatrine (HAAP) [140]. The measurement of gel fraction revealed a light difference between virgin and recycled LDPE, this discrepancy being duet to the accumulation of degrading compounds. PET/HDPE blends can be processed by γ-ray irradiation combined with using a cross-linking agent, TMPTA [68]. The specificity of this process consists of the new structure, HDPE-g-PET proved by FTIR records. When the weight ratio of PET/HDPE blend was 80/20, the content of TMPTA was 1 wt% and the absorbed dose was 30 kGy, the tensile strength, elongation at break and impact strength of irradiated blends were improved greatly compared with non-irradiated blends. Blends of high-density polyethylene (HDPE) can be crosslinked by radiation processing with recycled and pristine polyethylene terephthalate (PET) [141]. PET contains aromatic groups, which are effective at dissipation of the energy of the ionizing radiation during γ-radiolysis forming a copolymer capable of improving the compatibility of the blend HDPE/PET. The gel content differs for one composition to the other, when PET loading varies from 50 to 30 %. The initial lower amount of PET in the raw blend favorizes an advanced crosslinking level.

Radiation-induced compatibility behaviour of SBR–EPDM blends of different composition was studied for intimate compatibilizing components [66]. The results have shown that at a threshold dose of 10 kGy exists and good interaction between the components of the blends was achieved. Higher radiation doses lead to formation of crosslinked three-dimensional copolymer network, whose swelling depends on the radiation dose imparted. The anomalous diffusion of solvent into the gels was confirmed by rigorous treatment of the swelling data. Permeation data agreement with the series model indicated that in SBR–EPDM blends EPDM exists as continuous phase and SBR as dispersed phase.

The γ-processing of IIR gives the possibility of an efficient recycling of this material. The sharp drop of weight average molecular weight (Fig. 7) is an economical way for the manufacture of rubber ingredient used for toughening tire raw materials [17].

The EB treatment of the GTR containing LDPE and EVA matrix blends showed significant benefits. The 200 kGy EB absorbed dose (in air) resulted in a better tensile strength and increased elongation at break, without changing the tensile modulus, which provides more rubber-like properties [142]. The modest change in hardness proves the cross-linking effect caused by the EB treatment in all cases. The dynamic mechanical analysis confirmed the compatibilization effect of EVA and

Fig. 7 Change of weight average molecular weight in irradiated IIR

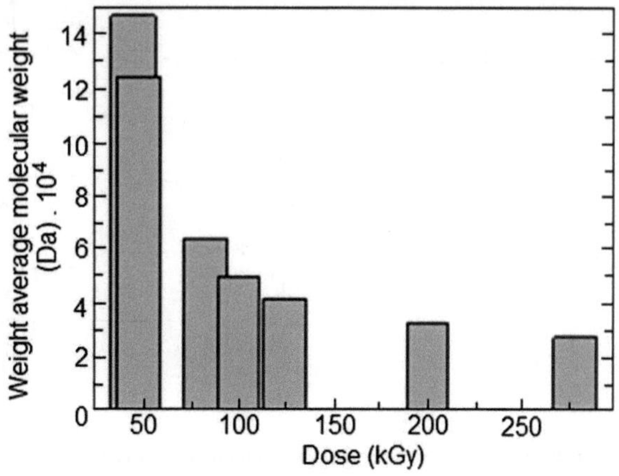

EB between the GTR and (thermoplastic) polymer matrix. Recycling of gamma irradiated inner tubes made of butyl rubber in butyl based rubber compounds was studied for the reclaming waste tires [143]. Gamma irradiated inner tube wastes and commercial butyl rubber crumbs devulcanized by conventional methods were replaced with butyl rubber up to 15 phr in the compound recipe. The rheological and mechanical properties and carbon black dispersion degree for both types of compounds were measured and then compared to those of virgin butyl rubber compound. The results lead to the procedure through which tires can be reinforced. γ-Irradiation is used for the in situ compatibilisation of blends of recycled high density polyethylene (rHDPE) and ground tyre rubber (GTR) powder [144]. The expected compatibilisation mechanism involves the formation of free radicals, leading to chain scission within rubber particles, crosslinking of polyethylene matrix and co-crosslinking between the two blend components at the interface. While uncompatibilised rHDPE/GTR blends show poor mechanical properties, especially for elongation at break and Charpy impact strength, irradiation leads to a significant increase of these mechanical performances.

The improvement in the quality of recycled polymers can be achieved by the accomplishment of high energy irradiation in hydrocarbon atmospheres [62, 145, 146], because the scission of π component of double bonds or the opening of cycles provides radicals, which crosslink polymer molecules. An example of mechanistic scheme applied in the irradiation of unsaturated structures is presented in Fig. 8.

There are other alternatives for the improvement of polymer stability starting with polymer wastes [147, 148]. The presence of sulphur in irradiated IIR increases the crosslinking level up to 100 kGy. The addition of stabilizers is a solution for the protection of materials for further oxidation [149].

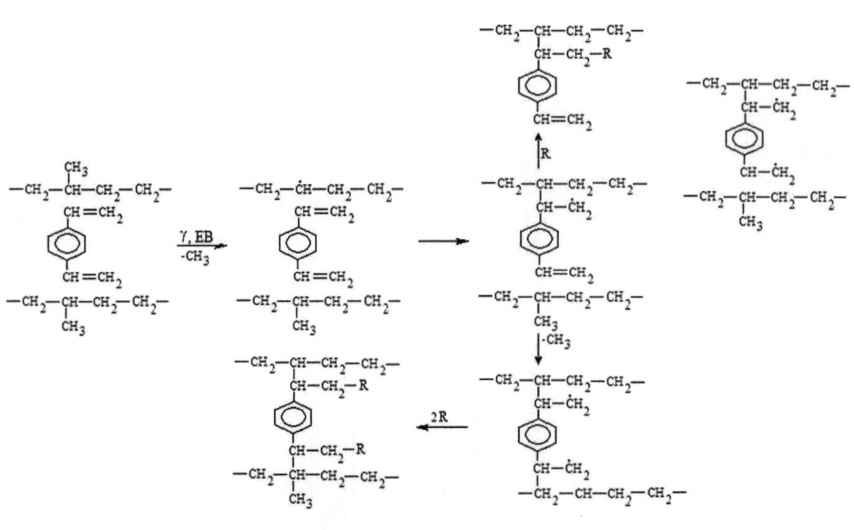

Fig. 8 Scheme for the reaction mechanism involving unsaturated monomer

2.2 Grafting

The association of new monomer onto macromolecules, which allows it, represents the grafting process [150]. The great advantages of radiation processing in grafting polymers are the initiation without catalyst and the simplicity of treatment, which can be conducted in connection with the reactivities of monomer and substrate. The connection of this outside structure brings different functional properties, with which the application area can be enlarged. The incidental radiation provides energy for the generation of reactive sites. Synthetic materials obtained by grafting are proper alternative to blended polymers, which may satisfy the practice requirements imposed in any certain service. High energy radiation substitute chemical reagents, which creates active positions on macromolecules. Some example of this kind of reactivity stimuli are ferrous ions in redox medium, cereous ions, manganate ions in acid environment and many others. Their involvement in redox processes causes the transfer of one electron for each ion from polymer molecule and, consequently, the formation of free radicals, with carry on the attachment of grafting entity.

Grafted copolymers can be obtained by several ways [1, 151–153]:

- the polymerization by addition of vinyl monomer M on polymer RR′,
- the reaction of two macroradicals appeared from two different polymer components RR′ and R_1R_2,
- polycondensation of two macromolecules RR′ and R_1R_2, which have functional groups that can react:
  - direct grafting of vinyl monomer on polymer macromolecule,
  - grafting catalysed by peroxyl structure appeared during radiolysis,
  - initial grafting of trapped radicals,
  - co-crosslinking of two different polymers by linking one type of macromolecule to other molecule from the second component.

Synthetic view on these routes is presented in Fig. 9. The application of grafting procedure, one polymer substrate gains supplementary features, which allow it to be used in particular conditions of operation. The new structure is regarded as a combination between the two initial configurations, but the basic material exhibits predominant characteristics. The grafted units correct some of undesirable properties, which are turned to the favourable behaviour. If grafting takes place in solution, the participation of solvent molecules increases the conversion yield, because they act as transporters of energy between radiation and irradiation environment [154]. The pre-irradiation plays also an important role, because it forms radicals, which become the process initiators.

The modification of polyethylene and cellulose by grafting of styrene and acrylonitrile simultaneously present in polymer bulk allows the attaining of maximum grafting percentage at about 30 kGy by γ-irradiation [155]. In the samples grafted with styrene no crosslinking were obtained. It could be concluded that when the yield of grafting is very high in acrylonitrile, its termination occurs via the

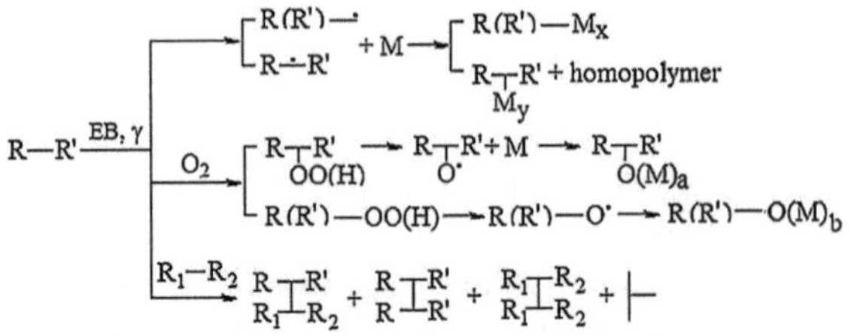

Fig. 9 View of possible grafting routes

combination of two growing chains of grafted branches and leads to a three-dimensional network. In contrast, cellulose samples after extraction with acetone showed decrease in the size of birling fragments that occurred at low dose rate and the cellulosic sheets broke into short fragments and particles. This could be explained by, polyethylene being a polymer of crosslinking type and cellulose being a polymer of degrading type.

The polymers compatibilization of different nonmiscible materials requires the grafting one of them with maleic structures. LLDPE can be modified by g-irradiation [156], when grafting degree increases as the concentration of DEM in the reaction mixture and the absorbed doses are increased up to 100 kGy. The evaluation of MFI values of irradiated samples up to 100 kGy proves the progress of grafting (Fig. 10), as well as spectroscopic records do. The highest grafting degree does not exceed 0.3 mol%.

The interaction between various compounding phases can be optimised by the grafting of base polymer with maleic anhydride ans other polar structures [157]. EB treatment is proper way for the modification of polyethylene, that be further subjected to addition of mineral filler, for example magnesium oxide (MgO).

Radiation-induced grafting of glycidyl methacrylate (GMA) onto high-density polyethylene (HDPE) and the radiation lamination of HDPE by bulk grafting of GMA can be accomplished for the production of medical items [158]. The γ-ray irradiation induced grafting of GMA onto HDPE occurs very easily the because of the existence of double bond in monomer (Fig. 10). The extent of grafting was higher in 2 *M* GMA than in 1 *M* GMA at the same irradiation dose. The extent of grafting initially increased quickly with the total irradiation dose and then remained almost constant. The maximum amount of GMA linked on polyethylene was obtained at 15 kGy. This low dose is explained by the radiation instability of monomer, which affects the radiation stability of grafted substrate. For various medical applications, the surface properties of low-density polyethylene (LDPE) can be modified radiochemically by the grafting of 2-hydroxyethyl methacrylate (HEMA) [159] (Fig. 10). LDPE films were grafted at 14.0 and 268.0 % using an initial concentration of $[HEMA]_i$ at 15 % (v/v) by irradiation in absence of air, during 10 and 30 h, respectively, at a dose-rate of 0.3 kGy h^{-1}.

The service of membranes for fuel cells asks grafting of appropriate monomers for assuring certain electrical conductivity and stability [160–166]. The radiation processing of these systems is accomplished by the initiation of radical generation at low

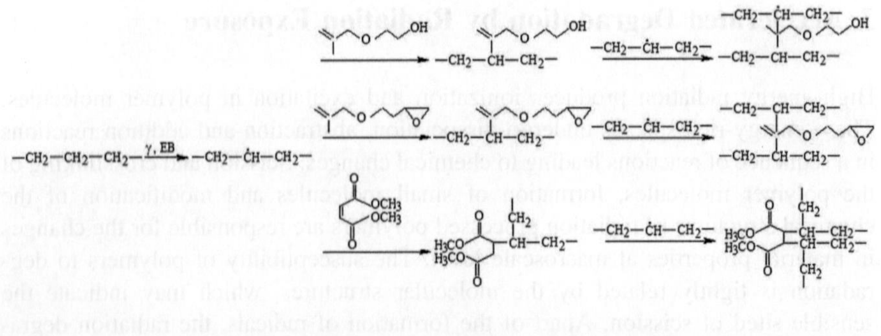

Fig. 10 Mechanistic scheme for the grafting of HEMA, GMA and DEM

and moderate doses, which interact with proton releasing units and provide suitable concentration of charge carriers. The increase in the polymer stability and functionality can be achieved by the grafting of various monomers, which induce a delocalization of electrons to create new structures with less reactivities. The grafting of styrene on the molecular configuration of polypropylene by gamma exposure [167] or EB treatment [168], starch [169], PTFE [170], cellulose [171] makes possible to enlarge application ranges onto the most aggressive conditions, because the benzene scavenge energy and does not release it on the entire molecules. The initial concentration of styrene influences the rate of grafting. The preirradiation or the presence of monomer capable to bind more efficiently styrene leads to high conversion of neat polymer, which must contain an optimal concentration of 30 % styrene. The localization of energy on benzene ring is also favourable to the compatibilization with different polymer structures that are less miscible with pristine polymer.

The high thermal stable polymers, fluorinated compounds, can be modified by grafting under irradiation [172, 173]. The modification of PTFE, FEP and PFA with styrene can be achieved at room temperature by the EB exposure in two steps. The first irradiation was accomplished at high temperatures (385 °C for PTF, 350 °C for PFA and 290 °C for FEP) for initiating grafting, where polymers were brought in liquid state in nitrogen atmosphere. The next step was EB irradiation at 50 kGy for the accomplishment of process. The maximum grafting yields were attained for high preirradiation dose (500 kGy). The order of the increasing efficiency for these polymers was PFA > PTFE > FEP because of their structures. For the proton exchange membranes (PEM) PTFE was grafted by styrene under γ-irradiation in air at room temperature. After the immersion in styrene, the grafting process was performed at different temperatures. The grafting yields reached different levels after 20 h of reaction: 76 % for 60 °C, 57 % at 80 °C, 22 % at 100 °C and 17 % at 120 °C. It seems that the lower temperature hinders the formation of homopolymer.

3 Accelerated Degradation by Radiation Exposure

High energy radiation produces ionization and excitation in polymer molecules. These energy-rich species undergo dissociation, abstraction and addition reactions in a sequence of reactions leading to chemical changes. Scission and crosslinking of the polymer molecules, formation of small molecules and modification of the chemical structures of radiation processed polymers are responsible for the changes in material properties at macroscale level. The susceptibility of polymers to degradation is tightly related by the molecular structures, which may indicate the sensible sited of scission. Apart of the formation of radicals, the radiation degradation is controlled by oxygen diffusion through which the feeding of molecular oxygen is defined. The distribution of final oxidation products has a parabolic shape because the gradient of oxygen concentration exists [63]. The degradation mechanisms under radiation are based on the processed model compounds, n-hexadecane and squalene [174], whose structures. The main criterion for the evaluation of radiation effects on polymers is the ratio between the two main radiochemical yields: scission and crosslinking. The values exceeding unity denotes stabilization of polymers by crosslinking. Its values less than unity define the degradable polymers [10]. The first process after scission is the formation of peroxyl radicals, which are the precursors of final oxidation products. The contribution of these structures may be found in the oxidation rates, which are reflected by the accelerated oxidation initiated even at moderate temperatures starting from 50 °C (Fig. 11) [99, 175].

The oxidation of high energy irradiated polymers progresses in relation with their basic component structure and formulation, which is initiated by the scission of backbones. The activation energy (E_a) required for oxidative degradation is the key of stability evaluation. The values of activation energies are placed on the range between 100 and 120 kJ mol^{-1} for different classes of insulation materials (polyethylene, ethylene-propylene copolymer, poly(vinyl chloride) used in the manufactures of cables for nuclear power plants [176]. The life time allows the prediction

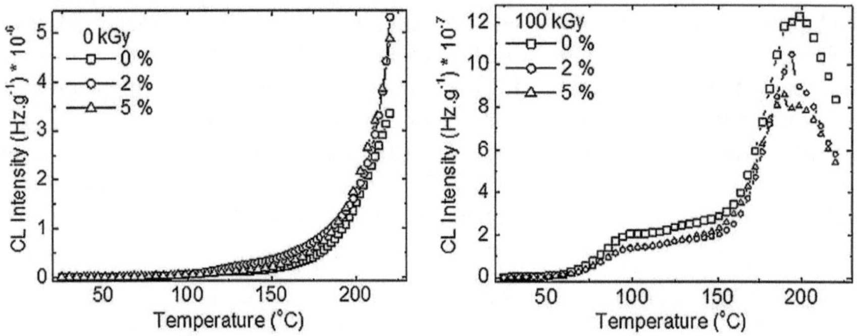

Fig. 11 The nonisothermal chemiluminescence spectra recorded on ethylene-propylene-diene terpolymer modified with silica

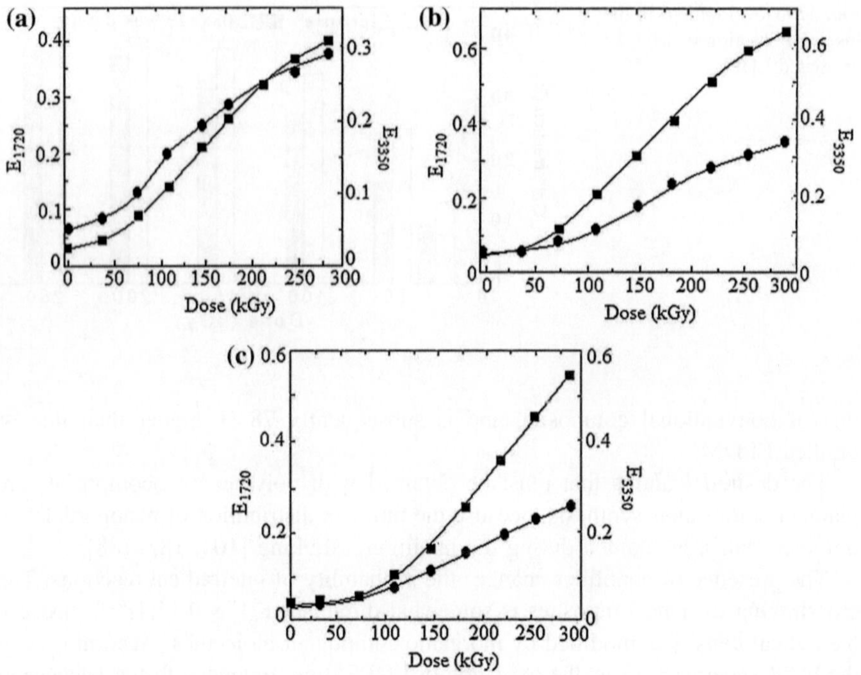

Fig. 12 Changes in absorbances of carbonyl () and hydroxyl () bands recorded for EPDM/PP blends. Blending ratios: **a** 80:20, **b** 60:40 and **c** 40:60

of the susceptibility to oxidation directly connected to the degradation mechanism and the service requirements [177]. The temperature regime of oxidation determines the ageing progress under different mechanisms, which modifies the diffusion rate of oxygen and, consequently, the Arrhenius dependency. The alteration of oxidation rate is also occurred due to the change in the molecular weight [178, 179], which characterizes the start of oxidation through the radical routes. The accumulations of carbonyl and hydroxyl containing compounds follow different rates because of the intermediate reactions involving hydroperoxides (Fig. 12) [180].

The evidences of existing differences between radiation degradation of polymers confirm the interaction between intermediates, which are the precursors of new structures [181]. The presence of additives like antioxidants or fillers turns the degradation onto specific alterations. An illustrative example for the contribution of nanostructured filled has been reported on EPDM/clay hybrids [97]. Apart of the improvement in the mechanical properties caused by radiation crosslinking, the irradiated nanocomposites show higher tensile strength than that of conventional composite and unfilled EPDM, inasmuch nanoclay dispersed in the matrix plays the role of reinforced filler which protects the matrix against degradation. The tensile strength of nanocomposites at 500 KGy irradiation dose is about 51 % higher than

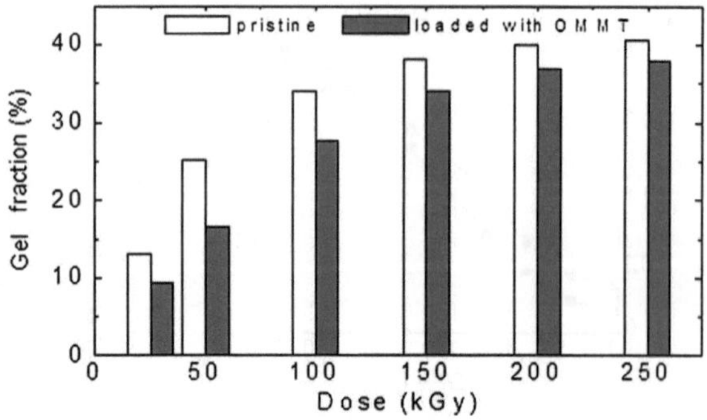

Fig. 13 Development in the insoluble fraction of EB-irradiated LDPE

that of conventional composite, and is subsequently 78 % higher than that of unfilled EPDM.

The desired features that must be obtained with polymer nanocomposites are planned before their synthesis, because the intimate distribution of nanoparticles in polymer matrix is attained during the in situ crosslinking [104, 182–188].

The presence of nanofiller changes the probability of interradical reactions. The crosslinking of nanocomposites is somewhat diminished (Fig. 13, [189]), because the radical density is modified by inorganic component molecules. According with the FTIR spectral records, the oxidation of LDPE runs somewhat faster, because in the presence of OMMT, some additional reactive sites come from the Hoffman elimination of the organo-modifier of the clay nano-particles. These results are related to different levels of modification in the values of polymer free volume accompanying to the intimate interaction between ionized molecules and organo-clay particles. The electron irradiation of PP/OMMT enhances the compatibility between the two components of the composite ensuring better cohesion in between [190].

The vicinity of polymer molecules and nanoparticles is an auspicious factor for the formation of bridges between components [92], so that the thermal and radiation stabilities are improved and their lifetime is significantly longer.

The mechanical properties of polymer nanocomposites are also influenced by the chemical treatment of nanoparticles due to the different neighborhood in the material. The free volume that characterizes the density of material is modified and, consequently, the penetration of fluids (solvents, oxygen) is rather favorable to degradation. The diffusion of xylene in ethylene-propylene diene terpolymer is unlike, if material presents different consistency (Fig. 14, [191]). The competitive radiochemical processes, crosslinking of polymer and degradation of covering layer are the most important reasons responsible for the different shapes of swelling curves.

Apart from the guiding the evolution of oxidative degradation by the adsorption of radicals on the enormous interphase surface, nanoparticle component of hybrid materials influences the modification induced in the molecular weights [192]. The ratios between weight and numerical average molecular weights are sharply increased and the macroscale properties are consequently modified. The activation

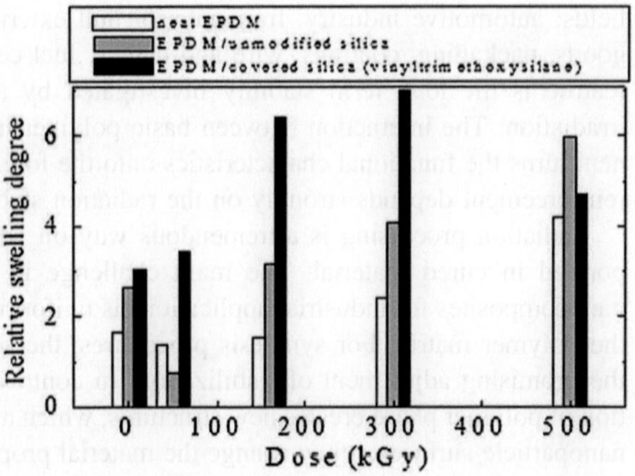

Fig. 14 Changes in the swelling degree of γ-irradiated of various EPDM samples

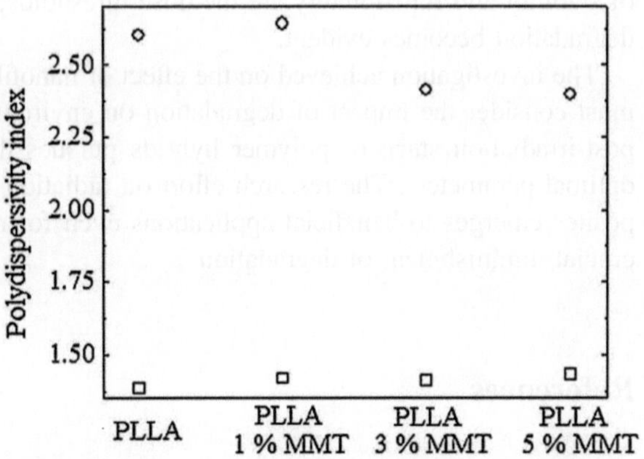

Fig. 15 Changes in the polydispersity of PLLA nanocomposites. (*Square*) before irradiation, (*Circle*) after exposure to 10 kGy

energies involved in radiation degradation follow similar trend, because the process enthalpies are directly correlated with molecular sizes (Fig. 15).

The addition of nanoparticles in the material formulations represents the key of polymer modification [193], which brings about the attaining upgraded performances simultaneously with the compatibilization of various blends. The features of polymer nanocomposites can be converted by high energy exposure and their application can be extended due to the favorable adjustment.

4 Conclusion

Polymeric nanocomposites show useful properties related to their stability. Their processing by exposure to high energy irradiation induces modifications, which enlarge the application limits. The potential applications can be found in various fields: automotive industry, from interior and exterior car parts to tyres, sporting goods, packaging, coatings, wire and cables, fuel cells, biomedical. The common feature is the long term stability investigated by accelerated degradation under irradiation. The interaction between basic polymer material and nanofiller component turns the functional characteristics onto the foreseen behavior, but the level of reinforcement depends strongly on the radiation stability of macromolecules.

Radiation processing is a tremendous way on which the filler is tightly incorporated in cured material. The main challenge in fabrication of these polymer nanocomposites for industrial applications is uniform dispersion of nanoparticles in the polymer matrix. For synthesis procedures, the scission of monomer promotes the promising adjustment of stabilization. In contrast, the decomposition/degradation of polymer phase creates new structures, which are able to be adsorbed onto the nanoparticle surface or they change the material properties to the unpredicted shift. However, the radiation treatment allows the proper preparation of ameliorated material in respect with conventional practice. During irradiation the basic aspects of stability and functionality are the dose thresholds, over which the proportion of degradation becomes evident.

The investigation achieved on the effect of nanofillers on the material durability must consider the impact of degradation on environmental health and safety. The post-irradiation stage of polymer hybrids pursues the long term operation under optimal parameters. The research effort on radiation processed polymer nanocomposites emerges to beneficial applications even for nuclear industry that develops crucial diminishment of degradation.

References

1. Chapiro, A.: Radiation Chemistry of Polymer Materials. Wiley Interscience Publishers, New York (1962)
2. Clough, R.L.: Radiation-resistant polymers. In: Encyclopedia of Polymer Science and Engineering, pp. 667–708, 2nd edn. Wiley, New York (1988)
3. Bhattacharya, A.: Radiation and industrial polymers. Prog. Polym. Sci. **25**, 371–401 (2000)
4. Clegg, D.W., Collyer, A.A. (eds.): Irradiation Effects on Polymers. Elsevier Applied Science, London (1999)
5. Woods, R.J.: Applied Radiation Chemistry: Radiation Processing. Wiley Interscience Publishers, New York (1994)
6. Clough, R.L.: High-energy radiation and polymers. A review of commercial processes and emerging applications. Nucl. Instrum. Meth. Phys. Res. B **185**, 8–33 (2001)
7. Spinks, J.W.T., Woods, R.J. (eds.): Introduction to Radiation Chemistry, 3rd edn. Wiley, New York (1990)
8. Dawes, K., Glover, L.C., Vroom, D.A.: The effects of electron beam and γ-irradiation on polymer materials. In: Mark, J.E. (ed.) Physical Properties of Polymer, Handbook, 2nd edn. Springer, New York (2007)
9. Makuuchi, K., Chang, S. (eds.): Radiation Processing of Polymer Materials and Its Industrial Applications. Wiley, New York (2012)

10. Zaharescu, T., Jipa S.: Radiochemical modifications in polymers. In: Arndt, K.F., Lechner M. D. (eds.) Landolt-Börnstein Series, vol. VIII/6 C2, Polymer Solids and Polymer Melts, pp. 95–184. Springer (2013)
11. Drobny, J.G.: Ionizing Radiation and Polymers: Principles, Technology, and Applications. Elsevier, PDL Handbook Series (2012)
12. Cleland, M.R., Park, L.A., Chang, S.: Applications for radiation processes of material. Nucl. Instrum. Meth. Phys. Res. B **208**, 66–73 (2003)
13. Zaharescu, T.: Irradiation effects on ethylene-propylene elastomers in aqueous environment. Polym. Test. **15**, 69–74 (1996)
14. Gillen, K.T., Clough, R.L.: Polymer ageing insights available from modulus profiling data. Polym. Eng. Sci. **29**, 29–35 (1989)
15. Ershov, B.G., Gordeev, A.V.: A model for radiolysis of water and aqueous solutions H_2, H_2O_2, O_2. Radiat. Phys. Chem. **77**, 928–935 (2008)
16. Zaharescu, T., Giurginca, M., Setnescu, R.: The radiation stability of the ethylene-propylene type rubbers during their gamma ray ageing. Rev. Roum. Chim. **40**, 181–190 (1995)
17. Zaharescu, T., Jipa, S., Setnescu, R., Setnescu, T.: Radiation processing of polyolefin blends. Part IV. Spectroscopic investigation on EPDM/PP blends. Nucl. Instrum. Meth. Phys. Res. B **265**, 260–264 (2007)
18. Gehring, J.: With radiation crosslinking of polyolefin engineering plastics into the next millennium. Radiat. Phys. Chem. **57**, 361–365 (2000)
19. Chmielewski, A.G., Haji-Saeid, M.: Radiation technologies: past, present and future. Radiat. Phys. Chem. **71**, 16–20 (2004)
20. Chmielewski, A.G., Haji-Saeid, M.: IAEA Program in the field of radiation technology. Nucl. Instrum. Meth. Phys. Res. B **236**, 38–43 (2005)
21. Haji-Saeid, M., Sampa, M.H., Ramamoorthy, N., Güven, O., Chmielewski, A.G.: The role of IAEA in coordinating research and transferring technology in radiation chemistry and processing of polymers. Nucl. Instrum. Meth. Phys. Res. B **265**, 51–57 (2007)
22. Haji-Saeid, M., de Sampa, M.H.O., Chmielewski A.G.: Radiation treatment for sterilization of packaging materials. Radiat. Phys. Chem. **76**, 1353 (2007)
23. Haji-Saeid M., Safrany, A., de Sampa, M.H.O., Ramamoorthy, N.: Radiation processing of natural polymers: the IAEA contribution. Radiat. Phys. Chem. **79**, 255–260 (2010)
24. Chmielewski, A.G., Al-Sheikhly, M., Berejka, A.J., Cleland, M.R., Antoniak, M.: Recent developments in the application of electron accelerators for polymer processing. Radiat. Phys. Chem. **94**, 147–150 (2014)
25. Berejka, A.J., Cleland, M.R., Walo, M.: The evolution of challenges for industrial radiation processing—2012. Radiat. Phys. Chem. **94**, 141–146 (2014)
26. Berejka, A.J.: Radiation response of industrial materials: dose rate and morphology implications. Nucl. Instrum. Meth. Phys. Res. B **261**, 86–89 (2007)
27. Nablo, S.V., Chrusciel, J., Cleghorn, D.A., Rangwalla, I.: Factors influencing equipment selection in electron beam processing. Nucl. Instrum. Meth. Phys. Res. B **208**, 90–101 (2003)
28. Miller, A.: Approval and control of radiation processing, EB and gamma. Radiat. Phys. Chem. **31**, 385–393 (1988)
29. Cleland, M.R., Park, L.A.: Medium and high-energy electron beam radiation processing for commercial applications. Nucl. Instrum. Meth. Phys. Res. B **208**, 74–89 (2003)
30. Barnard, J.W.: Factors influencing equipment selection in electron beam processing. Nucl. Instrum. Meth. Phys. Res. B **208**, 98–101 (2003)
31. Charlesby, A. (ed.): Atomic Radiation and Polymers. Pergamon Press, New York (1960)
32. Farah, K., Kuntz, F., Kadri, O., Ghedira, L.: Investigation of the effect of some irradiation parameters on the response of various types of dosimeters to electron irradiation. Radiat. Phys. Chem. **71**, 337–341 (2004)
33. Lavalle, M., Corda, U., Fuochi, P.G., Caminati, S., Venturi, M., Kovács, A., Baranyai, M., Sáfrány, A., Miller, A.: Radiochromic film containing methyl viologen for radiation dosimetry. Radiat. Phys. Chem. **76**, 1502–1506 (2007)

34. Abdel-Fattah, A.A., Ebraheem, S., Ali, Z.Y., Abdel-Rehim, F.: Ultraviolet and infrared spectral analysis of irradiated polyethylene films: correlation and possible application for large-dose radiation dosimetry. J. Appl. Polym. Sci. **67**, 1837–1851 (1998)
35. Castañeda Facio, A., Benavides, R., Martinez Pardo, M.E., Uribe, R.: Electron beam crosslinking of non-lead PVC formulations. Radiat. Phys. Chem. **76**, 1720–1723 (2007)
36. Peimel-Stuglik, Z., Fabisiak, S.: A comparison of the performance characteristics of four film dosimeters in a 10-MeV electron beam. Appl. Radiat. Isot. **66**, 346–352 (2008)
37. Whittaker, B., Watts, M.F.: Influence of dose rate, ambient temperature and time on the radiation response of Harwell PMMA dosimeter. Radiat. Phys. Chem. **60**, 101–110 (2001)
38. Seito, H., Ichikawa, T., Kaneko, H., Sato, H., Watanabe, Y., Kojima, T.: Characteristics study of clear polymethylmethacrylate dosimeter, Radix W, in several kGy range. Radiat. Phys. Chem. **78**, 356–359 (2009)
39. Khan, H.M., Ahmad, G., Sattar, A., Durrani, S.K.: Radiation dosimetry using clear PMMA and PVC in the range of 5-45 kGy. J. Anal. Nucl. Chem. **125**, 127–134 (1988)
40. Yang, B., Lu, Q., Wang, S., Townsend, P.D.: Studies on the thermoluminescence spectra and thermal stability of LiF:Mg, Cu, LiF:Mg, Cu, Na and LiF:Mg, Cu, Si. Nucl. Instrum. Meth. Phys. Res. B **239**, 171–178 (2005)
41. Hosni, F., Farah, K., Kaouach, H., Louati, A., Chtourou, R., Hamzaoui, A.H.: Effect of gamma-irradiation on the colorimetric properties of epoxy-resin film. Potential use in dosimetric application. Nucl. Instrum. Meth. Phys. Res. B **311**, 1–4 (2013)
42. de Magalhães, C.M.S., Macedo, Z.S., Valerio, M.E.G., Hernandes, A.C., Souza, D.N.: Preparation of composites of topaz embedded in glass matrix for applications in solid state thermoluminescence. Nucl. Instrum. Meth. Phys. Res. B **218**, 277–282 (2004)
43. Necmeddin Yazici, A., Bedir, M., Sibel Sökücü, A.: The analysis of dosimetric thermoluminescenty glow peak of CaF2: Mn after β-irradiation. Nucl. Instrum. Meth. Phys. Res. B **259**, 955–965 (2007)
44. McLaughlin, W., Desrosiers, M.F.: Dosimetric systems for radiation processing. Radiat. Phys. Chem. **46**, 1163–1174 (1995)
45. Thalacker, V.P., Simpson, O.T., Postma, N.B.: Electron beam dosimeters for radiation processing. Radiat. Phys. Chem. **31**, 473–479 (1988)
46. Cleland, M.R., Pageau, G.M.: Comparisons of X-ray and gamma-ray sources for industrial processes. Nucl. Instrum. Math. Phys. Res. B **24/25**, 967–972 (1987)
47. Saylor, M.C., Parks, L.A., Herring, C.H.: Technical and regulatory for radiation sterilization facilities using electron beam accelerators. Nucl. Instrum. Meth. Phys. Res. B **79**, 875–878 (1993)
48. Pilette, L.: Effects of ionizing treatments on packaging—food simulant combinations. Packag. Technol. Sci. **3**, 17–20 (1990)
49. Zimek, Z., Przybytniak, G., Nowicki, A., Mirkowski, K., Roman, K.: Optimization of electron beam crosslinking for cables. Radiat. Phys. Chem. 161–165 (2014)
50. Bartoníček, B., Plaček, V., Hnát, V.: Comparison of degradation effects induced by gamma radiation and electron beam radiation in two cable jacketing materials. Radiat. Phys. Chem. **76**, 857–863 (2007)
51. Charlesby, A.: Crosslinking of polyethylene. Proc. Royal Soc. (London), A **215**, 187–188 (1952)
52. Chapiro, A.: Chemical modifications in irradiated polymers. Nucl. Instrum. Meth. Phys. Res. B **32**, 111–114 (1988)
53. Cheng, S., Phillips, Ed, Parks, L.: Processability improvement of polyolefins through radiation-induced branching. Radiat. Phys. Chem. **79**, 329–334 (2010)
54. Deng, P.Y., Liu, M.H., Zhang, W.X., Sun, J.Z.: Preparation and physical properties of enhanced radiation induced crosslinking of ethylene-vinyl alcohol copolymer (EVOH). Nucl. Instrum. Meth. Phys. Res. **B 258**, 357–361 (2007)
55. Yoshii, F., Suhartini, M., Sagasawa, N., Mitomo, H., Kume, T.: Modification of biodegradable polymers by radiation crosslinking technique with polyfunctional monomers. Nucl. Instrum. Meth. Phys. Res. B **208**, 370–373 (2003)

56. Dadbin, S., Frounchi, M., Goudarzi, D.: Electron beam induced crosslinking of nylon 6 with and without the presence of TAC. Polym. Degrad. Stab. **89**, 436–441 (2005)
57. Mitomo, H., Kaneda, A., Quynh, T.M., Nagasawa, N., Yoshii, F.: Improvement in heat stability of poly(l-lactic acid) by radiation-induced crosslinking. Polymer **46**, 4695–4703 (2005)
58. Nagasawa, N., Kasai, N., Yagi, T., Yoshii, F., Tamada, M.: Radiation-induced crosslinking and post-processing of poly(l-lactic acid) composite. Radiat. Phys. Chem. **80**, 145–148 (2011)
59. Murray, K.A., Kennedy, J.E., McEvoy, B., Vrain, O., Ryan, D., Cowman, R., Higginbotham, C.L.: The effects of high energy electron beam irradiation in air on accelerated ageing and on the structure property relationships of low density polyethylene. Nucl. Instrum. Meth. Phys. Res. B **297**, 64–74 (2013)
60. Miličević, D., Trifunović, S., Popović, M., Vukašinović, T., Milić, Suljovrujić, E.: The influence of orientation on the radiation-induced crosslinking/oxidative behavior of different PEs. Nucl. Instrum. Meth. Phys. Res. B **260**, 603–612 (2007)
61. Yoshiga, A., Otaguro, H., Parra, D.F., Lima, L.F.C.P., Lugao, A.B.: Controlled degradation and crosslinking of polypropylene induced by gamma radiation in acetylene. Polym. Bull. **63**, 397–409 (2009)
62. Zaharescu, T., Feraru, E., Podină, C.: Thermal stability of ethylene propylene-diene monomer/divinylbenzene systems. Polym. Degrad. Stab. **87**, 11–16 (2005)
63. Gillen, K.T., Clough R.L.: In: Clough, R.L., Shalaby, S.W. (eds.) ACS Symp Series 475, ch. 28. ACS, Washington DC, (1991)
64. Zaharescu, T., Jipa, S., Setnescu, R., Setnescu, T.: Radiation processing of polyolefin blends. Part I. Crosslinking of EPDM/PP blends. J. Appl. Polym. Sci. **77**, 982–987 (2000)
65. Zaharescu, T., Jipa, S., Giurginca, M.: Radiochemical processing of EPDM/NB blends. J. Macromol. Sci. Pure Appl. Chem. A **35**, 1093–1102 (1998)
66. Dubey, K.A., Bhardwaj, Y.K., Chaudhari, C.V., Sabharwal, S.: Radiation-processed styrene–butadiene-*co*-ethylene– propylene diene rubber blends: compatibility and swelling studies. J. Appl. Polym. Sci. **99**, 3638–3649 (2006)
67. Chowdhury, R.: Electron-beam-induced crosslinking of natural rubber/acrylonitrile–butadiene rubber latex blends in the presence of ethoxylated pentaerythritol tetraacrylate used as a crosslinking promoter. J. Appl. Polym. Sci. **103**, 1206–1214 (2007)
68. Xiang, Z.I., Liu, H.R., Deng, P.Y., Liu, M.H., Yin, Y., Ge, X.W.: The effect of irradiation on morphology and properties of the PET/HDPE blends with trimethylolpropane trimethacrylete (TMPTA). Polym. Bull. **63**, 587–597 (2009)
69. Voit, W., Ware, T., Gall, K.: Radiation crosslinked shape-memory polymers. Polymer **51**, 3551–3559 (2010)
70. Banik, I., Bhowmick, A.K.: Effect of electron beam irradiation on the properties of crosslinked rubbers. Radiat. Phys. Chem. **58**, 293–298 (2000)
71. Haque, M.E., Dafader, N.C., Akhtar, F., Ahmad, M.U.: Radiation dose required for the vulcanization of narural rubber latex. Radiat. Phys. Chem. **48**, 505–510 (1996)
72. Kurtz, S.M., Muratoglu, O.K., Evans, M., Edidin, A.A.: Advances in the processing, sterilization and crosslinking of ultra-high molecular weight polyethylene for total joint arthroplast. Biomaterials **20**, 1659–1688 (1999)
73. Rezanejad, S., Kokab, M.: Shape memory and mechanical properties of cross-linked polyethylene/clay nanocomposites. Eur. Polym. J. **43**, 2856–2865 (2007)
74. Mahapram, S., Poompradub, S.: Preparation of natural rubber (NB) latex/low density polyethylene (LDPE) blown film and its properties. Polym. Test. **30**, 716–725 (2011)
75. Chattopadhyay, S., Chaki, T.K., Bhowmick, A.K.: Heat shrinkability of electron-beam-modified thermoplastic elastomeric films from blends of ethylene vinylacetate copolymer and polyethylene. Radiat. Phys. Chem. **59**, 501–505 (2000)
76. Zhu, G., Liang, G., Xu, Q., Yu, Q.: Shape-memory effects of radiation crosslinked poly(ε-caprolactone). J. Appl. Polym. Sci. **90**, 1589–1595 (2003)

77. Ware, T., Voit, W., Gall, K.: Effects of sensitizer length on radiation crosslinked shape-memory polymers. Radiat. Phys. Chem. **79**, 446–453 (2010)
78. Tikku, V.K., Biswas, G., Despande, R.S., Majali, A.B., Chaki, T.K., Bhowmick, A.K.: Electron beam initiated grafting of trimethylol propane trimethacrylate onto polyethylene—structure and properties. Radial. Phys. Chem. **45**, 829–833 (1995)
79. Gall, K., Dunn, M.L., Liu, Y.P., Finch, D., Lake, M., Munshi, N.A.: Shape-memory polymer nanocomposites. Acta Mater. **50**, 5115–5126 (2002)
80. Hu, J.L., Zu, Y., Huang, H.H., Lu, J.: Recent advances in shape-memory polymers: structure, mechanism, functionality, modeling and applications. Prog. Polym. Sci. **37**, 1720–1763 (2012)
81. Mishra, J.K., Chang, Y.W., Lee, B.C., Ryu, S.H.: Mechanical properties and heat shrinkability of electron beam crosslinked polyethylene-octene copolymer. Radiat. Phys. Chem. **77**, 675–679 (2008)
82. Rimnac, C.M., Kurtz, S.M.: Ionizing radiation and orthopaedic prostheses. Nucl. Instrum. Meth. Phys. Res. B **236**, 30–37 (2005)
83. Miguez Suarez, J.C., de Biasi, R.S.: Effect of gamma irradiation on the ductile-to-brittle transition in ultra-high molecular weight polyethylene. Polym. Degrad. Stab. **82**, 221–227 (2003)
84. Brunella, V., Bracco, P., Carpentieri, I., Paganini, M.C., Zanetti, M., Costa, L.: Lifetime of alkyl macroradicals in irradiated ultra-high molecular weight polyethylene. Polym. Degrad. Stab. **92**, 1498–1503 (2007)
85. Costa, L., Carpentieri, I., Bracco, P.: Post electron-beam irradiation oxidation of orthopaedic UHMWPE. Polym. Degrad. Stab. **93**, 1695–1703 (2008)
86. Chmielewski, A.G., Chmielewska, D.K., Michalik, J., Sampa, M.H.: Prospects and challenges in application of gamma electron and ion beams in processing of nanomaterials. Nucl. Instrum. Meth. Phys. Res. B **265**, 339–346 (2007)
87. Chmielewski, A.G., Michalik, J., Buczkowski, M., Chmielewska, D.K.: Ionizing radiation in naotechnology. Nucl. Instrum. Meth. Phys. Res. B **236**, 329–332 (2005)
88. Jung, C.H., Lee, D.H., Hwang, I.T., Im, D.S., Shin, J.W., Kang, P.H., Choi, J.H.: Fabrication and characterization of radiation-resistant LDPE/MWCNT nanocomposites. J. Nucl. Mater. **438**, 41–45 (2013)
89. Martínez-Morlanes, M.J., Castell, P., Martínez-Nogués, V., Martinez, M.T., Alonso, P.J., Puértolas, J.A.: Effects of gamma-irradiation on UHMWPE/MWNT nanocomposites. Compos. Sci. Technol. **71**, 282–288 (2011)
90. Huegun, A., Fernández, M., Muñoz, M.E., Santamaría, A.: Rheological properties and electrical conductivity of irradiated MWCNT/PP nanocomposites. Compos. Sci. Technol. **72**, 1602–1607 (2012)
91. Marković, G., Marinović-Cincović, M.S., Jovanović, V., Samaržija-Jovanović, S., Budinski-Simendić, J.: Gamma irradiation aging of NBR/CSM rubber nanocomposites. Compos. Part B- Eng. **43**, 609–615 (2012)
92. Wang, B.B., Song, L., Hong, N.N., Tai, Q.L., Lu, H.D., Hu, Y.: Effect of electron beam irradiation on the mechanical and thermal properties of intumiscent flame retarded ethylene-vinyl acetate copolymer/orcanically modified montmorillonite compositions. Radiat. Phys. Chem. **80**, 1275–1281 (2011)
93. Choi, J.H., Jung, C.-H., Kim, D.K., Suh, D.H., Nho, Y.C., Kang, P.H., Ganesan, R.: Preparation of polymer/POSS nanocomposites by radiation processing. Radiat. Phys. Chem. **78**, 517 (2009)
94. Lee, K.Y., Kim, K.Y., Hwang, I.R., Choi, Y.S., Hong, C.H.: Thermal, tensile and morphological properties of gamma-ray irradiated epoxy-clay nanocomposites toughened with a liquid rubber. Polymer. Test. **29**, 139–142 (2010)
95. Mohamed, R.M.: Radiation induced modification of NBR and SBR montmorillonite nanocomposites. J. Ind. Eng. Chem. **19**, 80–86 (2013)
96. Crăciun, E., Jitaru, I., Zaharescu, T., Jipa, S.: Qualification of epoxy resin by radiochemical ageing. Optoelectr. Adv. Mater. Rapid Commun. **4**, 1819–1822 (2010)

97. Ahmadi, S.J., Huang, Y.D., Ren, N.Q., Mohaddespour, A., Ahmadi-Brooghani, S.Y.: The comparison of EPDM/clay nanocomposites and conventional composites in exposure of gamma irradiation. Compos. Sci. Technol. **69**, 997–1003 (2009)
98. Planes, E., Chazeau, L., Vigier, G., Fournier, J., Stevenson-Royaud, I.: Influence of filler on mechanical properties of ATH filled EPDM during ageing by gamma irradiation. Polym. Degrad. Stab. **95**, 1029–1038 (2010)
99. Zaharescu, T., Pleşa, I., Jipa, S.: Kinetic effects of silica nanoparticles on thermal and radiation stability of polyolefins. Polym. Bull. **70**, 2981–2994 (2014)
100. Zaharescu, T., Jipa, S., Adrian, M., Supaphol, P.: Nanostructured isotactic polypropylene—TiO_2 systems. J. Optoelectr. Adv. Mater. **10**, 2205–2209 (2008)
101. Zaidi, L., Bruzaud, S., Kaci, M., Bourmaud, A., Gautier, N., Grohens, Y.: The effect of gamma irradiation on the morphology and properties of polylactyide/Cloisite 30B nanocomposites. Polym. Degrad. Stab. **98**, 348–355 (2013)
102. Ciuprina, F., Zaharescu, T., Pleşa, I.: Effect of γ-radiation on dielectric properties of LDPE-Al_2O_3 nanocomposites. Radiat. Phys. Chem. **84**, 145–150 (2013)
103. Khalid, M., Ismail, A.F., Ratnam, C.T., Faridah, Y., Rashmi, W., Al Khatib, M.F.: Effect of radiation dose on the properties of natural rubber nanocomposite. Radiat. Phys. Chem. **79**, 1279–1285 (2010)
104. Choi, J.H., Jung, C.H., Kang, D.W., Hwang, I.T., Choi, J.H.: Preparation and characterization of crosslinked poly(ε-caprolactone)/polyhedral oligomeric silsesquioxane nanocomposites by electron beam irradiation. Nucl. Instrum. Meth. Phys. Res. B **287**, 141–147 (2012)
105. Choi, J.H., Jung, C.H., Kim, D.K., Ganesan, R.: Radiation-induced grafting of inorganic particles onto polymer backbone: a new method to design polymer-based nanocomposite. Nucl. Instrum. Meth. Phys. Res. B **266**, 203–206 (2008)
106. Janowski, B., Pielichowski, K.: Thermo(oxidative) stability of novel polyurethane/POSS nanohybrid elastomers. Thermochim. Acta **478**, 51–53 (2008)
107. Abou Zeid, M.M.: Radiation effect on properties of carbonblack filled NBR/EPDM rubber blends. Eur. Polym. J. **43**, 4415–4422 (2007)
108. Thomas, J.K.: Fundamental aspects of the radiolysis of solid polymers, crosslinking and degradation. Nucl. Instrum. Meth. Phys. Res. B **265**, 1–7 (2007)
109. Olejniczak, J., Rosiak, J., Charlesby, A.: Gel/dose curves for polymers undergoing simultaneous crosslinking and scission. Radiat. Phys. Chem. **37**, 499–504 (1991)
110. Hill, D.J.T., Whittaker, A.K., Zainuddin, : water diffusion into radiation crosslinked PVA-PVP network hydrogel. Radiat. Phys. Chem. **80**, 213–218 (2011)
111. Kadłubowski, S., Henke, A., Ulański, P., Rosiak, I.: Hydrogels of poly(vinylpyrrolidone) (PVP) and poly(acrylic acid) (PAA) synthesized by radiation-induced crosslinking of homopolymers. Radiat. Phys. Chem. **79**, 261–266 (2010)
112. Adb El-Mohdy, H.L., Safrany, A.: Preparation of fast response superabsorbent hydrogel by radiation polymerization and crosslinking of isopropylacrylamide in solution. Radiat. Phys. Chem. **77**, 273–279 (2008)
113. von Sonntag, C., Bothe, E., Ulanski, P., Adhikary, A.: Radical transfer reactions in polymers. Radiat. Phys. Chem. **55**, 599–603 (1999)
114. Schmidt, T., Querner, C., Arndt, K.-F.: Characterization methods for radiation crosslinked pol(vinyl methyl ether) hydrogels. Nucl Instrum. Meth. Phys. Res. B **208**, 331–335 (2003)
115. Rosiak, J.M., Ulanski, I.P., Pajewski, L.A., Yoshii, F., Makuuchi, K.: Radiation formation of hydrogel for biomedical purposes. Some remarks and comments. Radiat. Phys. Chem. **46**, 161–168 (1995)
116. Varshney, L.: Role of natural polysaccharides in radiation formation of PVA-hydrogel wound dressing. Nucl Instrum. Meth. Phys. Res. B **255**, 343–349 (2007)
117. Zhou, Y., Zhao, Y.H., Wang, L., Xua, L., Zhai, M.L., Wei, S.C.: Radiation synthesis and characterization of nanosilver/gelatin/carboxymethyl chitosan hydrogel. Radiat. Phys. Chem. **81**, 553–560 (2012)

118. Lugão, A.B., Machado, L.D.B., Mirandal, L.F., Alvarez, M.R., Rosiak, J.M.: Study of wound dressing structure and hydration/dehydration properties. Radiat. Phys. Chem. **52**, 319–322 (1998)
119. Ajji, Z., Othman, I., Rosiak, J.M.: Production of hydrogel wound dressing using gamma radiation. Nucl. Instrum. Meth. Phys. Res. **B 229**, 375–380 (2005)
120. Şahiner, N., Pekel, N., Güven, O.: Radiation synthesis, characterization and amidoximation of *N*-vinyl-2-pyrrolidone/acrylonitrile interpenetrating polymer network. React. Funct. Polym. **39**, 139–146 (1999)
121. Razzak, M.T., Darwis, D., Zainuddin, Z., Sukirno, S.: Irradiation of polyvinyl alcohol and polyvinyl pyrrolidone blended hydrogel for wound dressing. Radiat. Phys. Chem. **62**, 107–113 (2001)
122. Kan, C.X., Wang, C.S., Zhu, J.J., Li, H.C.: Formation of gold and silver nanocomposites within polyvinylpyrrolidone (PVP) gel. J. Solid State Chem. **183**, 858–865 (2010)
123. Momesso, R.G.R.A.P., Moreno, C.S., Rogero, S.O., Rogero, J.R., Spencer, P.J., Lugão, A. B.: Radiation stability of resveratrol in immobilization on poly(vinyl pyrrolidone) hydrogel for dermatological use. Radiat. Phys. Chem. **79**, 283–285 (2010)
124. Ferraz, C.C., Varca, G.H.C., Lopes, P.S., Mator, M.B., Lugão, A.B.: Radiation-synthesized polyacrylamide hydrogels for proteins release. Radiat. Phys. Chem. **94**, 186–189 (2014)
125. Varca, G.H.C., Ferraz, C.C., Lopes, P.S., Mathor, M.B., Grasselli, M., Lugão, A.B.: Radiation-synthesized protein-based nanoparticles for biomedical purposes. Radiat. Phys. Chem. **94**, 181–185 (2014)
126. Abd El-Mohdy, H.L., Hegazy, E.A., El-Nesr, E.M., El-Wahab, M.A.: Metal sorption behavior of poly(*N*-vinyl-2-pyrrolidone)/(acrylic acid-co-styrene) hydrogels synthesized by gamma radiation. J. Environ. Chem. Eng. **1**, 328–338 (2013)
127. Radiation synthesis of stimuli-responsive membranes, hydrogels and adsorbents for separation purposes, IAEA-TECDOC 1465 (2005)
128. Gottlieb, R., Schmidt, T., Arndt, K.-F.: Synthesis of temperature-sensitive hydrogel blends by high-energy irradiation. Nucl. Instrum. Meth. Phys. Res. B **236**, 371–376 (2005)
129. Bhunia, T., Goswami, L., Chattopadhyay, D., Bandyopadhyay, A.: Sustained transdermal release of diltiazem hydrochloride through electron beam irradiated different PVA hydrogel membranes. Nucl. Instrum. Meth. Phys. Res. B **269**, 1822–1828 (2011)
130. Burillo, G., Briones, M., Adem, E.: IPN's of acrylic acid and *N*-isopropylacrylamide by gamma and electron beam irradiation. Nucl. Instrum. Meth. Phys. Res. B **265**, 104–108 (2007)
131. Casimiro, M.H., Gil, M.H., Leal, J.P.: Drug release assays from new chitosan/pHEMA membranes obtained by gamma irradiation. Nucl. Instrum. Meth. Phys. Res. B **265**, 406–409 (2007)
132. Safrany, A.: Macroporous gels with fast response prepared by e-beam crosslinking of poly (N-isopropylacrylamide) solution. Nucl. Instrum. Meth. Phys. Res. B **236**, 587–593 (2005)
133. Jipa, I.M., Stroescu, M., Stoica-Guzun, A., Dobre, T., Jinga, S., Zaharescu, T.: Effect of gamma irradiation on biopolymer films of poly(vinyl alcohol) and bacterial cellulose. Nucl. Instrum. Meth. Phys. Res. B **278**, 82–87 (2012)
134. Ahmed, E.M.: Hybrid composites prepared from industrial wastes: mechanical and swelling behavior. J. Adv. Res. doi:10.1016/j.jare.2013.12.002
135. Safrany, A.: Radiation processing: synthesis and modification of biomaterials for medical use. Nucl. Instrum. Meth. Phys. Res. **B 131**, 376–381 (1997)
136. Nho, Y.C., Lee, J.H.: Radiation of postsurgical adhesion formation with hydrogels synthesized by radiation. Nucl. Instrum. Meth. Phys. Res. B **236**, 277–282 (2005)
137. Burillo, G., Clough, R.L., Czvikovszky, T., Guven, O., Le Moel, A., Liu, W.W., Singh, A., Yang, J.T., Zaharescu, T.: Polymer recycling: potential application of radiation technology. Radiat. Phys. Chem. **64**, 41–51 (2002)
138. Żenkiewicz, M., Dzwonkowski, J.: Effects of electron radiation and compatibilizers on impact strength of composites of recycled polymers. Polym. Test. **26**, 903–907 (2007)

139. Dalai, S., Wenxiu, C.: Radiation effects on HDPE/EVA blends. J. Appl. Polym. Sci. **86**, 553–558 (2002)
140. Burillo, G., Galicia, M., del Pilar Carreo, M., Vázquez, M., Adem, E.: Crosslinking of recycled polyethylene by gamma irradiation in the presence of sensitizers. Radiat. Phys. Chem. **60**, 73–78 (2001)
141. Burillo, G., Herrera-Franco, P., Vazquez, M., Adem, E.: Compatibilization of recycled and virgin PET with radiation-oxidized HDPE. Radiat. Phys. Chem. **63**, 241–244 (2002)
142. Mészáros, L., Bárány, T., Czvikovszky, T.: EB-promoted recycling of waste tire rubber with polyolefins. Radiat. Phys. Chem. **81**, 1357–1360 (2012)
143. Karaağaç, B., Şen, M., Deniz, V., Güven, O.: Recycling of gamma irradiated inner tubes in butyl based rubber. Nucl. Instrum. Meth. Phys. Res. B **265**, 290–293 (2005)
144. Sonnier, R., Leroy, E., Clerc, L., Bergeret, A., Lopez-Cuesta, J.M.: Compatibilisation of polyethylene/ground tyre rubber blends by γ-irradiation. Polym. Degrad. Stab. **91**, 2375–2379 (2006)
145. Zaharescu, T., Feraru, E., Podină, C., Jipa, S.: Modifications of EPDM by gamma irradiation in hydrocarbon environment. Polym. Degrad. Stab. **89**, 373–381 (2005)
146. Jones, R.A., Groves, D.J., Ward, I.M., Taylor, D.J.R., Stepto, R.F.T.: Gel fraction and chain reactions in irradiated polyethylenes. Nucl. Instrum. Meth. Phys. Res. B **151**, 213–217 (1999)
147. Scagliusi, R.S., Cardoso, E.C.L., Lugão, A.B.: Radiation-induced degradation of butyl rubber vulcanized by three different crosslinking systems. Radiat. Phys. Chem. **81**, 991–994 (2012)
148. Scagliusi, R.S., Cardoso, E.C.L., Lugão, A.B.: Effect of gamma radiation on chlorobutyl rubber vulcanized by three crosslinking systems. Radiat. Phys. Chem. **81**, 1370–1373 (2012)
149. Zaharescu, T.: Degradation of ethylene-propylene copolymer in the presence of phenolic antioxidants. J. Mater. Sci. Lett. **14**, 923–925 (1995)
150. Bhattacharya, A., Misra, B.N.: Grafting: a versatile means to modify polymers: techniques, factors and applications. Prog. Polym. Sci. **29**, 767–814 (2004)
151. Garnett, J.L.: Grafting. Radiat. Phys. Chem. **14**, 79–99 (1979)
152. Stannett, V.T.: Radiation grafting—state-of-the art. Radiat. Phys. Chem. **35**, 82–87 (1990)
153. Moura, E., Somessari, E.S.R., Silveira, C.G., Paes, H.A., Souza, C.A., Fernandes, W., Manzoli, J.E., Geraldo, A.B.C.: Influence of physical properties on mutual polymer grafting by electron beam irradiation. Radiat. Phys. Chem. **80**, 175–181 (2011)
154. Ranogajec, F.: Effect of solvent on radiation grafting and crosslinking of polyethylene. Radiat. Phys. Chem. **76**, 1381–1384 (2007)
155. Hassanpour, S.: Radiation grafting of styrene and acrylonitrile to cellulose and polyethylene. Radiat. Phys. Chem. **55**, 41–45 (1999)
156. Catarí, E., Albano, C., Karam, A., Perera, R., Silva, P., González, J.: Grafting of a LLDPE using gamma irradiation. Nucl. Instrum. Meth. Phys. Res. B **236**, 338–342 (2005)
157. Legocka, I., Zimek, Z., Mirkowski, K., Nowicki, A.: Preliminary study on the application PE filler modified by radiation. Radiat. Phys. Chem. **57**, 411–416 (2000)
158. Li, Z.R., Wang, H.L.: Radiation-induced grafting of glycidyl methacrylate onto high density polyethylene (HDPE) and radiation lamination of HDPE. J. Appl. Polym. Sci. **96**, 772–779 (2005)
159. Ferreira, L.M., Falcão, A.N., Gil, M.H.: Modification of LDPE molecular structure by gamma irradiation for bioapplications. Nucl. Instrum. Meth. Phys. Res. B **236**, 513–520 (2005)
160. Gwon, S.J., Choi, J.H., Sohn, J.Y., An, S.J., Ihm, Y.E., Nho, Y.C.: Radiation grafting of methyl methacrylate onto polyethylene separators for lithium secondary batteries. Instrum. Meth. Phys. Res. B **266**, 3387–3391 (2008)
161. Abdel-Hady, E.E., El-Toony, M.M., Abdel-Hamed, M.O.: Grafting of glycidyl methacrylate/styrene onto polyvinyl fluoride membranes for proton exchange fuel cell. Electrochim. Acta **103**, 32–37 (2013)

162. Souzy, R., Ameduri, B.: Functional fluoropolymers for fuel cell membranes. Prog. Polym. Sci. **30**, 644–687 (2005)
163. Gubler, L., Slaski, M., Wallasch, F., Wokaun, A., Scherer, G.G.: Radiation grafted fuel cell membranes based on co-grafting of a-methylstyrene and methacrylonitrile into a fluoropolymer base film. J. Membr. Sci. **339**, 68–77 (2009)
164. Li, J.Y., Matsuura, A., Kakigi, T., Miura, T., Oshima, A., Washio, M.: Performance of membrane electrode assemblies based on proton exchange membranes prepared by pre-irradiation induced grafting. J. Power Sources **161**, 99–105 (2006)
165. Chen, J.H., Li, D.R., Koshikawa, H., Asano, M., Maekawa, Y.: Crosslinking and grafting of polyetheretherketone film by radiation techniques for application in fuel cell. J. Membr. Sci. **362**, 488–494 (2010)
166. Sherazi, T.A., Ahmad, S., Akram Kashmiri, S., Kim, D.S., Guiver, M.D.: Radiation-induced grafting of styrene onto ultra-high molecular weight polyethylene powder for polymer electrolyte fuel cell application. II. Sulfonation and characterization. J. Membr. Sci. **333**, 59–67 (2009)
167. Nho, Y.C., Chen, J., Jin, J.H.: Grafting polymerization of styrene onto preirradiated polypropylene fabric. Radiat. Phys. Chem. **54**, 317–322 (1999)
168. Vahdat, A., Bahrami, H., Ansari, N., Ziaie, F.: Radiation grafting of styrene onto polypropylene fibers by a 10 MeV electron beam. Radiat. Phys. Chem. **76**, 787–793 (2007)
169. Sheikh, N., Akhavan, A., Ataeivarjovi, E.: Radiation grafting of styrene on starch with high efficiency. Radiat. Phys. Chem. **85**, 189–192 (2013)
170. Yu, H.Y., Shi, H., Zeng, X.M., Bao, M., Zhao, X.Q.: A proton-exchange membrane prepared by the radiation grafting of styrene and silica into polytetrafluoroethylene films. Radiat. Phys. Chem. **78**, 497–500 (2009)
171. Barsbay, M., Güven, O., Davis, T.P., Barner-Kowollik, C., Barner, L.: RAFT-mediated polymerization and grafting of sodium 4-styrenesulfonate from cellulose initiated via gamma-radiation. Polymer **50**, 973–982 (2009)
172. Lappan, U., Geissler, U., Uhlmann, S.: Radiation-induced grafting of styrene into radiation modified fluoropolymer films. Nucl. Instrum. Meth. Phys. Res. B **236**, 413–419 (2005)
173. Sato, K., Ikeda, S., Iida, M., Oshima, A., Tabata, Y., Washio, M.: Studyon poly-electrolyte membrane of crosslinked PRFE by radiation grafting. Nucl. Instrum. Meth. Phys. Res. B **208**, 424–428 (2003)
174. Katsuma, Y.: Radiation induced degradation of polymers-an approach by using liquid paraffins as a model system. Angew. Makromol. Chem. **252**, 89–101 (1997)
175. Zaharescu, T., Jipa, S., Supaphol, P.: Thermal stability of isotactic polypropylene modified with $CaCO_3$ nanoparticles. Polym. Bull. **64**, 783–790 (2010)
176. Seguchi, T., Tamura, K., Ohshima, T., Shimada, A., Kudoh, H.: Degradation mechanism of cable insulation materials during radiation-thermal ageing in radiation environment. Radiat. Phys. Chem. **80**, 268–273 (2011)
177. Celina, M.: Review of polymer oxidation and its relationship with material performances and lifetime prediction. Polym. Degrad. Stab. **98**, 2419–2429 (2013)
178. Zaharescu, T., Cazac, C., Jipa, S., Setnescu, R.: Radiation processing of polyisobutylene. Nucl. Instrum. Meth. Phys. Res. B **185**, 360–364 (2001)
179. Burillo, G., Tenorio, L., Bucio, E., Adem, E., Lopez, G.P.: Electron beam irradiation effects on poly(ethylene terephtalate). Radiat. Phys. Chem. **76**, 1728–1731 (2007)
180. Zaharescu, T., Giurginca, M., Jipa, S.: Radiochemical oxidation of ethylene-propylene elastomers in the presence of some phenolic antioxidants. Polym. Degrad. Stab. **63**, 245–251 (1999)
181. Pandey, J.K., Raghunatha Reddy, K., Pratheep Kumar, A., Singh, R.P.: An overview on the degradability of polymer nanocomposites. Polym. Degrad. Stab. **88**, 234–250 (2005)
182. Kim, S.K., Kwen, H.D., Choi, S.H.: Radiation-induced synthesis of vinyl copolymer based nanocomposites filled with reactive organic montmorillonite clay. Radiat. Phys. Chem. **81**, 519–523 (2012)

183. Kharazmi, A., Saion, E., Faraji, N., Hussin, R.M., Yunus, W.M.M.: Structural, optical and thermal properties of PVA/CdS nanocomposites synthesized by radiolytic method. Radiat. Phys. Chem. **97**, 212–216 (2014)
184. Krklješ, A.N., Marinović-Cincović, M.T., Kacarević-Popović, Z.M., Nedeljković, J.M.: Radiolytic synthesis and characterization of Ag-PVA nanocomposites. Eur. Polym. J. **43**, 2171–2176 (2007)
185. Xu, X.L., Yin, Y.D., Ge, X.W., Wu, H.K., Zhang, Z.H.: γ-Radiation synthesis of poly(acrylic acid)—metal nanocomposites. Mater. Lett. **37**, 354–358 (1998)
186. Abou Taleb, M.F., Hegazy, D.E., Ismail, S.A.: Radiation synthesis, characterization and dye adsorption of alginate-organophilic montmorillonite nanocomposite. Carbohydr. Polym. **87**, 2263–2269 (2012)
187. Robinette, E.J., Palmese, G.R.: Synthesis of polymer-polymer nanocomposites using radiation grafting techniques. Nucl. Instrum. Meth. Phys. Res. **236**, 216–222 (2005)
188. Ni, Y.H., Ge, X.W., Zhang, Z.C.: Fabrication of CdS/polyacrylonitrile nanocomposites by γ-irradiation in an ethanol solution. Mater. Sci. Eng. B **130**, 61–65 (2006)
189. Dintcheva, N.Tz., Alessi, S., Arrigo, R., Przybytniak, G., Spadaro, G.: Influence of the e-beam irradiation and photo-oxidation aging on the structure and properties of LDPE-OMMT nanocomposite films. Radiat. Phys. Chem. **81**, 432–436 (2012)
190. Misheva, M., Djourelov, N., Zamfirova, G., Gaydarov, V., Cerrada, M.L., Rodríguez-Amor, V., Pérez, E.: Effect of compatibilizer and electron irradiation on free-volume and microhardness of syndiotactic polypropylene/clay nanocomposites. Radiat. Phys. Chem. **77**, 138–145 (2008)
191. Masellini-Varlot, K., Vigier, G., Vermogen, A., Gauthier, C., Cavaille, J.Y.: Quantitative structural characterization of polymer-clay nanocomposites and discussion of an ideal microstructure, leading to the highest mechanical reinforcement. J. Polym. Sci., Part B Polym. Phys. **45**, 1243–1251 (2007)
192. Yidirim, Y., Oral, A.: The influence of γ-ray irradiation on the thermal stability and molecular weight of poly(L-Lactic acid) and its nanocomposites. Radiat. Phys. Chem. **96**, 69–74 (2014)
193. Jovanović, Z., Jovanović, Ž., Krklješ, A., Stojkovska, J., Tomić, S., Obradović, B., Mišković-Stanković, V., Kačarević-Popović, Z.: Synthesis and characterization of silver/polyvinyl 2-pyrrolidone hydrogel nanocomposite obtained *in situ* radiolytic method. Radiat. Phys. Chem. **80**, 1208–1215 (2011)

Thermal Degradation of Synthetic Rubber Nanocomposites

Adali Castañeda Facio, Aide Saenz Galindo, Lorena Farias Cepeda, Lluvia López López and Ramón Díaz de León-Gómez

Abstract In recent years, synthetic rubbers nanocomposites have captured and held the attention of scientists because are the materials for the future, which have improved resistance to thermal degradation and stability of the nanocomposite. Commonly fillers like layered silicates, carbonaceous nanofillers (carbon nanotubes, carbon naanofibers and exfoliated nanographite), spherical particles (Silica, TiO_2, ZnO, $CaSO_4$, $CaCO_3$, $ZnFe_2O_4$) and polyhedral oligomeric silsesquioxane (POSS) are used for reinforcing elastomers. This new materials exhibit enhanced properties at very low filler level, usually ≤5 wt%. The properties of rubber nanocomposites strongly depend on the dispersion state of fillers and method of preparation. The effect to different nanoparticles on rubber properties is studied with thermal stability. This is mainly studied using TGA, TGA-MS, TGA-FTIR and other techniques. The thermal degradation mechanism of the rubber synthetic nanocomposites is generally considered to be related to the kind of used nanoparticles and its amount, the interactions between inorganic nanoparticles and polymer reactive group. Rubber synthetic nanocomposites play an important role in engineering, automotive, aerospace, construction, packaging and medical devices applications due to is possible to design new materials with unprecedented and improvements in their physical properties, particularly from the perspective of applications.

1 Introduction

The words nanocomposites and nanofillers are fairly recent, but have been in use since 1904 for example, carbon black is being used as a reinforcing filler in rubbers and apparently always existed in nature (in minerals and vegetation) [1, 2].

A.C. Facio (✉) · A.S. Galindo · L.F. Cepeda · L.L. López
Blvd. V. Carranza e Ing. José Cárdenas V., C.P. 25280, Saltillo, Coahuila, México
e-mail: adali.castaneda@uadec.edu.mx; adalif@hotmail.com

R.D. de León-Gómez
Blvd. Enrique Reyna No. 140, C.P. 25294, Saltillo, Coahuila, México

P.M. Visakh and Y. Arao (eds.), *Thermal Degradation of Polymer Blends, Composites and Nanocomposites*, Engineering Materials,
DOI 10.1007/978-3-319-03464-5_7

Nanocomposites represent a new development in the area of high thermal stability and offer significant improvement over conventional formulation where high filler loadings are often required [3]. Nanocomposites are multiphase materials containing two or more components mixed at nanometer scale. Such nanocomposites may contain organic (fullerenes, carbon nanotubes, graphene) and inorganic particles (Au, Ag, Cu, ZnO, TiO_2, layered silicates, etc.) obtained inorganic-inorganic, inorganic-organic or organic-organic nanocomposites. Organic-inorganic nanocomposites consisting of a polymeric matrix and various types of fillers with nanometric dimensions [4].

Polymer nanocomposite having inorganic particles within nanoscale dimensions have received considerable attention because of their much improved unique properties and numerous potential applications as in automotive, aerospace and construction industry, manufacture of tires and inner tubes. Other industrial rubber goods include various belts, oil seals, gasket and food packing.

Nanocomposites have been the exponentially growing field of research for developing the materials in the last few decades and have been mainly focusing on the structure-property relationships and their development. This is because polymer-nanocomposites can overcome the limitations of traditional micro-composites. With the rapid development of the nanotechnologies and nanomaterials the studies on polymer-based nanocomposites have been extensively carried out in order to find their promising alternatives to traditional composites, though mainly focused on multifunctional properties and filler dispersion. Properties of thermoplastics which have shown substantial improvement due to the incorporation of nanoparticles include: Thermal Stability and heat distortion temperature [5]; Mechanical properties, e.g., strength, modulus and dimensional stability [6]; Decrease permeability to gases, water and hydrocarbons; Flame retardancy and reduce smoke emissions; Chemical resistance; Surface appearance; Electrical and thermal conductivity; Optical clarity in comparison to conventionally filled polymers.

Major advantages that nanocomposites have over conventional composites are: lighter weight due to low filler loading, low cost due to fewer amounts of filler to be used.

Among various nanoparticles like clay minerals, carbon nanotubes and silica nanoparticles are more often used in enhancing physical, mechanical and thermal properties of polymers [7, 8]. Uniform dispersion of nanoparticles produces ultra-large interfacial area per volume between the nanoparticle and polymer.

Depending upon the number of dimensions of the dispersed particles in the nanometer range three types of nanocomposites can be distinguished as follows:

- Isodimensional nanofillers result when the three dimensions are in order of nanometers, such as spherical silica nanoparticles obtained by in situ solution-gel methods [9, 10] or by polymerization promoted from their surface [11].
- When two dimensions are in the nanometer scale while the third is large, an elongated structure results, as for example carbon nanotubes [12, 13], which are extensively, studied as reinforcing nanofillers yielding materials with exceptional properties.

- The third type of nanocomposite is characterized by only one dimension in the nanometer range. Here the fillers is in the form of sheets of one to a few nanometer thick to hundreds to thousands nanometers long. Clays and layered silicates belong to this family and the composites are known as polymer layered silicates nanocomposites (PLSNs).

Incorporation of clay or layered silicates to polymer matrix provides three different structures: conventional composites, intercalated and exfoliated nanocomposites and these are shown in Fig. 1. It is interdependent on the clay concentration the degree of clay layer separation and distribution in the composites [14]. In conventional composites the particles exist as aggregates with no insertion of polymer matrix.

In microcomposites or conventional composites the particles exist as aggregates with no insertion of polymers matrix. Hence it cannot impart any enhancement in properties. Intercalated nanocomposites consisting of a regular insertion of polymer in between the silicate layered in a crystallographically regular fashion [15–18]. In an exfoliated nanocomposite the individual 1 nm silicate layers are separated and dispersed in a continuous polymer matrix, with average distances between layers. Exfoliated nanocomposites exhibit better properties owing to the maximum polymer/filler interactions. In exfoliated structure, the entire surfaces of the layered are available for interactions with the polymer.

Due to attempts to establish morphology-property correlations in nanocomposites the discussion now focuses on the different types of nanofillers and reinforcing in synthetic rubbers nanocomposites.

New generation nano scale fillers are challenging the domination of traditional fillers such a as carbon blacks and silica in the rubbery industry. Nanoscaled fillers such as layered silicates, carbon nanotubes, carbon nanofibers (CNFs), exfoliated graphite, spherical particles and Polyhedral oligomeric silsesquioxane (POSS), etc. dispersed as a reinforcing phase in a rubber matrix are emerging as a relatively new form of useful material.

1.1 Nanofillers

Layered Silicates The structure of Layered silicates (LS)/clays consists of a 2-D layer of two fused silicate tetrahedral sheet with an edge-shared octahedral sheet of metal atoms, such as Al or Mg. This model was proposed by Hoffmann et al. [19].

The LS can be natural or synthetic clays, as well as phosphates of transition metals. The most widely used reinforcement is clay due to its natural abundance and its very high form factor.

Clay-based nanocomposites generate an overall improvement in physical performances. The most widely used ones are the phyllosilicates (smectites). They have a shell-shaped crystalline structure with nanometric thickness. Clays are classified according to their crystalline structures and also to the quantity and position of the ions within the elementary mesh. The elementary or primitive mesh

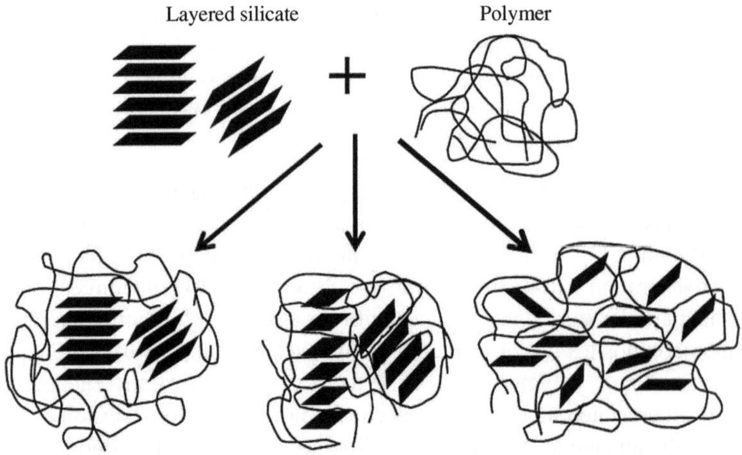

Fig. 1 Formation of conventional, intercalated and exfoliated nanocomposites

Table 1 Nanoclays identification

Family		Group	Formula
Phyllosilicates	TO(1:1)	Kaolinite	The thickness of the layer is about 0.7 nm. Kaolinite $Al_4Si_4O_{10}(OH)_8$
	TOT (2:1)	Smectite, (Talc, Mica, Mommorillonite), Sepiolite	The thickness of the layer is about 1 nm. The group includes many minerals that are major constituents of clays
	TOT:O (2:1:1)	Chlorite, Bentonite, Saponite	The reference plate is formed of three plates TOT and another isolated O plate. The thickness of the layer is about 1.4 nm. Chlorite di-tri $Al_2Mg_3Si_4O_{10}(OH)_8$
Polysilicate	Natural	Kenyaite, Magadiite, Kanemite, Ilerite, Silhydrite, Zeolite	Magadiite ($Na_2Si_{14}O_{29}H_2O$)
	Synthetic	FluoroHectorite, Zeolite	
Double lamellar hydroxide	Synthetic	Hydrotalcite	Hydrotalcites: $(Mg_6A_{12}(OH)_{16})(CO_3^{2-})4H_2O$

is the simplest atomic geometric pattern, which is enough for duplicating the crystalline network, by repeating itself indefinitely in the three directions. Table 1 presents the various natural and synthetic nanoclays available and used as fillers in rubbers.

The most common usage concerns organomodified Montmorillonite (MMT), a natural phyllosilicate extracted from Bentonite. Raw formula of Montmorillonite is $(Na,Ca)_{0,3}(Al, Mg)_2Si_4O_{10}(OH)_2{,}nH_2O$.

The layered system exhibits better properties due to higher polymer/LS interactions. The dispersion of inorganic fillers in organic polymers is like oil in water. The way to overcome this drawback is by transforming the organophobic galleries to organophilic. This is achieved by replacing the cation originally present in the galleries with organic cation, which has along organic chains with a positively charged end. It will act as compatibilising agent between organic and inorganic phase. Amino acids were the first compatibilising agents used for synthesis of polymers nanocomposites [20]. Alkyl ammonium ions are the most widely used promising compatibilizing agent. Lan et al. [21] observed that alkyl ammonium ions with a chain length larger than eight carbon atoms favored the synthesis of exfoliated nanocomposites whereas ammonium ions with shorter chains led to the formation of intercalated nanocomposite.

On the other hand Abdallah et al. [22] showed that bentonites modified by alkyl and aryl based phosphonium salts in poly(ethylene terephthalate) exhibited better dispersion than those unmodified. In 2010 Gu [23] showed that montmorillonite modified Octadecylammonium improves dispersion in natural rubber/butadiene rubber nanocomposites. Have also been studied in terms of particle size of styrene–butadiene rubber (SBR) nanocomposites showed remarkable improvement in thermal stability their compared to that of the pure SBR. Was also demonstrated that the increase in particle size is not beneficial in improving the thermal stability [5].

Carbonaceous Nanofillers Recently research efforts have focused on nano-scale variants of carbon black (carbon nanotubes, carbon nanofibers and exfoliated nanographite) as possible reinforcing fillers in elastomers. Among these, nanotubes are attracting the most attention.

In 1976 nano-filaments were discovered by Oberlin et al. [24, 25] however, Lijima had a huge impact on the scientific community with Iijima's paper on Nature, and his following discovery of single-wall carbon nanotubes (SWCNTs) [26].

The fiber could present a nanometric diameter and length of some orders of magnitude in comparison with its diameter. In general, three kinds of carbon nanotubes are considered:

- Single-wall carbon nanotubes (SWCNT). They present a diameter between 1 and 2 nm;
- Double-wall carbon nanotubes (DWCNT). Diameter is between 2 and 4 nm;
- Multi-wall carbon nanotubes (MWCNT). They present a diameter between 4 and 150 nm.

These nanotubes present a theoretical range of properties incredible (Young modulus up to 1 TPa, heat conductivity of 3000 W m^{-1} K^{-1}, electric conductivity of 107 S m^{-1}, etc.), but considering perfect nanotubes individually makes no sense. They nevertheless provide a wide range of new properties when used in nanocomposites, depending on their purity and dispersion in the matrix.

Sanchez studied the functionalization of oxidized SWCNTs and MWCNTs dispersed in thermoplastic elastomers based on poly(butylene terephthalate) (PBT)/ poly(tetramethylene oxide) (PTMO). These nanocomposites showed good dispersion and enhancement in thermo-oxidative stability [27]. 1 % of pristine multi-walled carbon nanotube (MWCNTs) were dispersed in silicon rubber. The SR nanocomposites showed 28 % better thermal stability and 100 % improvement in the ultimate tensile strength is achieved as compared with the pristine polymer matrix counterpart [28]. Also ionic liquids have been tested to improve the dispersion and thermal stability of MWCNTs in polychloroprene rubber (CR) showing improvement in these properties [29]. On the other hand the effect of carbon nanofiber on nitrile rubber was studied. It has been found that the nanofiber increase the thermal stability and decrease the flammability [4].

Xiong reported that both the thermal stability and the thermal conductivity of bromo-butyl rubber (BIIR) nanocomposites could be improved by incorporating the ionic liquids (ILs) modified graphene oxide (GO-Ils) using a solution compounding method [30]. Graphene has emerges as a subject of enormous scientific interest due to its exceptional electron transport, mechanical properties and high surface area. When incorporated, these atomically thin carbon sheets can significantly improve physical properties of polymers matrix at extremely small loading. Decomposition of graphene composites is substantially slower than neat polymer, which is attributed to restricted chain mobility of polymer near the graphene surface [31].

Spherical Particles Nanofiller with three dimensions in the nanometer regime are the spherical nanofillers obtained by sol-gel process [9, 10]. In sol-gel process the organic/inorganic hybrid material can be formed by the condensation reaction between the functionalized prepolymer and the metal alkoxides, leading to the formation of a chemical bond between the polymer and the inorganic filler. Therefore, the incorporation of filler particles in polymer through the sol-gel process avoids the aggregation of filler.

Peng et al. [32] have synthesized surface unmodified and modified silica nanoparticles. The strategy behind the surface modification of fillers is to enhance its interaction with the polymer matrix through hydrogen bonding with the reactive site of polymer matrix. Kickelbic [33] has extensively studied concept behind the nanoscale incorporation of inorganic building blocks in organic polymer. Silica, TiO_2, ZnO, $CaSO_4$, $CaCO_3$, $ZnFe_2O_4$ and so on, are the widely used spherical inorganic nanofillers in the polymer field [34–36].

Polyhedral oligomeric silsesquioxane (POSS) is a class of organic–inorganic hybrid nanomaterial constituted of an inorganic silica Rn(SiO1.5)n core cage structure where *n* is the number of silicon atoms of the cage (n = 8, 10, 12). The size of a POSS molecule is approximate 1.5 nm in diameter and about 1000D in mass—hence POSS nanostructures are nearly equivalent in size to most polymer dimensions and smaller than polymer radii of gyration. POSS chemical systems may be viewed as the smallest chemically particles of silica possible while the resins in which they are incorporated may be viewed as molecular nanocomposites which are intermediate between polymers and ceramics. Due to the great chemical flexibility of POSS molecules, POSS can be incorporated into polymers by copolymerization, grafting, or even blending using traditional processing methods.

Generally, the POSS are incorporated into the polymer matrix by copolymerization, grafting, or even blending using the traditional processing methods. The incorporation of POSS into polymers can lead to a successful improvement of the flammability, thermal or polymer mechanical properties. Besides, the problems associated with the polymer immiscibility are reduced. The thermo-mechanical properties of polymer-POSS nanocomposites systems are highly related to the resultant microstructure, which in turn depends on the type of interaction between the nanoPOSS and the matrix. POSS can be dispersed at a molecular scale or they can act like microdispersed fillers [37]. In general, to obtain molecular dispersion, some kind of chemical bonds between the POSS functional groups and the polymer matrix are required [38] For this reason the core structure can be modified by attaching functional groups to the Si atoms [39]. These functional groups allow POSS to be successfully incorporated into thermoplastics or thermosets, being possible to upgrade their final properties. Usually, positive reinforcement is obtained when good compatibility between the POSS and the polymer exists.

1.2 Nanocomposite Preparation

Use of nanoparticles as fillers in rubbers is highly relevant because end use applications of rubber compounds require filler reinforcement. Most of the literature on rubber nanocomposites is based on the use of nanoclay as the filler. It has been shown that incorporation of nanoclay in synthetic rubbers, like styrene butadiene rubber (SBR), chloroprene rubber (CR), nitrile rubber (NBR), ethylene propylene diene monomer (EPDM) rubber etc. enhances the mechanical, anti-ageing and barrier properties.

There are several methods of preparing elastomer nanocomposites like latex blending, solution mixing, internal mixing and open mill mixing [40]. These approaches have their own advantages and limitations. Most of the reported literature on elastomer nanocomposites is based on solution mixing technique, where a polymer is dissolved in a suitable solvent along with nanofiller followed by evaporation of solvent to obtain the nanocomposite. Solution mixing can seldom be used for bulk production of nanocomposites as dissolution of elastomer in the solvent and subsequent removal of the solvent pose engineering difficulties and environmental problems. Ganter et al. [41] found that the intergallery distance of the organoclay increased more than two times when it was incorporated in styrene/butadiene rubber (SBR) using toluene solvent. The solution technique has been used in the past for SBR, ethylene/vinylacetate [42], butadiene rubber (BR), polyepichlorohydrin [43], acrylonitrile/butadiene rubber (NBR) [44] and polyisoprene (IR) [45].

For preparation of elastomer based nanocomposites, mixing of latex and nanoclay followed by coagulation and drying is a viable method in the case of rubbers that are available in latex form. Most rubbers are available in latex form. Latex is a more or less stable aqueous dispersion of fine rubber particles (in the submicron-micron range). The related particle size distribution strongly depends on the manufacturing. So, mixing of the latex with layered silicates followed by coagulation (coprecipitation) is a promising way to produce rubber nanocomposites. This technique has been used for NR [46], SBR [47], carboxilated NBR [48], NBR [49], polyurethane (PUR) [50] and polychloroprene (CR).

Use of internal mixers offers a direct and environment friendly technique for preparation of nanocomposites. A two-step mixing in an internal mixer followed by addition of curative on a two roll mill for preparation of elastomer-layered silicate nanocomposites [51]. In the study on the effect of processing conditions (mixer type, mixing temperature) and formulations on the properties of EPDM nanocomposites, it has been shown that open two roll mill mixing results in inadequate dispersion of the nanofiller in the elastomer matrix compared to compounding in an internal mixer. Melt compounding with layered silicates has been reported for NR [52], ENR [46], SBR [53], NBR [54], EPDM [55] and PU [56] systems.

2 Thermal Degradation of Synthetic Rubber Nanocomposite

The exposure of synthetic rubber nanocomposites during service against environmental influences like oxygen, temperature, static and dynamic mechanical load or UV-light effects aging processes.

Rubber materials undergo both physical and chemical changes when heat is applied; this will usually result in undesirable changes to the properties of the material. A clear distinction needs to be made between thermal decomposition and thermal degradation. Thermal decomposition is "a process of extensive chemical species change caused by heat." Thermal degradation is "a process whereby the action of heat or elevated temperature on a material, product, or assembly causes a loss of physical, mechanical, or electrical properties" [57].

The chemical processes are responsible for the generation of flammable volatiles while physical changes, such as melting and charring, can markedly alter the decomposition and burning characteristics of a material.

The various physical processes that occur during thermal decomposition can depend on the nature of the material.

The thermal decomposition of polymers may proceed by oxidative processes or simply by the action of heat. In many polymers, the thermal decomposition processes are accelerated by oxidants (such as air or oxygen). In that case, the minimum decomposition temperatures are lower in the presence of an oxidant. Of particular interest is the fact that the effect of oxygen (or air) on thermal decomposition depends on the mechanism of polymerization: free-radical polymerization leads to a neutralization of the effect of oxygen. A study on poly(vinylidene fluoride) indicated that the effect of oxygen can lead to changes in both reaction rate and kinetic order of reaction [58]. Inabi's work in particular has resulted in the development of models for the kinetics of general random chain scission thermal decomposition [59], as well as for the thermal decomposition of thermoplastics [60]. There are a number of general classes of chemical mechanisms important in the thermal decomposition of polymers: 1 random-chain scission, in which chain scissions occur at apparently random locations in the polymer chain. The chain reaction is terminated by recombination reactions with the formation of stable compounds, such as, for instance, the constitution of C–C or C–O–C bonds from two radicals tantamount to an increase in crosslinking density; 2 end-chain scission, in which individual monomer units are successively removed at the chain end; 3 chain-stripping, in which atoms or groups not part of the polymer chain (or backbone) are cleaved; and 4 cross-linking, in which bonds are created between polymer chains.

At the same time the thermo-oxidation generally occurs quite slowly at ambient temperatures, but may have serious consequences as the temperature is increased. Oxidation of rubbers can involve the reaction of free radicals present in the rubber with molecular oxygen. The free radicals can be formed by decomposition of hydroperoxides, which are present in the rubber in minute amounts after processing. These reactions then lead to chain scission and/or additional crosslinking, depending on the type of rubber. For instance, oxidation of synthetic rubber is initially dominated by chain scission, which causes the rubber to soften. This softening is followed by an increase in crosslinking, which then leads to hardening. Both processes results in a weaker, brittle polymer. At ambient temperatures the main degradation process involves the presence of oxygen from the atmosphere. The reaction with oxygen leads to chain scission and the formation of polymeric peroxides, which can also increase the rate of thermo-oxidation [61] The presence of double bonds and branching in the macromolecular chains of the polymer increases the oxidation rate, even if a stabilization agent is present [62]. Changes in the properties of polymers depend on the extent of such reactions [63].

Curing systems can influence the thermal degradation of rubber nanocomposites. Accelerated sulphur curing systems are commonly used in vulcanizing the elastomers. The crosslink density and type of crosslink networks significantly influence the mechanical properties of the rubber vulcanisates. In general, the vulcanisates cured with efficient and conventional systems predominantly contain monosulphide, disulphide networks and polysulphide networks [64]. As an alternative, the peroxide curing system has been used to overcome this problem. The thermal stability of C–C linkages improves the ageing properties of elastomer compounds [65].

Rubber is material with important applications in different areas, highlighting the automotive area, due to chemical and physical property, especially their structure composition, this material is considered an elastomer within the classification of polymeric materials.

Synthetic rubbers are a highly elastic material that enhances the quality of modern life. It is a unique material polymeric in that it stretches many times it is length without breaking, and it recovers to its original shape, "memory", is indispensable for many key applications such as automobiles tire, conveyor belts, gloves, and many more [66].

Nowadays, synthetic rubber, has the characteristics of being able to be combined, mixed or modified with different type of fillers, organic and or inorganic, may be of different size, morphology and chemical nature. An important area to explore in the implementation and use of nanocomposites is degradation, which may be by different methods: chemical, enzymatic [67], thermal, etc. highlighting the thermal degradation as an important method that could occur during the use of polymeric nanocomposites. Below are some examples of polymer nanocomposites degradation.

With respect to UV degradation, Arantes et al., reported the photodegradation of styrene–butadiene rubber (SBR)/TiO_2 nanocomposites. Nanocomposites films were obtained by solution mixing. The study was conducted by monitoring cross-linking reactions between the polymer chains during the degradation process by nuclear magnetic resonance spectroscopy (NMR) in the solid state with magic angle spinning (MAS), when the material becomes crosslinked becomes more rigid and less soluble in organic solvents [68].

The electron beam irradiation is other type of energy that could produce degradation, in 2012 Tonny and Nemr reported the degradation of acrylonitrile/butadiene/silica nanocomposite, the materials were electron beam irradiated at 25 and 100 kGy, finding that during the electron beam the thermal and thermo oxidative degradation are the principal reaction that occur confirmed by the fragments C–C of the polymer and volatiles as CO, CO_2, CH_4, H_2O, etc. [69].

The thermal degradation has been amply studied by researches, meanwhile Chayan et al. in 2012, reported the study of Nitrile rubber/silica nanocomposites prepared with surface modified silica nanoparticles. They observed that the first weight loss occur at the temperature range 350–490 °C is due to degradation of the rubber component. The next weight loss at the temperature range 520–640 °C is due to the decomposition of carbonaceous residue. The temperature at maximum weight loss (Tmax) was determined from the peaks obtained in Derivative Thermogravimetry (DTG). It is observed that the value is higher for the composites containing surface modified silica particles indicating greater thermal stability for them over that of composite containing unmodified silica [70, 71]. Presence of silane coupling agents in silica nano particles are detected Improvement in thermal stability [70].

In 2011 Mohomane et al., prepared nanocomposites from polychloroprene rubber (PCP) as the matrix and organically modified montmorillonite clays (Cloisite 15A, 20A, 25A, 10A and 93A) as fillers by two-roll mill, with clays contents of 2.5, 5 and 10 phr., The results for Cloisite 93A and Cloisite 15A depicted an exfoliated structure and a well-dispersed morphology in the polymer matrix at all filler contents, while complete exfoliation was not observed for the other clays, especially at higher clay contents. The initial stage of thermal degradation was accelerated with the incorporation of organoclays. The TGA results show that degradation occurs in two stages. The first weight loss takes place around 300 °C and was attribute to the elimination of HCl molecule, leaving behind the polyene chains. This is followed by a gradual decrease in weight between 300 and 450 °C. Also condensation reactions occur above 500 °C, demonstrating the stability by incorporating organically modified montmorillonite and that clays have a significant influence on the PCP degradation mechanism, even at low clay contents.

Introduction of the organic modified clay generally reduces the onset of the decomposition temperature of the PCP rubber. On the other hand, the degradation of the nanocomposite samples is slowed down in the first stage of decomposition. The former effect was most pronounced in the case of the Cloisite 93A and Cloisite 15A nanoclays, where it is observed at low clay contents. As the filler content increased, this effect was also observed in the other PCP nanocomposites [72].

Is important to emphasize that the fillers in the polymer matrix have as effect an increase in the thermal stability of the nanocomposite. Interactions between polymer and inorganic filler materials are powerful and increase the molecular cohesion energy and hence the heat needed to activate the mechanism of thermal degradation [73].

Synthetic rubbers are characterized by good thermal stability due to its structure, for this reason the rubbers have very important applications. The nitrile rubber (NBR) is a polymer used in applications requiring resistance to high temperatures of oil. Balachandran et al. reported behavior NBR-nanoclay nanocomposites, showing that the temperature at which degradation occurs is increased by adding the nanoclays [74].

3 Thermal Stability of Synthetic Rubber Nanocomposites

The addition of nanoparticles to synthetic rubber resulting in enhancement in thermal, stiffness and resistance to fracture is one of the most important phenomena in material science technology. The commonly used white filler in rubber industry are clay and silica. The polymer/clay nanocomposites offer enhanced thermo mechanical properties. Bourbigot et al. observed that the thermal stability of polystyrene (PS) is significantly increased in presence of nanoclay [75]. Thermal and mechanical properties of clays multiwalled carbon nanotubes reinforced ethylene vinyl acetate (EVA) prepared through melt blending showed synergistic effect in properties [76].

Thermogravimetric analysis (TGA) is one of the most commonly used techniques to study the primary reactions of the decomposition of polymers and other materials. TGA is also useful for the characterization and evaluation of polymer thermal stability. Although synthetic rubber nano composites have excellent mechanical properties, these properties may interfere with its low thermal stability. This may cause the polymer chain to be more susceptible to degradation. Degradation usually starts from a head-to-head structure, a site of unsaturation or a tertiary carbon atom [77].

Attenuated total reflectance-Fourier transform infrared (ATR-FTIR) method can be used for spectral characterization of the physical properties of the synthetic rubber nanocompounds, as well as other complex samples such as polymer composites. The ATR-FTIR techniques allow analysis of specimens with minimum sample preparation and without solvent casting, grinding or pressing, as in the case in transmission experiments.

Recently, TGA equipment is often being connected to Fourier transform infrared spectrometers (FTIR) for a complete chemical identification and analysis of the gases evolved at each stage, making the technique even more powerful.

In the area of polymers (especially rubbers) one is always searching for improved thermical mechanical and properties to fit certain specific applications. Reinforcement of rubber is normally achieved by adding nanofillers, such as carbon black and silica.

The synthetic rubber is widely used, this type of elastomer have important properties and numerous application due to stability. These properties are favored when the synthetic rubber is reinforced with organic nanoparticle as: carbon nanotubes, nanofibers, graphene, and fullerene, among other [78–80]. These may or may not be modified nanofillers surface for subsequent incorporation in the matrix, promoting greater interaction between inorganic nanofillers and polymeric matrix, thus improving the thermal stability of the nanocomposite etc. [81, 82].

The content of the nanoparticle with respect to synthetic rubber may be typically 60 %, however there is an interrogation concerning the relevance of incorporating nanoparticles at such loadings: making a calculation of an ideal dispersion [83].

Malas et al. reported (SBR/BR)/expanded graphite (EG) and black carbon (CB) nanocomposites by melt blending, this study demonstrated that the presence of EG improvement thermo-mechanical properties and the presence of CB are a factor important to improve such properties [84].

Recently carbon nanotubes are the organic nanoparticle more attractive to be incorporate in different polymeric matrix, due to their excellent electrical and conductive properties. The CNs can be surface modified using simple and economical methods [85], the modifications of the CNs are important for good dispersion in synthetic rubber nanocomposites.

In 2011 Inuaki et al. analyzed the polystyrene-block-polybutadiene-block-polystyrene-composite with multiwall carbon nanotubes (MWCNT). They showed a significant improvement in flexibility and thermal properties, indicating that this type of nancomposite, is attractive for applications requiring high temperature. The increase of the properties can be explained by the formation of a three-dimensional structure at the interfacial phase of the synthetic rubber matrix along the MWCNTs, which they designated as a "cell structure". However were detecting secondary reactions (oxidative degradation and crosslinking) when the temperature were higher than 200 °C [86].

An important factor in study of thermal stability of synthetic rubber nanocomposites is the dispersion of the nanoparticle on the matrix. The presence of nanoparticles exfoliated influences the thermal or fire behavior of nanocomposite [83].

Gerorge et al., reported about dispersion of nanographite in EVA (60 % vinyl acetate), used two type of: graphite: natural (NG) and expanded graphite (EG) characterized by different technical, microscopy type TEM, found that NG formed agglomerates inside the rubber matrix at 4 phr as well 8 phr, on the contrary to EG platelets which were homogeneously distribute in the EVA matrix. They explained that EG presented greater surface area can lead to higher interaction between the platelets and rubber matrix [87, 88]. The main parameter responsible for the good dispersion quality in the nanocomposite is the compatibility of the nanoparticle with the polymeric matrix [83].

At the same time inorganic nanoparticles are important component in synthetic rubber nanocomposites, this type of nanoparticles are less reactive than organic nanoparticles, because they are not involved in thermal degradation reactions. Inorganic nanoparticles can be subjected to high temperature, due of natural chemical, are considered an excellent choice as reinforcement in polymer matrix, has been reported that using this type of nanoparticles improve the thermal stability of nanocomposite.

The group of Gu Z. in the 2009, reported the behavior of styrene butadiene/ rubber/organo-bentonite nanocomposite prepared from latex dispersion, content was lower than 12 mass%. The results showed were that presence of organo-bentonite in the nanocoposite affects direct in the thermo stability, mechanical properties and swelling behavior, which was attribute to the good barrier properties of the dispersed nanoparticles. The dispersion is an important factor that can affect various properties such as thermal stability [81].

In 2012 Rybinski et al., reported the study of butadiene-styrene-rubber (SBR) nanocomposite with montmorillonite, found that the nanofillers used do not explicitly influence in the thermal stability of the nanocomposite but they decrease the thermal decomposition rate of these material under thermo-oxidative conditions [89].

Other interesting kind of nanoparticle are silice, Li et al. studied the properties of vulcanized rubber nanocomposite filled with nanokaolin and precipitated silica, demonstrating that the presence of nanokaolin improve the thermal stability, this due to the good interfacial interaction between the nanokaolin and the chains of rubber in the nanocomposite [90].

4 Different Step for Thermal Degradation of Synthetic Rubber Nanocomposites

There are various chemical mechanisms important in the thermal degradation of polymers. It is sufficient here to note that thermal decomposition of a polymer generally involves more than one of these classes of reactions.

The general mechanisms common in polymer decomposition are illustrated in Fig. 2. The reactions can be divided into those involving atoms in the main polymer chain and those involving principally side chains or groups. While the decomposition of some polymers can be explained by one of these general mechanisms, others involve combinations of these general mechanisms. Nonetheless, these categorizations are useful in the identification and understanding of particular decomposition mechanisms.

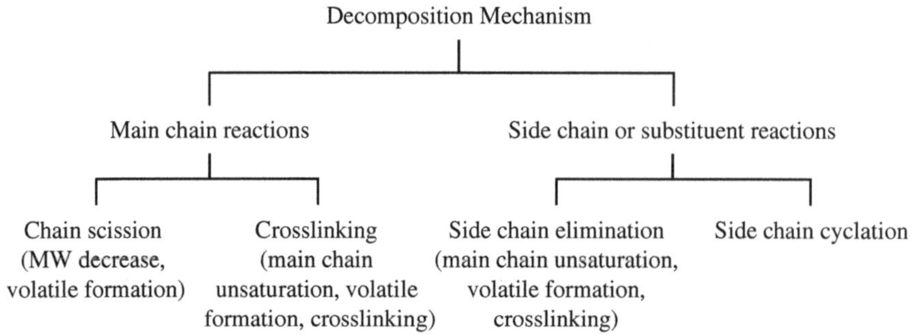

Fig. 2 General decomposition mechanisms

The most common reaction mechanism involves the breaking of bonds in the main polymer chain. These chain scissions may occur at the chain end or at random locations in the chain. End chain scissions result in the production of monomer, and the process is often known as unzipping. Random-chain scissions generally result in the generation of monomers and oligomers (polymer units with 10 or fewer monomer units) as well as a variety of other chemical species. The type and distribution of volatile products depend on the relative volatility of the resulting molecules.

Cross-linking is another reaction involving the main chain. It generally occurs after some stripping of substituents and involves the creation of bonds between two adjacent polymer chains. This process is very important in the formation of chars, since it generates a structure with a higher molecular weight that is less easily volatilized. The main reaction types involving side chains or groups are elimination reactions and cyclization reactions.

In elimination reactions, the bonds connecting side groups of the polymer chain to the chain itself are broken, with the side groups often reacting with other eliminated side groups. The products of these reactions are generally small enough to be volatile. In cyclization reactions, two adjacent side groups react to form a bond between them, resulting in the production of a cyclic structure [91].

Synthetic rubber or polyisoprene decomposes by random-chain scission with intramolecular hydrogen transfer. This, of course, gives small yields of monomer. Other polydienes appear to decompose similarly though the thermal stability can be considerably different. The average size of fragments collected from isoprene decomposition is 8–10 monomer units long. This supports the theory that random-chain scission and intermolecular transfer reactions are dominant in the decomposition mechanism. In nitrogen, decomposition begins at 200 °C. At temperatures above 400 °C, increases in monomer yield are attributable to secondary reaction of volatile products to form monomer. Between 200 and 300 °C, low molecular weight material is formed, and the residual material is progressively more insoluble and intractable. Preheating at between 200 and 300 °C lowers the monomer yield at higher temperatures. Decomposition at less than 300 °C results in a viscous liquid and, ultimately, a dry solid. The monomer is prone to dimerize to dipentene as it cools. There seems to be little significant difference in the decomposition of natural rubber and synthetic polyisoprene.

However polybutadiene is more thermally stable than polyisoprene due to the lack of branching. Decomposition at 320 °C can lead to monomer yields of up to 60 %, with lower conversions at higher temperatures. Some cyclization occurs in the products. Decomposition in air at 250 °C leads to a dark impermeable crust, which excludes further air. Continued heating hardens the elastomer.

Besides polychloroprene decomposes with initial evolution of hydrochloric acid at around 341 °C and subsequent breakdown of the residual polyene. The sequences of the polyene are typically around three (trienes), much shorter than PVC. Polichloroprene melts at around 50 °C.

ABS copolymer is a popular engineering thermoplastic because of its unique properties, which include an excellent mechanical response, chemical resistance, fine surface appearance, and easy processing characteristics. Its unique properties. It consists of a styrene–acrylonitrile (SAN) continuous phase partially grafted onto a dispersed polybutadiene phase of an elastomeric nature. ABS resin is its inherent flammability and lower thermal stability when it is exposed to heat, mechanical stress, and ionizing or ultraviolet radiation in the presence of oxygen because of the formation of reactive intermediates such as free radicals and hydroperoxides.

ABS/clay nanocomposites have been the subject of many investigations and very promising results have been obtained especially in reducing the potential flammability of ABS and thermal stability enhancement. Jang et al. studied ABS/MMT nanocomposites prepared by direct intercalation through one step emulsion polymerization. The onset temperature of decomposition shifted to the higher temperature region as much as 40–50 °C when the MMT content reached 33.3 wt%, though the direct intercalation was not accompanied by delamination of the clay layers [92].

Choi et al. [93] synthesized the ABS/clay nanocomposites by using two clays (sodium montmorillonite and laponite). Both colloidal stability and mechanical properties of the nanocomposites depended on aspect ratios of clays. As was found by TGA the onset temperatures of thermal decomposition of nanocomposites moved towards a higher temperature compared with neat ABS. The neat ABS showed 20 % weight loss at 420 °C, but ABS clay nanocomposites decomposed at 13 °C higher temperature. This may result from the thermal barrier property of clay platelets in the ABS copolymer matrix. In similar ABS/clay nanocomposites it was found that ABS samples with higher acrylonitrile contents produced a greater increase in the decomposition temperature, which was attributed to better clay intercalation/exfoliation due to the higher polarity brought about by the higher acrylonitrile content [94]. Except thermal stability, it was found that clay and its content in the nanocomposites also change the kinetic parameters of the thermal degradation reactions [95]. Reduction in activation energy Ea and reaction order n, were observed by increasing the clay content in ABS nanocomposites. Even though the thermal property enhancement effect of clay nanofillers observed in many of the polymeric nanocomposites has been attributed to the barrier property and char promoting effect of the clay, which protects the inner materials from further decomposition or combustion, the reduced Ea values by increasing clay content showed their accelerating effect on thermal degradation reactions. It was concluded that clay particles or layers take part in the thermal degradation mechanisms and they must not be regarded as inert materials.

Wang et al. [96, 97] prepared intercalated-exfoliated structure of ABS nanocomposites by melt blending ABS and organic-modified clay. In the TGA curves, it can be seen that there are two steps in the degradation of ABS and its nanocomposite. The first step is roughly from 300 to 500 °C and the second step is about from 500 to 650 °C. The first step (300–500 °C) of degradation is attributed to the main chain pyrolysis, commencing at about 340 °C with the evolution of butadiene followed closely by aromatics from the degradation of the styrenic portion and then the evolution of the acrylonitrile at about 400 °C, leading to about 75–85 % weight loss and the formation of a 15–25 wt% charred residue. The second step occurs above 500 °C. It may be assigned to the degradation of the carbonaceous residue formed during the first step. Thus, the first step is the main step of ABS resin degradation. Thermal degradation of the ABS/OMMT hybrids was studied using TGA. Although the onset of weight loss of nanoABS occurs at a lower temperature, which is partly caused by the degradation of OMMT at lower temperature, than that of pure ABS, nanoABS gives a larger residue from the first step of degradation. In the first step of degradation, ABS has 85.9 wt% mass loss, while, nanoABS has 9.4 wt% less mass loss than that of pure ABS. After pyrolysis, the nanocomposite forms a char with a multilayered carbonaceous–silicate structure. The type of organic modifier can affect also the thermal stability of ABS [98]. The onset degradation temperatures increased from 364.5 °C for pure ABS resin to 365.2 °C for ABS/OMMT-P16 nanocomposites (P16 is hexadecyl triphenyl phosphonium bromide) and 374.3 °C for ABS/OMMT-CPC nanocomposites (CPC is cetyl pyridium chloride), respectively. The maximum of thermal decomposition of the ABS nanocomposites was also shifted towards a higher temperature range compared to the pure ABS. This is because from the morphology study it was found that the OMMT dispersed well in the ABS resin and intercalated structure for ABS/OMMT-P16 nanocomposites and intercalated-exfoliated structure for ABS/OMMT-CPC nanocomposites were respectively formed. Except of clays other types of nanoparticles have been also used for the preparation of ABS nanocomposites with enhanced properties such as SiO_2 and SWCNTs [99, 100]. Ma et al. [101] reported that the coexistence of clay and CNTs could improve the thermal stability of ABS nanocomposites as clay and CNTs can form a more effective confine space. This synergistic effect between CNTs and organoclay in thermal stability and flame retardancy in polymers was also reported in literature in polymer nanocomposites [102, 103] The synergistic mechanism was generally explained that the coexistence of clay and CNTs could enhance the clay-CNTs network structure which can hinder the movement of polymer chains and thus improve flame retardancy.

Barick et al. investigate the thermal stability of thermoplastic polyurethane (TPU) nanocomposite based on organically modified layered silicate (OMLS) prepared by intercalation process followed by compression molding, founding that thermal stability of TPU nanocomposite shift to higher temperature compared to pure TPU because of the barrier effect produced by nanofiller [104]. Mohamane et al. reported that the initial stage of thermal degradation in polychloroprene rubber (PCP) as the matrix and organically modified montmorillonite clays, was accelerated with the incorporation of organoclays. The presence of quaternary ammonium anion in the nanocomposite systems is probably responsible for the acceleration of the polymer decomposition [72].

The thermal stability of calcium carbonate ($CaCO_3$) nanoparticles on polybutadiene rubber (PBR) were studied by Shimpi and Mishra [105]. They observed that the incorporation of nano $CaCO_3$ in PBR shows better thermal stability. At 12 wt% of nano $CaCO_3$ (21, 15, and 9 nm) filled in PBR shows decomposition temperature at 491, 483, and 472 °C, respectively. At 4 wt% loading of filler, decomposition temperature is observed to be 488, 480, 450 °C for nano $CaCO_3$ (21, 15, and 9 nm), respectively. This enhancement in thermal stability is due to uniform dispersion of nano $CaCO_3$ throughout the matrix that keeps the rubber chains intact on crosslinking, which prevent out diffusion of the volatile decomposition product [106]. The presence of nanoinorganic particles in between the rubber chains is responsible for preventing the diffusion of the volatile decomposition products from the rubber nanocomposites at same time. It is clear that nanoinorganic filler provides better thermal stability as compared with commercial micron size filler.

5 Effect of Reinforcement on Thermal Stability on Synthetic Rubber Nanocomposites

Addition of small amount of nanofillers may improve the properties of rubber and thermoplastics. In the polymer industry, polymer-filler nanocomposites are a promising class of material that offers the possibility of developing new hybrid materials with desired set of properties. Properties of rubbers and thermoplastics which have shown substantial improvements due to the incorporation of nanoparticles, are mechanical properties, decreased permeability to gases, water and hydrocarbons, thermal stability and heat distortion temperature, flame retardancy and reduced smoke emissions, chemical resistance, surface appearance, electrical and thermal conductivity, optical clarity in comparison to conventionally filled polymers [107].

Polychloroprene is a family of synthetic rubbers, exhibits good chemical stability, and maintains flexibility over a wide temperature range. It is used in a wide variety of applications, such as laptop sleeves, orthopedic braces, electrical insulation, liquid and sheet applied elastomeric membranes or flashings, and automotive fan belts, and others. Subramaniam et al. studied the thermal degradation of polychloroprene rubber composites based on unmodified and ionic liquid modified multi-walled carbon nanotubes (MWCNTs), in aerobic and anaerobic conditions. They found that the polychloroprene rubber and its composite exhibit three and four stage degradation in nitrogen and air respectively [29]. The second and third stages of degradation were predominant as the main degradation of the elastomer occurs. They found that the presentence of unmodified CNTs alone does not improve the thermal stability of the nanocomposite, however when modified CNTs ionic liquid a reasonable enhancement is observed, they attributed to the interfacial interactions of ionic liquid modified CNTs with polychloroprene rubber, also they found better dispersion of the modified CNTs than the unmodified ones. The activation energy was calculated form degradation kinetics studies, founding to be high for modified CNTs–polychloroprene rubber nanocomposites.

Recently Iqbal et al. [28] incorporated four diverse loadings of pristine multi-walled carbon nanotubes (MWNTs) in a silicon rubber (SR) matrix using dispersion kneader and two roller mixing mill to disperse the nanotubes evenly in the polymer matrix. The SR nanocomposite (1 wt% filler contents) has 28 % better thermal stability and 100 % improvement in the ultimate tensile strength is achieved as compared with the pristine polymer matrix counterpart. TGA of their nanocomposite showed thermal degradation with temperature evolution up to 850 °C. They observed % weight loss at 300, 500, 650, and 800 °C during the TGA for all SR nanocomposite and demonstrated an improved thermal stability. They found that the maximum thermal degradation in the range 400–600 °C for all nanocomposites as the polymer pyrolysis of SR matrix is usually analyzed within this temperature range. Thermal stability and heat quenching ability of the SR nanocomposites were gradually augmented with increasing filler loadings in the polymer matrix. Li et al. [108] also, investigated the effect of the interfacial interaction on the thermal oxidative stability of chemically modified carbon nanotubes (CNTs)/SR composites and of nanoparticle-attached CNTs/SR composites, poly(dimethylsiloxane)-grafted CNTs and iron(III)-oxide-attached CNTs were prepared and embedded into a SR matrix, respectively. They found that the effect of interfacial interactions of poly (dimethylsiloxane)-grafted CNTs/SR composite was damaged at 250 % strain. The interfacial interaction of iron (III)-oxide-attached CNTs/SR composites was damaged at 100 % strain. After the thermal oxidative aging, the mechanical properties of the poly(dimethylsiloxane)-grafted CNTs/SR composites decreased sharply due to the damaged interface and also the thermal stability was reduced. However, similar damages did not affect the thermal stability of the iron (III)-oxide-attached CNTs/SR composites.

In addition, SR/GO nanocomposites has been prepared for Wang et al. [109] via a solution intercalation method. The thermal behavior of the nanocomposites characterized by TGA showed that that the addition of graphite oxide is beneficial to improve the thermal stability of the nanocomposites at high temperature. They found that the initial decomposition temperature of the filled rubber was decreased compared with the pure SR. They argued that the decreasing of the initial decomposition temperature with the GO may be due to the degradation of GO. When further heated, they observed that the incorporation of GO into SR improves thermal stability remarkably. The degradation of siloxane chains occurs in two ways: degradation of siloxane chains initiated by the terminal group and degradation of the main chains of siloxane. GO contains polar OH and COOH groups, and silicone rubber backbones consist of polar Si–O units. The graphite oxide easily absorbs polar Si–O bond, resulting in formation of physical crosslinking points. This increases the rigidity of siloxane chains, and thus hinders the degradation of siloxane chains. In addition, SR filled with conductive carbon black has better thermal stability than that of SR without any filler. Also, SR filled with conductive carbon black has better thermal stability than that of silicone rubber filled with the same amount of silica [5].

Also, it is reported in the literature [110] that Cellulose nanocrystals (CNs) can be used to obtain nitrile rubber (NBR)/CNs nanocomposites, by mixing a water suspension of CNs and the NBR latex directly. In this study is reported that mechanical performance showed that CNs have a good reinforcing effect on NBR. The results showed of DMA demonstrated that the glass transition temperature (Tg) of the composites was shifted from 10.8–17.2 °C with CNs content increasing to 20 phr and the storage modulus increased simultaneously. TGA result shows that the degradation corresponding to CNs in NBR/CNs nanocomposites is much higher than the degradation temperature of pure CNs. The thermogram of NBR/CNs nanocomposites show two degradation stages: the first region at a temperature range of 340–390 °C is probably due to the degradation of CNs, and the weight loss is in accordance with the CNs content; the second stage occurred at around 400–480 °C and is due to the degradation of NBR. Cao et al. [110] found that the degradation corresponding to CNs in NBR/CNs nanocomposites is much higher than the degradation temperature of pure CNs. This significant enhancement of thermal resistance by the presence of NBR matrix can be attributed to the formation of a confined structure in the NBR/CNs nanocomposites. That is, the CNs is protected by NBR matrix, indicating the strong interaction between the CNs and NBR molecules.

The effect of organoclay (OC) on the performance of styrene–butadiene rubber (SBR)/phenolic resin (PH) blend prepared by two-roll mill was investigated by Shojaei and Faghihi [111]. They showed through the XRD analysis that dispersion state of OC is improved by incorporation of PH possibly due to the better interaction of PH (present in SBR) with OC. Analysis of cure behavior revealed that both OC and PH increase the cure rate and crosslinking density. In addition, tensile and DMTA results showed that the mechanical properties increase by addition of OC in both SBR and SBR/PH systems, however, the extent of enhancement is higher for the SBR compared to SBR/PH blend. This was attributed to the pronounced reinforcing efficiency of PH in the system, which dominates the overall performance of the ternary hybrid and reduces the flexibility of the SBR chains by enhancing the crosslinking density. Their TGA revealed that the OC increases the thermal stability of SBR. The incorporation of OC along with PH results in higher value of thermal degradation rate, which can be indicative of catalytic role of OC in presence of PH.

In order to minimize the oxidative degradation of SBR at high temperature, He et al. [112] prepared SBR/clay/carbon black (CB) nanocomposites, then the effects of nano-clay on the properties of SBR nanocomposites were investigated. They found that the two fillers were uniformly dispersed in the SBR matrix at nano-scale, and the nano-dispersed clay layers can provide the SBR/clay/CB nanocomposites excellent thermal aging resistance and heat resistance. In addition, they proposed a heat resistant mechanism; the incorporation of the layered silicate could hinder the rubber from generating oxygen-containing groups in its surface during high temperature aging. Then the changes of Tg between the rubber outer layer and inner layer is lowered down, which keep the rubber macromolecular chains flexible, so as to improve the bending resistance of the high temperature aging nanocomposites. They predicted that the rubber/clay nanocomposites could be exposed to high temperature occasions and most likely be used to produce heat-resistant products, for example, heat-resisting conveyor belt.

Hwang et al. [113] synthesized via in situ polymerization high-impact polystyrene (HIPS)/organically modified montmorillonite (organoclay) nanocomposites. X-ray diffraction and TEM experiments revealed that intercalation of polymer chains into silicate layers was achieved, and the addition of nanoclay led to an increase in the size of the rubber domain in the composites. In comparison with neat HIPS, they found that the HIPS/organoclay nanocomposites exhibited improved thermal stability as well as an increase in both the complex viscosity and storage modulus, and they may have been influenced by a competition between the incorporation of clay and the decrease in the molecular weight of the polymer matrix.

The bromobutyl rubber (BIIR), modified from butyl rubber (IIR) with bromine, is known to have faster cure reactivity and better adhesion than unmodified IIR. Meanwhile, BIIR retains the excellent impermeability and mechanical property of IIR. It has been used in a wide variety of commercial products, such as tire inner liners, pharmaceutical closures and protection materials. Nevertheless, the stability of BIIR has always been a major concern. BIIR is known to be thermally unstable due to the formation of double bonds, which could be generated by HBr elimination from its backbone [114]. Recently Xiong et al. demonstrated that both the thermal stability and the thermal conductivity of BIIR nanocomposites could be improved by incorporating the ionic liquids (ILs) modified graphene oxide (GO-ILs) using a solution compounding method. They found by X-ray diffraction analysis of GO-ILs that ILs had been effectively intercalated into the interlayer of GO, which be able to raise the exfoliation degree of GO. The increased exfoliation degree facilitated a good dispersion of GO-ILs in the BIIR matrix observed bay SEM images. The successful incorporation of GO-ILs in BIIR matrix was found to be able to raise the Tg of GO-ILs/BIIR nanocomposites due to the attractive interactions of the polymer chains with the large surface area of the GO-ILs. Most importantly, the thermal stability of BIIR nanocomposites was enhanced by adding GO-ILs. Their calculated the thermal degradation kinetic parameters of BIIR nanocomposites demonstrate that the increased activation energy (E_a) with the addition of GO-ILs was observed in both methods. In addition, they found that adding 4 wt% GO-ILs, the thermal conductivity of GO-ILs/BIIR nanocomposites had 1.3-fold improvement compared with that of the unfilled BIIR [30].

6 Effect of Nanoparticle on Thermal Stability on Synthetic Rubber Nanocomposites

Hydrogenated nitrile rubber (HNBR) has been developed to improve thermal stability of nitrile butadiene rubber (NBR). During hydrogenation of NBR, a small number of double bonds and all the cyano groups are kept unhydrogenated for subsequent sulphur vulcanization and oil resistance. This elastomer is known for its physical strength and retention of properties after long term exposure to heat, oil, and chemicals. Addition of small amount of nanofillers may improve the properties of this rubber. Choudhury and coworkers incorporate different kind of clays into HNBR by solution dispersion in order to study the effect of these nanofilleres in the thermal, mechanical and dynamical mechanical proprieties of the nanocomposites. They used organically modified and unmodified layered silicates such as Cloisite NA, Cloisite 30B and Cloisite15A, rod-like sepiolite and spherical nanosilica. They found that on addition of only 4 phr of silica nanofiller to neat HNBR, the temperature at which maximum degradation took place (T_{max}) increased by 4–16 °C, this is attributed to the high thermal stability of A300 particles. These particles have a tendency to form agglomeration because of the presence of active silanol group on its surface for which the mechanical properties drop, in contrast, they found that 30B has undergone full exfoliation within the bulk of the matrix, and thus provides highest tensile strength, excellent storage modulus and very good thermal stability. In addition, they observed that Sepiolite because of its fibrous nature, high aspect ratio and uniform dispersion gives rise to excellent improvement in modulus at 100 % elongation and storage modulus and provides very good thermal stability to the matrix. The nanocomposite based on 15A clay exhibited intercalated structure, they claimed that is responsible for intermediate thermal, mechanical and dynamic mechanical property to the matrix. On other hand an unmodified NA particles, they observed form XDR and TEM microphotograph agglomeration on addition to the rubber matrix, and thus NA has the least effect on the thermal and mechanical properties of the elastomer. They increased up 8 phr the nannofiller lading in the nanocposite, these increase the thermal as well as mechanical properties of HNBR-filler nanocomposites [6].

Kong et al. [115] synthesized by melt-intercalation silicone rubber (SR)/clay nanocomposites using synthetic Fe-montmorillonite (Fe-MMT) and natural Na-MMT which were modified by cetyltrimethylammoniumbromide, surfactant. They obtained exfoliated and intercalated nanocomposites. With TGA and mechanical performance found that with the presence of iron significantly increased the onset temperature of thermal degradation in SR/Fe-MMT nanocomposites. In addition, the thermal stability, gel fraction and mechanical property of SR/Fe-MMT were different from the SR/Na-MMT nanocomposites, so the iron not only in thermal degradation but also in the vulcanization process acted as an antioxidant and radicals trap. A new flame-retardant system, SR/Fe-OMT based on an EVA matrix, was examined Fang et al. [116]. The experimental analyses showed that the exfoliated Fe-OMT had better dispersion in the EVA matrix than Na-OMT, and it was more effective in improving the flame retardancy, thermal stability, and mechanical properties of EVA nanocomposites. Their TGA also showed that EVA/SR/Fe-OMT had the maximum amount of char residue at 700 °C and had a higher thermal stability on the deacetylated polymer than EVA/SR/Na-OMT. From the tensile tests, they found that the addition of OMT, especially Fe-OMT, improved the tensile strength and elongation at break of the nanocomposites. They claimed that OMT could serve as a compatibilizer of EVA and SR.

Liu et al. [117] present a very interesting research about different kind of rubbers filled with nanokaolin or precipitated silica. The fillers were introduced in styrene butadiene rubber (SBR), natural rubber (NR), butadiene rubber (BR) and ethylene–propylene diene methylene (EPDM). Their results showed that nanokaolin could greatly improve the vulcanizing process by shortening the time to optimum cure and prolonging the setting-up time of cross-linked rubbers, which improves production efficiency and operational security. In addition, they found that the rubber composites filled with nanokaolin have good mechanical properties, they suggest that these result are by the suitable modification of nanokaolin can make kaolinite platelets dispersed in rubber matrix in directional parallel arrangements, with 20–50 nm in thickness. The tightly linked interface at the nanoscale range between the kaolin sheets and rubber molecules is the main reason for the improvements of mechanical properties and thermal stability. As compared with the rubber composites filled with precipitated silica, they found an increase of onset weight loss temperature of that filled with nanokaolin is between 3.5 and 126.8 °C in thermogravimetric tests, and were especially good in SBR and BR. In addition, the stable weight loss temperature for nanokaolin composite can increase 50–90 °C over that of a precipitated silica composite. Their shows that nano kaolin has not ably improved thermal the stability of the rubber composites.

In other hand, Shimpi et al. [105] synthesized by solution spray calcium carbonate ($CaCO_3$) nanoparticles of three different diameters 9, 15, and 21 nm to form polybutadiene rubber (PBR) nanocomposites with variations in wt% loading (4, 8, and 12 %). Their results of PBR nanocomposites were compared with commercial $CaCO_3$ (40 lm) and fly ash (75 lm) filled PBR microcomposites. They observed that the incorporation of nano $CaCO_3$ in PBR with reduced size shows better thermal stability as compared to commercial $CaCO_3$ and fly ash filled PBR. They found at 12 wt% of nano $CaCO_3$ (21, 15, and 9 nm) filled in PBR shows decomposition temperature at 491, 483, and 472 °C, respectively, while for commercial $CaCO_3$ and fly ash decomposition temperature were observed to be 459 and 461 °C for same wt% of filler loading. At 4 wt% loading of filler, decomposition temperature were observed to be 488, 480, 450 °C for nano $CaCO_3$ (21, 15, and 9 nm), respectively. They argued that this enhancement in thermal stability is due to uniform dispersion of nano $CaCO_3$ throughout the matrix that keeps the rubber chains intact on crosslinking, which prevent out diffusion of the volatile decomposition product [106]. The presence of nanoinorganic particles in between the rubber chains is responsible for preventing the diffusion of the volatile decomposition products from the rubber nanocomposites at same time. From the above results of rubber nanocomposites, it is clear that nanoinorganic filler provides better thermal stability as compared with commercial micron size filler.

7 Recent Studies on Synthetic Rubber Nanocomposites

In recent years, synthetic rubber nanocomposites are of great interest for both academic researches and industrial applications because they frequently exhibit unexpected properties synergistically derived from the two or more components. The properties of elastomeric materials as thermal stability and chemical resistance are of relevant importance considering both economic and ecologic impacts in many different areas.

There are two types of rubber polymers: natural rubber and synthetic rubber. Synthetic rubber is one type of artificial elastomer mainly synthesized from petroleum byproducts. It has good mechanical property, thermal stability, and compatibility with petroleum products.

Acrylonitrile rubbers are elastomers of butadiene and acrylonitrile copolymers of special applications. They are used in applied hydraulics and pneumatics, the percentage proportion determines their resistance to oil and freeze. Nitrile vulcanizates show high elasticity and tensile strength, oil resistance and a low permanent compressive strain. Prochon et al. reported the influence of keratin, recovered from the tanning industry, on the thermal and mechanical properties of vulcanizates with synthetic rubber acrylonitrile-butadiene rubber NBR. The addition of waste protein to NBR vulcanizates influences the improvement of resistance at high temperatures and mechanical properties like tensile strength and hardness. The introduction of keratin to the mixes of rubber previously blended with zinc oxide (ZnO) before vulcanization process leads to an increase in the cross-linking density of vulcanizates. It is present a clear influence of protein on the formation of spatial lattice of elastomers between protein, zinc oxide, and the elastomer matrix. The polymer materials received including addition of proteins will undergo biodecomposition in natural conditions. In that aerobic environment, microorganisms, bacteria, and fungus digested better polymer materials containing natural additives [118].

The development of nanoparticles with different sizes and shapes, including spherical particles such as silica, platelets such as layered silicates, and carbon nanotubes have been used as fillers in elastomeric matrices to improve their properties like mechanical and barrier. The interaction of the polymer chains with the filler particles is also crucial in controlling the performance of nanocomposites.

Moreover Ponomarenko et al. reported the melt-compounded composites of synthetic styrene–co–butadiene rubber (BUNA SL18) and silica particles (Silica VN3, Degussa). They showed that at low elongations, silica particles provided a considerably weaker reinforcement effect of the rubber matrix when compared to organoclay nanoparticles. The overall thermoelastic behavior of both the rubber/silica composites and the rubber/organoclay nanocomposites could be quantitatively accounted for by a model that explicitly assumed the contributions of local strain amplification for the rubber matrix and of successive decay of a spatial network of filler particles with increasing strain, generating the exothermal effects of external friction between particles. Significantly more intensive heat generation, concomitant to higher amplitudes of stress relaxation in the rubber/silica composites at high uniaxial extensions as compared to the rubber/organoclay nanocomposites, was regarded as evidence for a lower mechanical strength of a spatial network of silica particles [119].

Furthermore Li et al. reported the use of a novel method (Two-Step Method, TSM) to investigate the modification process in preparation of rubber, silica, and bis(3-triethoxysilylpropyl) tetrasulfide (TESPT). The TSM modification indicated that the TESPT hydrolyzed firstly to generate the silanol (Si–OH), and the silanol reacted with the hydroxyl groups on the surface of silica. The properties of modified silica were studied and results exhibited the advantages of TSM, and also revealed that 8 % TESPT amount was suitable than 12 and 15 % TESPT amount [120].

Mean while Scotti et al. reported the preparation of silica/styrene butadiene rubber (SBR) nanocomposites by blending method using shape-controlled spherical and rod-like nanoparticles with different aspect ratios as filler for the rubber reinforcement. The influence of the particle morphology on the reinforcing effect independently of the silica surface chemistry and considering the aspect ratio as the only geometrical variance. Spherical and anisotropic rod-like particles, dispersed in the nanocomposites, formed a network of particles bridged by thin rubber layers throughout the SBR matrix [121].

At the same time Bala et al. studied the filler effects of organomodified dodecyl alkylammonium intercalated montmorillonite (12C-MNT) for natural rubber (NR) and styrene butadiene rubber (SBR). The results showed that tensile strength, modulus, and elongation at break increased significantly as compared to pure vulcanized NR and SBR when 4 wt% of 12C-MNT was used as filler. In the case of NR, these properties decreased drastically for 1 wt% of pure Na-MNT [122].

In recent years, lamellar nanofillers have been established as the most important filler type for barrier and mechanical reinforcement. Dal Point et al. reported a novel nanocomposite series based on styrene–butadiene rubber (SBR latex) and alpha-zirconium phosphate (α-ZrP) lamellar nanofillers. The use of surface modified nanofillers improvement the mechanical properties. However, no modification of the gas barrier properties is observed. The addition of bis(triethoxysilylpropyl) tetrasulfide (TESPT) as coupling agent in the system is discussed on the nanofiller dispersion state and on the filler-matrix interfacial bonding. Simultaneous use of modified nanofillers and TESPT coupling agent is found out with extraordinary reinforcing effects on both mechanical and gas barrier properties [123].

Polymer nanocomposites are one of the highly discussed research topics in recent time. Yehia et al. reported the preparation and the properties of different nanoclays based on sodium montmorillonite (bentonite) and some organic amines of varying chain lengths (dodecylamine, hexadecylamine and octadecylamine) beside amine-terminated butadiene-acrylonitrile copolymer (ATBN) The nanocomposite clays were incorporated in natural and synthetic rubbers (NR, SBR and

NBR). The modified organo-clay (MMT-ATBN) can act as a reinforcing agent in the natural and synthetic rubber vulcanizates with low concentrations (4 phr) in the same level as carbon black (40 phr, HAF). The modified organo-clay appears to be more compatible in the NBR compounds, which can be attributed to the chemical nature of the alkylamines [82].

In 2013, cellulose nanocrystals (CNs) were fabricated from sulfuric acid hydrolysis of cottonseed linter. The crystals were then utilized to prepare nitrile rubber (NBR)/CNs nanocomposites by mixing a water suspension of CNs and the NBR latex directly. CNs formed a strong filler-filler network in the NBR matrix which resulted in an obvious "Payne effect" [110].

Songmin et al. reported the preparation of chitosan hydrochloric acid salt and the effect in improving the dispersion of carbon nanotubes in different solvents and silicone rubber. It was also found that treated carbon nanotubes could be dispersed in the silicone rubber homogeneously based on SEM and XRD analysis. The incorporation of carbon nanotubes enhanced the thermal stability of the silicone rubber. They also significantly improved the mechanical properties of the silicone rubber matrix. From the FTIR spectra, it was believed that chitosan hydrochloric acid salt adsorbed on the surface of the carbon nanotube and then interacted with silicone rubber matrix, resulting in the enhancement of dispersion of carbon nanotubes, improving thermal and mechanical properties in the resulting nanocomposites [124].

On the other hand, Xiubin et al. reported the preparation of ethylene propylene diene methylene rubber composites filled with pretreated multi-walled carbon nanotubes by a double-roll milling dispersion method. The results showed that the pretreated multi-walled carbon nanotubes can improve the thermal conductivity and increase the thermal decomposition temperature. However, different filling content of multi-walled carbon nanotubes showed different effects on the mechanical performance of ethylene propylene diene methylene [125].

At the same time, Peddini et al. described the preparation of nanocomposites from styrene–butadiene rubber (SBR) and multiwall carbon nanotubes (MWCNT). MWCNT are important nanostructures due to the exceptionally high modulus and aspect ratios there has been much interest in using them as reinforcing agents for polymer composites. Styrene–butadiene rubber (SBR), commonly used as a tread stock for tires, is employed here as the matrix for creation of a masterbatch with oxidized MWCNT (12.3–15 wt%). These materials do not show a high level of electrical conductivity as might be expected from a percolation concept, signifying excellent tube dispersion and formation of a bound rubber layer on the discrete MWCNT [126].

Investigated the effect of reinforcing styrene butadiene rubber (SBR) with functionalized carbon nanotubes on the mechanical and thermal properties. MWCNT were functionalized with phenol functional group to enhance their dispersion in SBR matrix. This study demonstrated the capability of MWCNTs as reinforcing filler as demonstrated by the substantial improvement in Young's Modulus, tensile strength and energy of absorption of the nanocomposites. The tensile strength increased from 0.17 (SBR) to 0.48 MPa while the Young's Modulus increased from 0.25 to

0.83 MPa when 10 wt% functionalized CNTs was added. With the addition of 1 wt% reinforcement, a peak value of 4.1 kJ energy absorption was obtained. The homogenous dispersion of CNT-Phenol is thought to be responsible for the considerable enhancement in the reported properties [127].

As can be seen an important aspect in material development is to achieve good combination of properties. The performance of a composite material depends on several factors including formulation, dispersion of filler in the polymer matrix and adhesion between the filler and the matrix. The properties of nanocomposites depend on the type of nanofiller used and its interaction with the polymer matrix.

8 Thermal Application of Synthetic Rubber Nanocomposites

Practical application of the vulcanization of silicone rubbers using UV light is opening up new areas of application for silicone elastomers, since UV vulcanization enables extremely rapid curing to take place at low temperatures. This process makes it possible to combine silicone elastomers with heat-sensitive plastics, components, chemicals or medicinal products, which would previously have led to the destruction of the material or product because of the high curing temperatures [128].

Asphalt binder is a thermoplastic material that behaves as an elastic solid at low service temperatures or during rapid loading; and as a viscous liquid at high temperatures or slow loading. This double behavior creates a need to improve the performance of the asphalt binder to minimize stress cracking, which occurs at low temperatures, and permanent deformation, which occurs at high service temperatures addition of crumb rubber to sulfur asphalt enhanced the temperature resistance of the binder. Utilization of waste crumb rubber and sulfur in asphalt modification proved to enhance asphalt pavement [129].

Styrene butadiene rubber (SBR) composites filled with SBR industrial scraps (SBR-r) were devulcanized by microwave. For further application in automotive profiles, the set of results indicated that the best performance was achieved by the composite containing SBR-r devulcanized for 2 min in a microwave [130].

Tri-block copolymer styrene–butadiene–styrene (SBS) composites with different butadiene/styrene ratios and multi-walled carbon nanotubes (MWCNTs) can be used for the development of electro-mechanical sensors for large strains applications (Costa) [131]. Butyl rubber–$Ba_{0.7}Sr_{0.3}TiO_3$ composites (BR–BST) were prepared by sigma mixing followed by hot pressing. The stress-strain studies show the good flexibility of the composite. The dielectric properties of the composites were investigated at both radio and microwave frequencies. This type of composite is proposed for electronic applications [132].

Recent and important application of synthetic rubber polymers is the use in transdermal drug delivery as pressure sensitive adhesives. Thus, transversal drug delivery systems can improve the disadvantage of orally taken drugs. Commonly, many drugs taken via the oral route are ineffective because of the first pass metabolism and drug degradation in the gastrointestinal tract. There must be significant controlled drug release into the systemic blood circulation to target organs via the skin. Suksaeree et al. reported the use of rubber polymers for transdermal drug delivery systems, including materials for transdermal drug delivery systems such as styrene-isoprene-styrene (SIS) block copolymer, acrylic rubber hybrid, PIB, and silicone rubber. For example, capsaicin is commonly used in topical skin preparations to relieve pain. The SIS block copolymer, in combination with resin and plasticizer, is used in transdermal drug delivery systems to produce capsaicin-matrix patches Silicone rubber, a rubber-like material, is an elastomer composed of silicon together with carbon, hydrogen, and oxygen. The matrix-type patches are made from silicone rubber containing lidocaine hydrochloride. These matrix type patches can be applied to control the rapid release of active principle. The authors concluded that rubber polymers as important drug carriers have made a great impact on the advancement of drug delivery science and technology in pharmaceutical sciences and medical applications [133].

9 Future Trends

Rubber nanocomposites represent a new class of multiphase materials containing dispersion of nano-sized filler materials such as nanoparticles, nanoclays, nanotubes, nanofibers etc. within the polymer matrices. Owing to their nanoscale size features and very high surface-to-volume ratios, they possess unique combination of multifunctional properties not shared by their more conventional composite counterparts reinforced with micro-sized fillers. These multifunctional nanocomposites exhibit excellent mechanical, thermal, and other physico-chemical properties. It is believed that the molecular level interactions between the nanoparticles and polymer matrices along with the presence of very high nanoparticle-polymer interfacial area play a major role in influencing the thermal, physical and mechanical properties of nanocomposites. Another advantage of nanocomposites is that the strength, shrinkage, warpage, viscosity and optical properties of the polymer matrix are not significantly affected.

The enhanced properties are attributed to the structure and morphology of the nanocomposite, as they contain organically treated clays and synthetic mica. These nanomaterials have a large aspect ratio and each one is approximately 1 nm thick and hundreds or thousands of these layers are stacked together with weak Van der Waals forces to form a clay particle, resulting in subsequent exfoliation in which the individual layers are peeled apart and then dispersed throughout the polymer matrix. The excellent degree of exfoliation, which results in smaller particle sizes and provides the greater surface area to interact with the polymer, results in improved performance. The simultaneous use of organoclays and coupling agent like silanes, titanates, etc. is also promising strategy.

There are several challenges in the area of rubber nanocomposites. Complete exfoliation and uniform dispersion of nanofillers in rubber matrix still remains to be a challenge. Efficient surfactants have to be designed for the excellent dispersion. In the case of clay filled rubber nanocomposites, the extent of exfoliation-intercalation has not been quantified yet. The orientation of nanoplatelets in rubber matrix by special extrusion is also a major challenge.

Currently, researchers are using other nanofillers besides clays, nanomaterials like graphene, carbon nanofibers, nanofoams, multiscale hybrid reinforcement and graphene-enabled rubber nanocomposites. The advantages of graphene are challenging carbon nanotubes in polymer nanocomposites applications due to its intrinsic properties and it is predicted that a single, defect-free graphene platelet could have an intrinsic tensile strength higher than that of any other material. Initial successes have demonstrated that graphite nanoplates provide dramatic reinforcement and multifunction to rubber, the potential of rubber material loaded graphite nanoplates is undoubtedly tremendous and promising. One on the biggest challenges in the research of rubber/graphite nanocomposites is to develop composition technologies for improving the exfoliation and dispersion of graphitic fillers in matrix beside the enhancement of graphite matrix interfacial adhesion. However, there are still many limitations and challenges for nanocomposites production.

Moreover the combinations of organoclays with other nanofiller like carbon nanotubes, graphene, quantum dots to obtain multifunctional materials with enhanced mechanical thermical and electrical properties are also envisaged.

10 Conclusions

Rubber based nanocomposites is one of the many composites materials in which scientist and engineers are interested and find good scope for possible exploitation of commercial applications.

Composites that exhibit a change in composition and structure over a nanometer scale have shown remarkable property enhancements relative to conventional composites.

Because of their nanometer size, rubber clay nanocomposites exhibit markedly improved properties when compared to the pure polymers or their traditional composites. Most notable properties are increased tensile strength, tear strength, modulus, gas barrier properties, thermal stability.

An important aspect in material development is to achieve good combination of properties. The performance of a composite material depends on several factors including formulation, dispersion of filler in the polymer matrix and adhesion between the filler and the matrix. The properties of nanocomposites depend on the type of nanofiller used and its interaction with the polymer matrix.

The study of the thermal degradation of synthetic rubber nanocomposites plays an important role for different application, knowing the degradation temperature of nanocomposites is a useful tool to predict lifetimes and applications. However it is important reviser type, chemical natural, morphology of nanoparticle used for the preparation the nanocomposite with synthetic rubber, it is important to be subjected to high temperature, leading to different reaction secondary factor.

The designing and compounding are key factors in obtaining the intercalated–exfoliated structure of the clay minerals in the rubber matrix, which ultimately governs the final physical properties of the composites.

In conclusion the addition of nanoparticles can lead to thermal stabilisation effect of polymer during their decomposition. However in this sense the effect of nanoparticles content is very crucial due at higher concentration nanoparticles can form aggregates and the effective area of nanoparticles in contact with polymer macromolecular is lower. The thermal stability enhancement takes place at low loading of nanoparticle.

References

1. Donnet, J.B.: Nano and microcomposites of polymers elastomers and their reinforcement. Compos. Sci. Technol. **63**, 1085–1088 (2003)
2. Yehia, A.A., Akelah, A.M., Rehab, A., El-Sabbagh, S.H., Nashar, D.E., Koriem, A.A.: Evaluation of clay hybrid nanocomposites of different chain length are reinforcement agent for natural and synthetic rubbers. Mater. Des. **33**, 11–19 (2012)
3. Periadurai, T., Vijayakumar, C.T., Balasubramanian, M.: Thermal decomposition and flame retardant behaviour of SiO_2-phenolic nanocomposite. J. Anal. Appl. Pyrol. **89**, 244–249 (2010)
4. Rybiński, P., Janowska, G., Jóźwiak, M., Jóźwiak, M.: Thermal properties and flammability of nanocomposites based on nitrile rubbers and activated halloysite nanotubes and carbon nanofibers. Thermochimica Acta **549**, 6–12 (2012)
5. Zhang, Y., Liu, Q., Xiang, J., Frost, R.L.: Thermal stability and decomposition kinetics of styrene-butadiene rubber nanocomposites filled with different particles sized kaolinites. Appl. Clay Sci. (2014) http://dx.doi.org/10.1016/j.clay.2014.04.002
6. Choundhury, A., Bhowmick, A.K., Ong, C.: Effect of different nanoparticles on thermal, mechanical and dynamic mechanical properties of hydrogenated nitrile butadiene rubber nanocomposites. J. Appl. Polym. Sci. **116**, 1428–1441 (2010)
7. Sahoo, N.G., Rana, S., Chob, J.W., Li, L., Chan, S.H.: Polymer nanocomposites based on functionalized carbon nanotubes. Prog. Polym. Sci. **35**, 837–867 (2010)
8. Paul, D.R.; Robeson, L.M.: Polymer nanotechnology: nanocomposites. Polymer **49**, 3187–3104 (2008)
9. Mark, J.E.: Ceramic-reinforced polymers and polymer-modified ceramics. Polym. Eng. Sci. **36**(24), 2905–2920 (1996)
10. Wen, J., Wilkes, G.L.: Organic/inorganic hybrid network materials by the sol-gel approach. Chem. Mater. **8**(8), 1667–1681 (1996)
11. von Werne, T., Patten, T.E.: Preparation of structurally well-defined polymer nanoparticle hybrids with controlled/living polymerization. J. Am. Chem. Soc. **121**, 7409–7410 (1999)
12. Calvert, P.: Carbon Nanotubes, Ebbesen, T.W. (ed.) CRC Press, Boca Raton (1992)
13. Dresselhaus, M.S., Dresselhaus, G., Avouris, P.: Carbon Nanotubes: Synthesis, Structure, Properties and Applications, Topics of Applied Physics, vol. 80. Springer, Heidelberg (2001)
14. Luo, J.J., Daniel, I.M.: Characterization and modeling of mechanical behavior of polymer/clay nanocomposites. Compos. Sci. Technol. **63**, 1607–1616 (2003)

15. Messermith, P.B., Giannelis, E.P.: Polymer layered silicate nanocomposites: in situ intercalative polymerization of ε-caprolactone in layered silicates. Chem. Mater. **5**, 1064–1066 (1993)
16. Giannelis, E.P.: Polymer layered silicate nanocomposites. Adv. Mater. **8**, 29–35 (1996)
17. Komarneni, S.: Nanocomposites. J. Mater. Chem. **2**, 1219–1230 (1992)
18. Ruiz-Hitzky, E.: Conducting polymer intercalated in layered solids. Adv. Mater. **5**, 334–340 (1993)
19. Hoffman, U., Endell, K., Wilm, D.: Kristallstruktur und Quellung von Montmorillonit. (Das Tonmineral der Bentonite). Zeitschrift fur Kristallografie Mineral Petrografie Abteilung A **86**, 340–348 (1993)
20. Okada, A., Kawasumi, M., Usuki, A., Kujima, Y.: Nylon 6-clay hybrid. Mater. Res. Soc. Symp. Proc. **171**, 45–50 (1990)
21. Lan, T., Kaviratna, P.D., Pinnavaia, T.J.: Mechanism of clay tactoid exfoliation in epoxyclay nanocomposite. Chem. Mater. **7**, 2144–2150 (1995)
22. Abdallah, W., Yilmazer, U.: Preparation and Characterization of thermally stable phosphonium organoclays and their use in poly (ethylene terephthalate) nanocomposites. J. Appl. Polym. Sci. **128**, 4283–4293 (2013)
23. Gu, Z., Gao, L., Song, G., Liu, W., Li, P., Shan, C.: Octadecylammonium montmorillonite/natura rubber/cis-1,4-polybutadiene nanocomposites. Appl. Clay Sci. **50**, 143–147 (2010)
24. Oberlin, A., Endo, M., Koyama, T.: Filamentous growth of carbon through benzene decomposition. J. Cryst. Growth **32**(3), 335–349 (1976)
25. Endo, M., Koyama, T., Hishiyama, Y.: Structure improvement of carbon fibers prepared from benzene. Jpn. J. Appl. Phys. **15**(11), 2073–2076 (1976)
26. Iijima, S.: Helical microtubules of graphitic carbon. Nature **354**(6348), 56–58 (1991)
27. Prado, LASdA, Kopyniecka, A., Chandrasekaran, S., Broza, G., Roslaniec, Z., Schulte, K.: Impact of filler functionalization on the crystallinity, thermal stability and mechanical properties of thermoplastic elastomer/crbon nanotube nanocomposites. Macromol. Mater. Eng. **298**, 359–370 (2013)
28. Iqbal, N., Bilal, M., Sagar, A., Maqsood, A.: Fabrication and characterization of multiwalled carbon nanotubes/silicon rubber composites. J. Appl. Polym. Sci. **128**(4), 2439–2446 (2013)
29. Subramaniam, K., Das, A., Häußler, L., Harnisch, C., Werner, K.S., Heinrich, G.: Enhanced thermal stabity of polycloroprene rubber composites with ionic liquid modified MWCNTs. Polym. Degrad. Stab. **97**, 227–287 (2012)
30. Xiong, X., Wang, J., Jia, H., Fang, E., Ding, L.: Structure, thermal conductivity, and thermal stability of bromobutyl rubber nanocomposites with ionic liquid modified graphene oxide. Polym. Degrad. Stab. **98**, 2208–2214 (2013)
31. Kim, H., Abdala, A., Macosko, C.: Graphene/polymer nanocomposites. Macromolecules **43**, 6515–6530 (2010)
32. Peng, C.C., Göpfert, A., Drechsler, M., Abetz, V.: Smart silica-rubber nanocomposites in virtue of hydrogen bonding interaction. Polym. Adv. Technol. **16**, 770–782 (2005)
33. Kickelbick, G.: Concepts for the incorporation of inorganic building blocks into organic polymers on a nanoscales. Polymer **28**, 83–114 (2003)
34. Piña, H., Flores, V., del Castillo, L., Dominguez, O.: Processing and mechanical properties of natural rubber-$ZnFe_2O_4$ nanocomposites. J. Mater. Eng. Perform. **16**, 470–476 (2007)
35. Sahoo, S., Bhowmick, A.K.: Influence of ZnO nanoparticles on the cure characteristics and mechanical properties of carboxylate nitrile rubber. J. Appl. Polym. Sci. **106**, 3077–3083 (2007)
36. Mishra, S. Shimpi, N.G.: Effect of the variation in the weight percentage of the loading and reduction in the nanosizes of $CaSO_4$ on the mechanical and thermal properties of styrene-butadiene rubber. J. Appl. Polym. Sci. **104**, 2018–2016 (2007)
37. Zhang, Y., Lee, S., Yoonessi, M., Liang, K., Pittman, C.U.: Phenolic resin trisilanolphenyl polyhedral oligomeric silsesquioxane (POSS) hybrid nanocomposites: structure and properties. Polymer **47**, 2984–2996 (2006)

38. Iyer, S., Schiraldi, D.A.: Role of specific interactions and solubility in the reinforcement of bisphenol-A polymers with polyhedral oligomeric silsesquioxanes. Macromolecules **40**, 4942–4952 (2007)
39. Hanssen, R.W.J.M., van Santen, R.A., Abbenhuis, H.C.L.: the dynamic status quo of polyhedral silsesquioxane coordination chemistry. Eur. J. Inorg. Chem. **2004**, 675–683 (2004)
40. Sengupta, R., Chakraborty, S., Bandyopadhyay, S., Dasgupta, S., Mukhopadhyay, R., Auddy, K., Deuri, A.S.: A short review on Rubber/Clay nanocomposites with emphasis on mechanical properties. Polym. Eng. Sci. **47**(11), 1956–1974 (2007)
41. Ganter, M., Gronski, W., Semke, H., Zilg, T., Thomann, R., Mülhaupt, R.: Surface-compatibilized layered silicates—a novel class of nanofillers for rubbers with improved mechanical properties. Kautsch. Gummi Kunstst. **54**(4), 166–171 (2001)
42. Pramanik, M., Srivastava, S.K., Samantaray, B.K., Bhowmick, A.K.: Rubber-clay nanocomposite by solution blending. J. Appl. Polym. Sci. **87**, 2216–2220 (2003)
43. Lim, Sk, Lim, S.T., Kim, H.B., Chin, I., Choi, H.J.: Preparation and physical characterization of polyepichlorohydrin elastomer/clay nanocomposites. J. Macromol. Sci. B Phys. B **42**(6), 1197–1199 (2003)
44. Wu, C.M., Hwang, W.G., Tien, K.C., Chang, Y.C., Fu H.L.: 11th National Conference on Science and Technology of National Defense, Taipei, Taiwan (2003)
45. Jeon, H.S., Rameshwaram, J.K., Kim, G.: Structure-property relationships in exfoliated polyisoprene/clay nanocomposites. J. Polym. Sci. B Polym. Phys. **42**, 1000–1009 (2004)
46. Varghese, S., Karger-Kocsis, J.: Natural rubber-based nanocomposites by latex compounding with layered silicates. Polymer **44**, 4921–4927 (2003)
47. Zhang, L., Wang, Y., Wang, Y., Sui, Y., Yu, D.: Morphology and mechanical properties of clay/styrene butadiene rubber nanocomposites. J. Appl. Polym. Sci. **78**, 1873–1878 (2000)
48. Wu, Y., Jia, Q., Yu, D., Zhang, L.: Structure and properties of nitrile rubber (NBR)-clay nanocomposites by co-coagulating NBR latex and clay aqueous suspension. J. Appl. Polym. Sci. **89**, 3855–3858 (2003)
49. Wu, Y.P., Zhang, L.Q., Wang, Y.Q., Liang, Y., Yu, D.S.: Structure of carboxylated acrylonitrile-butadiene rubber (CNBR)–clay nanocomposites by co-coagulating rubber latex and clay aqueous suspension. J. Appl. Polym. Sci. **82**(11), 2842–2848 (2001)
50. Varghese, S., Gatos, K.G., Apostolov, A.A., Karger-Kocsis, J.: Morphology and mechanical properties of layered silicate reinforced natural and polyurethane rubber blends produced by latex compounding. J. Appl. Polym. Sci. **92**, 543–551 (2004)
51. Balachandran, M., Bhagawan, S.S.: Studies on acrylonitrile—butadiene copolymer (NBR) layered silicate composites: mechanical and viscoelastic properties. J. Compos. Mater. Sept. **45**, 2011–2022 (2011)
52. Varghese, S., Karger-Kocsis, J.: Melt-compounded natural rubber nanocomposites with pristine and organophilic layered silicates of natural and synthetic origin. J. Appl. Polym. Sci. **91**, 813–819 (2004)
53. Mousa, A., Karger-Kocsis, J.: Rheological and thermodynamical behavior of styrene/butadiene rubber-organoclay nanocomposites. Macromol. Mater. Eng. **286**, 260–266 (2001)
54. Kojima, Y., Fukumori, K., Usuki, A., Okada, A., Kurauchi, T.: Gas permeabilities in rubber clay hybrid. J. Mater. Sci. Lett. **12**, 889–890 (1993)
55. Zheng, H., Zhang, Y., Peng, Z., Zhang, Y.J.: Influence of the clay modification and compatibilizer on the structure and mechanical properties of ethylene-propylene-diene rubber/montmorillonite composites. J. Appl. Polym. Sci. **92**, 638–646 (2004)
56. Mishra, K., Kim, Il, Chang-Sik., Ha: New millable polyurethane/organoclay nanocomposite: preparation characterization and properties. Macromol. Rapid Commun. **24**, 671–675 (2003)
57. ASTM E 176, Standard Terminology of Fire Standards. In: Annual Book of ASTM Standards. American Society for Testing and Materials. 4.07 West Conshohocken, PA
58. Hirschler, M.M.: Effect of oxygen on the thermal decomposition of poly(vinylidene fluoride). Eur. Polym. J. **18**, 463–467 (1982)

59. Inabi, A., Kashiwagi, T.: A calculation of thermal degradation initiated by random scission, unsteady radical concentration. Eur. Polym. J. **23**(11), 871–881 (1987)
60. Steckler, K.D., Kashiwagi, T., Baum, H.R., Kanemaru, K.: Analytical model for transient gasification of noncharring thermoplastic materials. In: Cox, G., Langford, B. (eds.) Proceedings of Third International Symposium on Fire Safety Science, London (1991)
61. Sommers, A.E., Bastow, T.J., Burgar, M.I., Forysth, M., Hill, A.J.: Quantifying rubber degradation using NMR. Polym. Degrad. Stab. **2000**(70), 31–37 (2000)
62. Ranby, B., Rabek, J.F.: Photodegradation, photooxidation and photostabilization of polymers. Wiley, London (1975)
63. Stephen, R., Jose, S., Joseph, K., Thomas, S., Ommen, Z.: Thermal stability and ageing properties of sulphur and gamma radiation vulcanized natural rubber (NR) and carboxylated styrene butadiene rubber (XSBR) latices and their blends. J. Polym. Degrad. Stab. **91**, 1717–1725 (2006)
64. Choi, S., Han, D.H., Ko, S.W., Lee, H.S.: Thermal aging behaviors of elemental sulfur-free polyisoprene vulcanisates. Bull. Korean Chem. Soc. **26**, 1853–1855 (2005)
65. Akiba, M., Hashim, A.S.: Vulcanization and crosslinking in elastomers. Prog. Polym. Sci. **22** (3), 475–521 (1997)
66. Gehani, R.: National innovation system and disruptive innovations in synthetic rubber and tire technology. J. Thechnol. Manag. innov. **2**, 55–72 (2007)
67. Bhatt, R., Shah, D., Patel, K.C., Trivedi, U.: PHA-rubber blends: synthesis, characterization and biodegradation. Bioresour. Technol. **99**, 4615–4620 (2008)
68. Arantes, T., Leao, K., Tavares, M., Ferreira, A.: NMR study of styrene-butadiene rubber (SBR) and TiO_2 nanocomposites. Polym. Testing **28**, 490–494 (2009)
69. Tonny, M.M., Nemr, K.F.: Application of acrylonitrile butadiene rubber for management of industrial waste silica. Mater. Sci. Eng. **1**, 1–6 (2012)
70. Chayan, Das, Bharat, K.: Preparation and studies of nitrile rubber nanocomposites with silane modified silica nanoparticle. Research Journal of Recent Sciences **1**, 357–360 (2012)
71. Chaichua, B., Prasassaraki, P., Poompradub, S.: In situ silica reinforcement of natural rubber by sol-gel process via rubber solution. J. Sol-Gel Sci. Technol. **52**, 219–227 (2009)
72. Mohomane, S.M., Djokovic, V., Thomas, S., Luyt, A.S.: Polychloroprene nanocomposite filled with different organically modified clays: Morphology, thermal degradation and stress relaxation behavior. Polym. Testing **30**, 585–593 (2011)
73. Popovic, I., Katsikas, L.: The characterization of polymer composite by thermogravimetry. Mater. Compos. **40**, 7–12 (2006)
74. Balachandran, M., Bhagawan, S.: Mechanical, thermal and transport properties of nitrile rubber (NBR)-nanoclay composite. J. Polym. Res. **19**, 2–10 (2012)
75. Bourbigot, S., Gilman, J.W., Wilkie, C.A.: Kinetic analysis of the thermal degradation of polystyrene montmorillonite nanocomposite. Polym. Degrad. Stab. **84**(3), 483–492 (2004)
76. Peeterbroeck, S., Lepoittevin, B., Pollet, E., Benali, S., Broekaert, C., Alexandre, M., Bonduel, D., Viville, P., Lazzaroni, R., Dubois, P.: Polymer layered silicate/carbon nanotube nanocomposites: the catalyzed polymerization approach. Polym. Eng. Sci. **46**, 1022–1030 (2006)
77. Gamlin, C., Markovic, G., Dutta, K., Choudhury, R., Matisons, J.G.: Structural effects on the decomposition kinetics of EPDM elastomers by high-resolution TGA and modulated TGA. J. Therm. Anal. Calorim. **591**, 319–336 (2000)
78. Carretero, G.J., Verdejo, R., Arroyo, M., López, M.M.A.: Nuevos avances en el desarrollo de nanocompuestos elastoméricos. Suplemento de la Revista Latinoamericana de Metalurgía y Materiales **S2**(1), 33–34 (2009)
79. Zhang, J., Wang, L., Zhao, Y.: Improving performance of low-temperature hydrogenated acrylonitrile butadiene rubber nanocomposites by using nano-clays. Mater. Des. **50**, 322–331 (2013)
80. Sadaka, F., Campistron, I., Laguerre, A., Pilard, J.F.: Telechelic oligomers obtained by metathetic degradation of both polyisoprene and styrene-butadiene rubbers. Appl. Recycl. Waste Tire Rubber **98**, 736–742 (2013)

81. Gu, Z., Song, G., Liu, W., Li, P., Gao, L., Li, H., Hu, X.: Preparation and propertied of styrene butadiene rubber /natural rubber/ organo-bentonite nancomposite prepared from latex dispersion. Appl. Clay Sci. **46**, 241–244 (2009)
82. Yehia, A.A., Akelah, A.M., Rehab, A., Sabbagh, S.H., Nashar, D.E., Koriem, A.A.: Evaluation of clay hybrid nanocoposite of different chain length as reinforcing agent for natural and synthetic rubbers. Mater. Des. **33**, 11–19 (2012)
83. Cerin, O., Fontaine, G., Duquesne, S., Bourbigot, S.: Thermal stability of synthetic rubber nanocomposites. In: Mittal. V. (ed.) Recent Advance in Elastomeric Nanocomposites. Springer, Berlin, Heidelberg (2011)
84. Malas, A., Pal, P., Das, ChK: Effect of expanded graphite and modified graphite flakes on the physical and thermo-mechanical properties of styrene butadiene rubber/polybutadiene rubber (SBR/BR) blends. Mater. Des. **55**, 664–673 (2014)
85. Cabello, Ch., Saénz, A., López, Ll, Pérez, C., Barajas, L., Ávila, C.: Modificación superficial de (MWCNT) con H_2SO_4/HNO_3 mediante ultrasonido. Afinidad **68**, 370–374 (2012)
86. Inukai, S., Niihara, K.-I., Noguchi, T., Ueki, H., Magario, A., Yamada, E., Inagaki, S., Endo, M.: Preparation and properties of multiwall carbon nanotubes/ polystyrene-block-polybutadiene-block-polystyrene-composite. Ind. Eng. Chem. Res. **50**, 8016–8022 (2011)
87. Gerorge, J.J., Bhowmick, A.K.: Influence of matrix polarity on the properties of ethylene vinyl acetate-carbon nanofiller nancomposite. Nanoscale Res. Lett. **4**, 655–664 (2009)
88. Gerorge, J.J., Bhowmick, A.K.: Ethylene vinyl acetate/expanded graphite nancomposite by solution intercalation: preparation, characterization and properties. J. Mater. Sci. **43**, 702–708 (2009)
89. Rybinski, P., Janowska, G., Jozwiak, M., Pajak, A.: Thermal stability and flammability of butadiene-styrene rubber nanocomposites. J. Therm. Anal. Calorim. **109**, 561–571 (2012)
90. Liu, Q., Zhang, Y., Xu, H.: Properties of vulcanized rubber nanocomposite filled with nanokaolin and precipitated silica. Appl. Clay Sci. **42**, 232–237 (2008)
91. Madorsky, S.L.: Thermal Degradation of Polymer, Reprinted by Robert E. Kreiger, New York (1976)
92. Jang, L.W., Kang, C.M., Lee, D.C.: A new hybrid nanocomposite prepared by emulsion copolymerization of ABS in the presence of clay. J. Polym. Sci. Part B Polym. Phys. **39**, 719–727 (2001)
93. Choi, Y.S., Xu, M.Z., Chung, I.J.: Synthesis of exfoliated acrylonitrile–butadiene–styrene copolymer (ABS) clay nanocomposites: role of clay as a colloidal stabilizer. Polym. Degrad. Stab. **46**, 531–538 (2005)
94. Karahaliou, E.-K., Tarantili, P.A.: Preparation of poly(acrylonitrile–butadiene–styrene)/ montmorillonite nanocomposites and degradation studies during extrusion reprocessing. J. Appl. Polym. Sci. **113**, 2271–2281 (2009)
95. Pourabbas, B., Azimi, H.: Indirect synthesis of ABS/clay nanocomposites, comparison and thermal properties. J. Compos. Mater. **42**, 2499–2522 (2008)
96. Wang, S.F., Hu, Y., Song, L., Wang, Z.Z., Chen, Z.Y., Fan, W.C.: Preparation and thermal properties of ABS/montmorillonite nanocomposite. Polym. Degrad. Stab. **77**, 423–426 (2002)
97. Wang, S., Hu, Y., Lin, Z., Gui, Z., Wang, Z., Chen, Z., Fan, W.: Flammability and thermal stability studies of ABS/montmorillonite nanocomposite. Polym. Int. **52**, 1045–1049 (2003)
98. Cai, Y., Huang, F., Xia, X., Wei, Q., Tong, X., Wei, A., Gao, W.: Comparison between structures and properties of ABS nanocomposites derived from two different kinds of OMT. J. Mater. Eng. Perform. **19**, 171–176 (2010)
99. Ying, G.H., Fung, J.L.: Organic–inorganic composite materials from acrylonitrile–butadiene–styrene copolymers and silica through an in situ sol–gel process. J. Appl. Polym. Sci. **75**, 275–283 (2000)
100. Yang, S., Castilleja, J.R., Barrera, E.V., Lozano, K.: Thermal analysis of an acrylonitrile–butadiene–styrene/SWNT composite. Polym. Degrad. Stab. **83**, 383–388 (2004)
101. Ma, H., Tong, L., Xu, Z., Fang, Z.: Synergistic effect of carbon nanotube and clay for improving the flame retardancy of ABS resin. Nanotechnology **18**, 375602 (2007)

102. Liu, L., Grunlan, J.C.: Clay assisted dispersion of carbon nanotubes in conductive epoxy composites. Adv. Funct. Mater. **17**, 2343–2348 (2007)
103. Tang, C., Xiang, L., Su, J., Wang, K., Yang, C., Zhang, Q., Fu, Q.: Largely improved tensile properties of chitosan film via unique synergistic reinforcing effect of carbon nanotube and clay. J. Phys. Chem. B **112**, 3876–3881 (2008)
104. Barick, A.K., Tripathy, D.K.: Thermal and dynamic mechanical characterization of thermoplastic polyurethane/organoclay nanocomposites prepared by melt compounding. Mater. Sci. Eng. A **527**, 812–823 (2010)
105. Shimpi, N.G., Mishra, S.: Synthesis of nanoparticles and its effect on properties of elastomeric nanocomposites. J. Nanopart. Res. **12**, 2093–2099 (2010)
106. Chen, G., Liu, S., Chen, S., Qi, Z.: FTIR spectra, thermal properties, and dispersibility of a polystyrene/montmorillonite nanocomposites. Macromol. Chem. Phys. **202**, 1189–1193 (2001)
107. Chrissafis, K., Bikiaris, D.: Can nanoparticles really enhance thermal stability of polymers? Part I: an overview on thermal decomposition of addition polymers. Thermochim. Acta **523**, 1–24 (2011)
108. Li, H., Wang, H., Wu, Y., Zhang, X., Zheng, J.: Effect of the interfacial interaction on thermal oxidative stability of carbon nanotubes/silicone rubber composites. Sci. Adv. Mater. **5**, 453–461 (2013)
109. Wang, X., Dou, W.: Preparation of graphite oxide (GO) and the thermal stability of silicone rubber/GO nanocomposites. Thermochim. Acta **529**, 25–28 (2012)
110. Cao, X., Xu, C., Wang, Y., Liu, Y., Liu, Y., Chen, Y.: New nanocomposite materials reinforced with cellulose nanocrystals in nitrile rubber. Polym. Testing **32**, 819–826 (2013)
111. Shojaei, A., Faghihi, M.: Physico-mechanical properties and thermal stability of thermoset nanocomposites based on styrene-butadiene rubber/phenolic resin blend. Mater. Sci. Eng., A **527**, 917–926 (2010)
112. He, S.-J., Wang, Y.-Q., Xi, M.-M., Lin, J., Xue, Y., Zhang, L.-Q.: Prevention of oxide aging acceleration by nano-dispersed clay in styrene-butadiene rubber matrix. Polym. Degrad. Stab. **98**, 1773–1779 (2013)
113. Hwang, S.J., Joo, Y.L., Lee, S.J.: Properties of high-impact polystyrene/organoclay nanocomposites synthesized via in situ polymerization. J. Appl. Polym. Sci. **110**, 1441–1450 (2008)
114. Malmberg, S.M., Parent, J.S., Pratt, D.A., Whitney, R.A.: Isomerization and elimination reactions of brominated poly(isobutylene-co-isoprene). Macromolecules **43**, 8456–8461 (2010)
115. Kong, Q., Hu, Y., Song, L., Wang, Y., Chen, Z., Fan, W.: Influence of Fe-MMT on crosslinking and thermal degradation in silicone rubber/clay nanocomposites. Polym. Adv. Technol. **17**, 463–467 (2006)
116. Fang, S., Hu, Y., Song, L., Wu, J.: Preparation and investigation of ethylene-vinyl acetate copolymer/silicone rubber/clay nanocomposites. J. Appl. Polym. Sci. **113**, 1664–1670 (2009)
117. Liu, Q., Zhang, Y., Xu, H.: Properties of vulcanized rubber nanocomposites filled with nanokaolin and precipitated silica. Appl. Clay Sci. **42**, 232–237 (2008)
118. Prochon, M., Przepiórkowska, A.: Innovative application of biopolymer keratin as a filler of synthetic acrylonitrile-butadiene rubber NBR. J. Chem. **2013**, 1–8 (2013)
119. Ponomarenko, S.M., Privalko, E.G., Privalko, V.P., Schön, F., Gronski, W.: Structure and thermoelasticity of synthetic rubber/silica composites. J. Macromol. Sci. Part B **43**, 1231–1242 (2004)
120. Li, Y., Han, B., Liu, L., Zhang, F., Zhang, L., Wen, S., Lu, Y., Yang, H., Shen, J.: Surface modification of silica by two-step method and properties of solution styrene butadiene rubber (SSBR) nanocomposites filled with modified silica. Compos. Sci. Technol. **88**, 69–75 (2013)
121. Scotti, R., Conzatti, L., D'Arienzo, M., Di Credico, B., Giannini, L., Hanel, T., Stagnaro, P., Susanna, A., Tadiello, L., Morazzoni, F.: Shape controlled spherical (0D) and rod-like (1D) silica nanoparticles in silica/styrene butadiene rubber nanocomposites: role of the particle morphology on the filler reinforcing effect. Polymer **55**, 1497–1506 (2014)

122. Bala, P., Samantaray, B.K., Srivastava, S.K., Nando, G.B.: Organomodified montmorillonite as filler in natural and synthetic rubber. J. Appl. Polym. Sci. **92**, 3583–3592 (2004)
123. Dal Pont, K., Gérard, J.-F., Espuche, E.: Microstructure and properties of styrene-butadiene rubber based nanocomposites prepared from an aminosilane modified synthetic lamellar nanofiller. J. Polym. Sci. B Polym. Phys. **51**, 1051–1059 (2013)
124. Songmin, S., Lu, G., Chun-Wah, Y.M.: Improvement of carbon nanotubes dispersion by chitosan salt and its application in silicone rubber. Compos. Sci. Technol. **86**, 129–134 (2013)
125. Xiubin, Z., Haiyan, Z., Jin, L., Liping, L., Qiguang, W.: Thermal conductivity and thermal stability enhancement of ethylene propylene diene methylene with carbon nanotube. J. Reinf. Plast. Compos. **33**, 767–774 (2014)
126. Peddini, S.K., Bosnyak, C.P., Henderson, N.M., Ellison, C.J., Paul, D.R.: Nanocomposites from styrene-butadiene rubber (SBR) and multiwall carbon nanotubes (MWCNT) part 1: morphology and rheology. Polymer **55**, 258–270 (2014)
127. Laoui, T.: Mechanical and thermal properties of styrene butadiene rubber-functionalized carbon nanotubes nanocomposites. Fuller. Nanotub. Carb. Nanostruct. **21**, 89–101 (2013)
128. Ganter, B., Boßhammer, Irmer, U.: UV-curable silicone rubbers open up new fields. Int. Polym. Sci. Technol. **40**, 1–4 (2013)
129. Anwar Parvez, M., Al-Mehthel, M., Al-Abdul, W.H.I., Hussein, I.A.: Utilization of sulfur and crumb rubber in asphalt modification. J. Appl. Polym. Sci. **131**, 1–11 (2014)
130. Zanchet, A., Carli, L.N., Giovanela, M., Brandalise, R.N., Crespo, J.S.: Use of styrene butadiene rubber industrial waste devulcanized by microwave in rubber composites for automotive application. Mater. Des. **39**, 437–443 (2012)
131. Costa, P.P., Silvia, C.C., Viana, J.C., Lanceros Mendez, S.S.: Extruded thermoplastic elastomers styrene-butadiene-styrene/carbon nanotubes composites for strain sensor applications. Compos. B Eng. **57**, 242–249 (2014)
132. Chameswary, J., Sebastian, M.: Butyl rubber–Ba0.7Sr0.3TiO3 composites for flexible microwave electronic applications. Ceram. Int. **39**, 2795–2802 (2013)
133. Suksaeree, J., Pichayakorn, W., Monton, C., Sakunpak, A., Chusut, T., Saingam, W.: Rubber polymers for transdermal drug delivery systems. Ind. Eng. Chem. Res. **53**, 507–513 (2014)

Outdoor Exposure Degradation of Ethylene-Vinyl-Acetate Encapsulant Material for Photovoltaic Application

K. Agroui and G. Collins

Abstract Apart from chemical structure, thermal properties of polymers depend mainly on polymer morphology, which is affected by the processing conditions. Outdoor exposure of the materials leads to changes in polymer morphology. The main objective of this experimental investigation was to better understand the changes due to thermal transitions and the molecular organizations of the cross-linked Ethylene-Vinyl-Acetate (EVA) encapsulant material after aging in outdoor exposure. EVA samples are characterized by various thermal and spectroscopic analysis techniques like thermally stimulated current, which is a very sensitive toll to detect polymer degradation. Results show a significant decrease in most properties of EVA in natural field exposure due principally to the specificity of the exposure site. For the aged EVA samples, the distinctive feature of these results is that there are two different endothermic processes due to the recrystallization phenomenon. Furthermore, the difference of the magnitude of peak current by TSC technique suggests increased crosslinking exposure occurring selectively in the high temperature phase as a result of outdoor exposure.

Keywords EVA encapsulant · Photovoltaic module · Outdoor aging · Thermal analysis

1 Introduction

Qualification test sequences, such as those described in IEC standards, are an established a pass/fail indicator, whether the product is qualified for sale. However, qualification testing is not a method for a quantitative prediction of the products

K. Agroui (✉)
Semiconductors Technology for Energetic Research Center (CRTSE),
2, Bd. Dr. Frantz Fanon, BP 140 Alger 7 Merveilles, Algiers, Algeria
e-mail: kagroui@yahoo.fr

G. Collins
Department of Biomedical Engineering, New Jersey Institute of Technology,
Newark, NJ 07102, USA

P.M. Visakh and Y. Arao (eds.), *Thermal Degradation of Polymer Blends, Composites and Nanocomposites*, Engineering Materials,
DOI 10.1007/978-3-319-03464-5_8

service life in general [1]. Accelerated aging such as damp-heat, UV-irradiation and corrosive atmosphere add to degradation and form a complex field of influencing factors.

One important application of EVA (Ethylene Vinyl Acetate) material is the encapsulation of solar cells in photovoltaic (PV) modules, where the material has to fulfill several basic functions. Unfortunately, the IEC standard does not provide any information on the changes in the internal structure of the EVA encapsulant material as a consequence of encapsulation process or environmental exposure.

Generally, PV modules degradation is investigated by using analysis methods such as I–V curve measurements, electroluminescence, thermo-imaging or more sophisticated techniques [2, 3]. Investigating the mechanism causing degradation of the components in a PV module is an approach to get information about the degradation progress. To assess the long term performance and durability of EVA encapsulant material, it was necessary not only to measure the deterioration of macroscopic physical properties, but also to gain information about degradation processes taking place at a molecular level.

Many studies have been conducted, utilizing various technical analysis to assess the conditions of the EVA. These methods require the destructive extraction of samples of the polymer from the module [4, 5]. Extraction of EVA samples from field deployed PV module is very difficult and contributes to the whole destruction of the PV module. For this reason, the durability in outdoor exposure of EVA encapsulant under the conditions simulating those in PV module is often used to assess the weathering of polymeric materials [6].

In several papers, the thermal behavior of EVA material is discussed in detail. The changes in thermal properties due to storage at room temperature and annealing at elevated temperatures are investigated and explained by characterizing the changes in polymer morphology [7, 8]. Nevertheless, these findings were not linked to Thermally Stimulated Current (TSC), which defines the transitions and relaxations that depend on changes in the mobility of molecular scale dipolar structural units. The high interest of TSC technique is due to the very low equivalent frequency and its capability to resolve complex dielectric transitions. TSC is based on the ability of polar molecules to be moved by an electrostatic field and also has been frequently employed to investigate the molecular motions in polymeric materials [9, 10]. TSC complements perfectly the thermal analysis methods like as Differential Scanning Calorimetry (DSC) to determine fundamental properties of EVA and also has been frequently employed to investigate the molecular motions in polymeric materials [11].

The aim of this work is to analyze the changes in the relevant thermal behavior of the crosslinked EVA encapsulant material due to outdoor exposure in Mediterranean climate by TSC and DSC thermal analysis. The changes in thermal properties should be discussed and special interest will be focused on the specific TSC relaxation parameters like activation energy and relaxation frequency determination by using the initial rise method.

2 Sample Preparation

2.1 Material and Crosslinking Conditions

EVA encapsulant test uses EVA A9918 (33 % of vinyl acetate) standard cure from STR (Specialized Technology Resources) manufacturer and is referred to as original EVA [12]. The original EVA films were inserted between two pieces of Teflon and placed on the lower chamber of the laminator in order to produce the correct thermal conditions during the cure. The EVA has to be crosslinked chemically in order to obtain the required thermo-mechanical properties. The cure cycle is executed using the SPIIRE-LAMINATOR (Fig. 1) at temperature of 160 °C for 15 min according to the standard module lamination conditions.

Original EVA film submitted to the cure step cycle is referred to as cured or crosslinked EVA. The gel content can be determined by a solvent (toluene) extraction method as a standard procedure described by STR. The gel content measurement during the cure cycle, is 87.4 % according to STR procedure.

2.2 Outdoor Exposure Conditions

Natural weathering involves placing samples on inclined racks orientated at the sun at an angle of 45° in a southerly direction as described in ASTM standard [6]. The laminate configuration was used is: glass/cured EVA encapsulant/TPT back sheet polymer with Aluminum frame and without the solar cell circuit as shown in Fig. 2. This was done to simulate the similar situation of EVA material when the PV modules operate outdoors. The outdoor measurements were performed at two different sites. The first site is located at Renewable Energy Development Centre (CDER) located at Algiers as Mediterranean climate and the second site is the Applied Renewable Energy Research Unit (URAER) located at Adrar (South-west of Algiers) as desert climate. In order to investigate the physical aging of the EVA material, specimens were exposed for 18 months and 8 years in natural weathering of CDER and URAER site respectively.

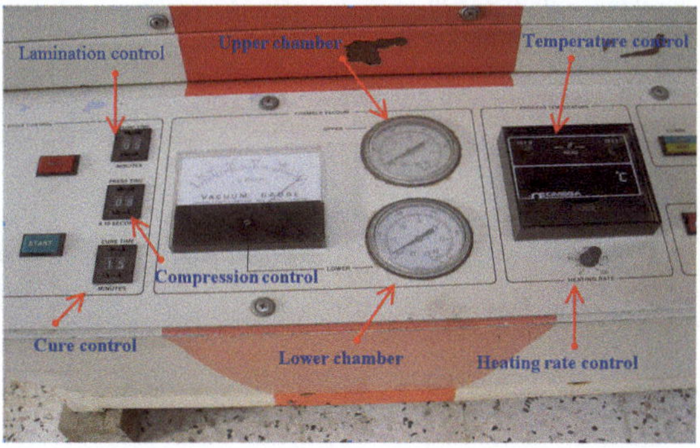

Fig. 1 PV module laminator description

Fig. 2 EVA outdoor exposure at CDER site

Special instrumentation has been provided for the measurement of the meteorological parameters of CDER site, like horizontal global irradiance-humidity and maximal-minimal ambient temperatures as illustrated in Figs. 3 and 4 respectively. The ambient temperature and relative humidity of the exposure location site is in the range of 18–40 °C and 20–95 % respectively.

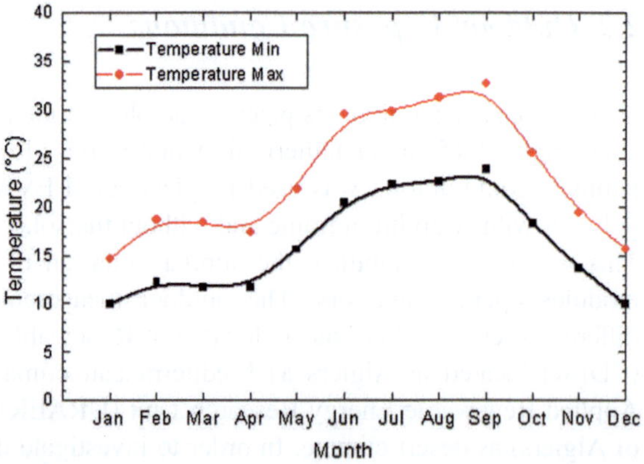

Fig. 3 Variation of ambient temperature at CDER exposure site during year 2011

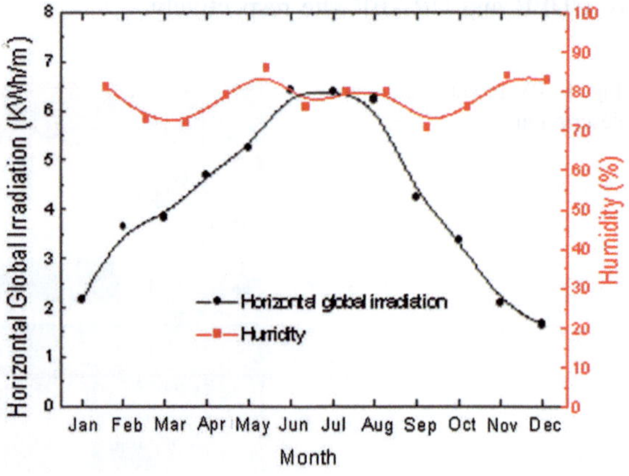

Fig. 4 Average variation of global horizontal irradiation and relative humidity at CDER exposure site during year 2011

3 Technical Analysis Description

3.1 Thermally Stimulated Current (TSC)

3.1.1 TSC Principle

TSC is a technique for detecting the transitions that depend on changes in the mobility of molecular scale dipolar structural units. TSC is based on the ability of polar molecules to be moved by an electrostatic field. Two types of current are generated: thermally stimulated polarization current (TSPC) and thermally stimulated depolarization current (TSDC). TSPC is generated when dipolar structures orient in a static electric field with increasing in the temperature. TSDC is generated because of the relaxation of the previously polarized molecules as reported in Fig. 5.

In the TSC technical literature, the most frequently described experiment is TSDC. When performing the experiment measurements by TSC technique it was noted that the thermostimulated current is greatly influenced by the choice of the heating rate value, due to the fact that the detected current peak is very small (in the order of pA).

3.1.2 TSC Experimental Set-up Description

Several prior investigations were done in order to favor optimal sensibility of current peak relaxation measurements, for this reason the heating rate was fixed at 7 °C/min. In the present study, the experimental parameters of TSC are summarized as follow: An applied voltage of 200 V DC, heating rate of 7 °C/min and a temperature range of −80 to 60 °C, because the glass transition temperature (T_g) of EVA lies within this range. Figure 4 illustrates the profile of a TSDC at 0 °C polarization temperature. The instrument used for the study is TSC/RMA Model

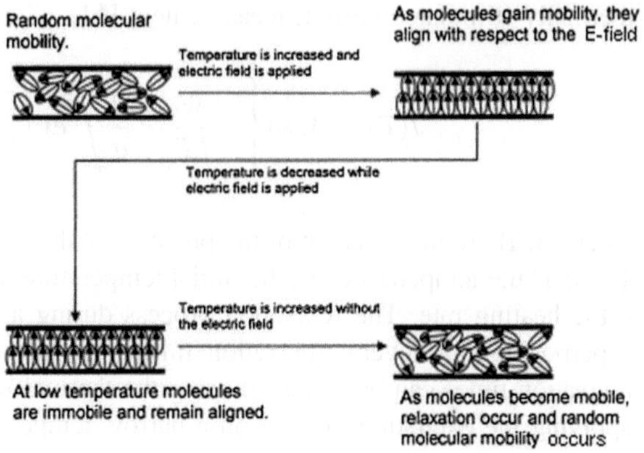

Fig. 5 Description of a TSC technique in TSDC mode

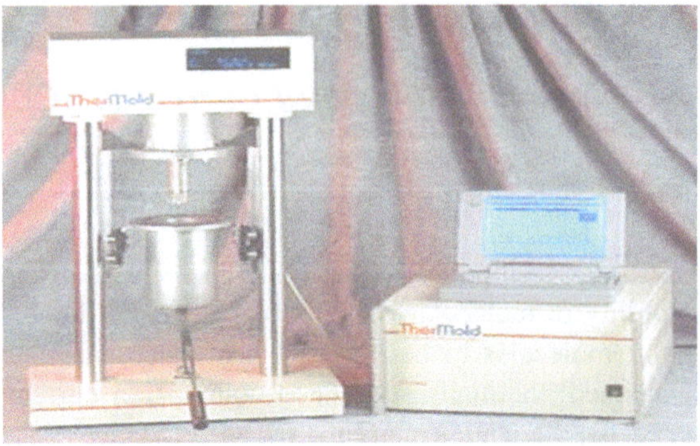

Fig. 6 Apparatus used for TSC test (TSC/RMA Model 9000)

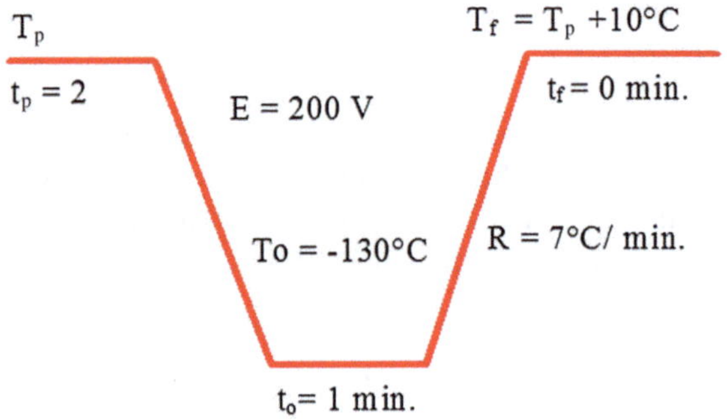

Fig. 7 TSDC experimental profile at 0 °C polarization temperature

9000 from Stamford Instrumentation as represented by Fig. 6. The global TSC experiments were carried out following the procedures using the parameters schematised in diagram of Fig. 7.

3.1.3 Theoretical Background of TSC Principle

Equation (1) shows the current density J(T) at temperature T released during thermally stimulated current measurement [11].

$$J(T) = Aexp\left\{ -\frac{W}{kT} - \frac{B}{q}\int_{T_o}^{T} exp\left[-\frac{W}{kT} \right] dT' \right\} \tag{1}$$

where A, B are the constant of the process, W the activation energy of the process, T and T′ the temperature, T_o the initial temperature, k the Boltzman's constant and q the heating rate. The relaxation process during a glass transition is usually the superposition of several relaxation times. The separation of this distribution of relaxation times can be achieved using the thermal windowing (TW) technique by applying the excitation voltage on a narrow temperature window, the response of the material can be modeled using a single relaxation time. Shifting the polarization window across the entire temperature range of a transition generates the thermal windowing spectra corresponding to a map of that distribution of the relaxation times for the transition. For the determination of the characteristic parameters of the process i.e. the activation energy and the relaxation time factor, the initial rise method is mostly used. This method is based on the fact that the integral term in Eq. (1) is low for $T \ll T_{max}$ and consequently the first exponential term dominates. In this situation the current density during a thermal stimulated measurement has an Arrhenius expression. Therefore the activation energy can be determined by the slope of the plot ln(J(T)) versus 1/T as shown in Eq. (2).

$$J(T) = J_0 \exp\left[-\frac{W}{kT}\right] \tag{2}$$

The relaxation time τ(T) at temperature T can be extracted from the elementary spectrum by using Eq. (3).

$$\tau(T) = \frac{P(T)}{J(T)} \tag{3}$$

where P(T) is the polarization at temperature T as described in Eq. (4):

$$P(T) = \frac{1}{q}\int_{T}^{\infty} J(T')dT' \tag{4}$$

Equation (5) shows the relaxation frequency $\alpha(T_{max})$ at the peak temperature T_{max}.

$$T_{max} = \sqrt{\frac{qW}{k\alpha(T_{max})}} \tag{5}$$

The Debye formalism that is represented in Eqs. (1–5) strictly only applies for a single relaxation process. That is to say, the equations can only be applied for an isolated relaxation. The global polarization experiments that were performed produce a depolarization current that represents the entire distribution of relaxations. In order to isolate a relaxation process and properly apply the equations, it would be necessary to carry out thermal windowing experiments. In our study we applied the equations to the low temperature relaxation process in order to get a first approximation of the magnitude of the activation energy.

The TSC measurements provide information on the molecular scale mobility in the internal structure of a solid material and is based on the motions of dipoles in response to temperature and a static electric field. TSC is well suited to study the degree of curing and the determination of the glass transition temperature. The combined change in TSC peak parameters of EVA encapsulant is correlated with the degree of curing.

3.2 Differential Scanning Calorimetric (DSC)

All DSC experiments were conducted in a nitrogen atmosphere by using a TA Q100 instrument. A heat-cool-heat temperature programming protocol was used in DSC experiments to evaluate the behavior of the cured EVA sample. The first heating was done at 7 °C/min; the cooling was done at −40 °C/min; the final heating was done at 7 °C/min. The temperature heating rate of 7 °C/min was chosen (as special case for this study) because it was the same rate as used in the TSC technique, in order to compare DSC and TSC in the same temperature program. Furthermore the high value of cooling rate was chosen in order to avoid crystallization of the polymer sample. The heating program was a conventional temperature range from −50 to 85 °C, as the expected thermal events lie within this range temperature.

3.3 X-Ray Diffraction

X-ray diffraction pattern measurements were carried out using a BRUKER D8 ADVANCE diffractometer. The accelerating voltage was 40 kV and the current was 40 mA. The diffraction intensities were recorded as a function of increasing scattering angle 2θ, which was in the range 2°–88° with a step angle of 0.05 and the scan time of 0.4 s per step.

3.4 Dielectric Analysis as Function of Temperature and Frequency

For evaluating the dielectric properties of EVA an experimental set-up was installed according to the NFC 26 200 et NFC 26 230 recommendations system. The main component is the ring guard capacitor type 16451B (Hewlett-Packard). EVA samples are contained between the upper and lower metal plate electrodes within a confinement ring soldered to the lower plate. The upper electrode is connected to high voltage terminal bridge type 4284A (Hewlett-Packard). The electrode are connected to heating resistances to allow measurements to be taken at several temperatures from −50 to 65 °C at a rate of 2 °C/min in the frequency range from 1 Hz to 100 kHz. Capacitance (C) and dissipator factor (tgδ) were measured at 0.05 % degree of precision. Dielectric constant (ε) were deduced from the measurements [13]. The dielectric properties of EVA samples are measured as function of temperature at various frequencies. In the first time EVA sample is kept between electrodes to permit capacitance and dissipator factor measurements as function of temperature at various frequencies. In the second time EVA sample is removed from test cell and vacuum capacitance of sample is determined. Permittivity of EVA may therefore be defined as the ratio of sample and vacuum capacitance respectively.

3.5 *Dielectric Analysis as Function of Temperature at Fixed Frequency*

For evaluating the electrical properties of EVA an experimental set-up was installed according to the IEC recommendations system. Figure 8 shows a schematic representation of the test facility. The main component is the ring guard capacitor type 2904 (TETTEX). EVA samples are contained between the upper and lower metal plate electrodes within a confinement ring soldered to the lower plate. The electrodes are connected to heating resistances to allow measurements to be taken at several temperatures up to 150 °C which are stabilised with a temperature regulator type 2965 (TETTEX). The electrode pressure is kept constant and equal to 2.5×10^4 pa. The applied voltage is adjusted to a value of 500 VAC and frequency of 50 Hz to favour optimal sensibility of measurements. Insulating resistance (R) was measured with a Megohmeter type MOM11 WTW, capacitance (C) and dissipator factor (tgδ) were measured with a precision bridge type 2891 (TETTEX). The resistivity (ρ) and dielectric constant (ε) were deduced from the measurements.

3.6 *UV-Visible Spectrophotometer*

To monitor UV-stabilization and discoloration (yellowing) during EVA weathering, UV-Visible spectroscopy is applied. UV-Visible transmittance measurements were carried out with a spectrophotometer type Carry 500 from Varian with integrate sphere. All UV-Visible spectra were recorded at 600 nm/min at 1 nm.

3.7 *Attenuated Total Reflection by Infrared Spectroscopy Spectrometer (ATR)-IR*

To identify EVA formulation, degradation and even stabilizer diffusion and consumption, the Fourier Transform Infrared (FTIR) spectroscopy in ATR mode is a simple and satisfying method. The FTIR measurements were carried out with a

Fig. 8 Experimental set-up for measuring electrical properties of EVA

spectrometer type Hyperion 2000 Nexus from Thermo Nicolet manufacturer. The spectrometer is equipped with a DTGM KBr detector in transmission mode. All FTIR spectra were recorded at 4 cm^{-1} resolution in the 400–4000 cm^{-1}.

3.8 Tensile Properties

Tensile tests were carried out with a screw driven universal test machine of the type Instron 4505 (Instron International Ltd., High Wycombe, UK) at 23 °C. The test samples were prepared according to ASTM D638-76 IV within maximum dimensions size of 15 × 75 mm, the standard dumbbell-shaped test samples had a gauge length of 30 mm and a width of 12 mm [14]. The average speed value of 50 mm/min was used for plotting the stress–strain curve. From a total of at least five samples for each test series were tested and their average plotted where the stress-at-break (σ_b) and strain-at-break (ε_b) were deduced.

3.9 DMTA

In order to investigate the viscoelastic behavior of crosslinked EVA, rheological measurements were made to determine at what temperatures the phase transitions occur and their effect on the dynamic mechanical modulus. The complex dynamic modulus E expression is given by Eq. (6).

$$E = E' + jE'' \tag{6}$$

where: E′ is the storage modulus, E″ is the loss modulus. The loss factor tgδ expression is given by Eq. (7).

$$\mathrm{tg}\delta = \frac{E''}{E'} \tag{7}$$

Thermo-mechanical properties of the crosslinked EVA were measured using a Rheometric DMTA IV in a nitrogen atmosphere. Crosslinked EVA specimens were cut from the samples and mounted in the rectangular tension fixture. The typical sample dimensions were 10 mm long, 3.3 mm wide and 0.4 mm thick. During the dynamic temperature ramp experiment, the heating rate was 7 °C/min and the frequency was 1 Hz. The static force was maintained at 20 % higher than the dynamic force.

4 Results and Discussion

4.1 *Effect of Outdoor Exposure on Thermal Relaxations*

Figure 9 shows the comparison of TSDC and TSPC results, after second run at polarization temperature of 85 °C, of the EVA sample before and after aging for 6 months in outdoor exposure. TSC experiments on EVA show that a low temperature peak located at −24 °C is the T_g for a polyethylene rich phase, and the high temperature peak located at +31 °C is the T_g for a poly(vinyl acetate) rich phase. The high temperature peak appears also to have a shoulder above 50 °C, characteristic of the change in molecular crystal melting at it will be revealed by DSC analysis. It is noted that depolarization is the relaxation of dipoles away from the direction of the field, and polarization is the motion of the dipoles into the direction of the field. For this reason, TSDC and TSPC experiments generate current peaks in opposite directions; in this case, TSDC will generate positive peaks and TSPC will generate negative peaks.

It would be expected that if the identical dipoles in the same number were relaxing in the case of TSDC and polarizing in the case of TSPC, that the peaks

Fig. 9 TSDC and TSPC spectra of unaged and aged EVA under polarization temperature of 85 °C at CDER site

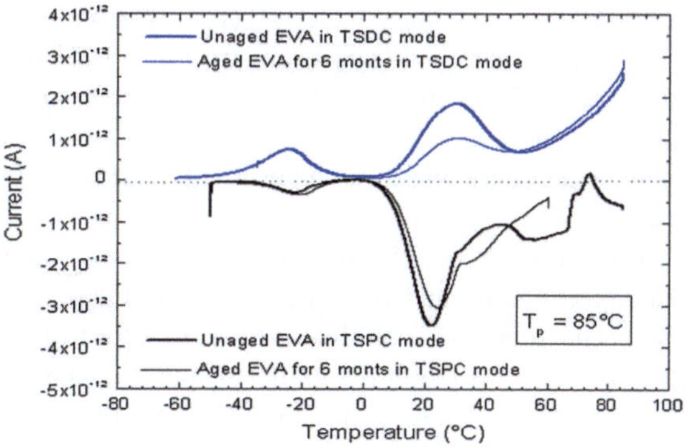

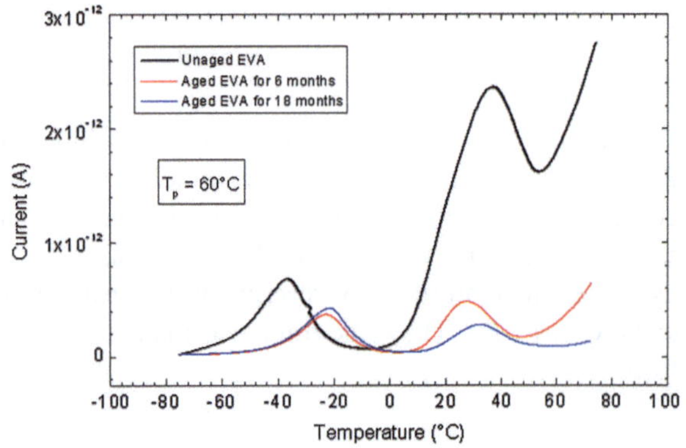

Fig. 10 TSDC spectra of unaged and aged crosslinked EVA under polarization temperature of 60 °C at CDER site

would nominally be mirror images of each other. Also, it is noted that while there is agreement in terms of the positions of the peaks, there is substantial disagreement in terms of their shape and magnitude. In all cases, the low temperature TSDC peak is observably larger than the low temperature TSPC peak. Again, this is consistent with the crystalline melting at the polarization temperature and generation of a higher fraction of mobile amorphous material as has been discussed above.

The high temperature TSDC peak, has a smooth regular shape that would be expected from a dipole relaxation process. In all cases, however, the high temperature TSPC peak has the expected shape on the low temperature side, and develops irregular shape on the high temperature side. These irregularities in shape are not expected for dipole relaxations and likely reflect the non-systematic effect of dipole motion as a consequence of crystal melting.

Figures 10 and 11 show the TSDC curves at 60 °C polarization temperature of the EVA before and after exposure at CDER and URAER sites respectively. There are differences in relative magnitudes of the low and high temperature relaxation peaks. These results suggest that prolonged exposure selectively affects the poly (vinyl acetate) rich phase, with much less impact on the polyethylene rich phase. This is due to the progress of EVA crosslinking reaction such as temperature increase by long-term exposure. We found also, that the aged EVA after exposure showed considerable decrease in current intensity for the high temperature polarization due to secondary melting peaks as it will be revealed by DSC technique.

TSC peak parameters according to Fig. 11 are illustrated in Tables 1 and 2. The difference of the magnitude of peak current by TSC technique is presumably attributed to post-crosslinking and recrystallization, occurring selectively in the high temperature phase as a result of outdoor exposure under different climatic parameters.

For the determination of the characteristic parameters of the process i.e. the activation energy and the relaxation time factor, the initial rise method is mostly used. This method is based on the fact that for restricted conditions [15], the current density during a thermal stimulated measurement can be expressed by Arrhenius equation. From TSDC curves in Fig. 12, under polarization temperature of 0 °C, the

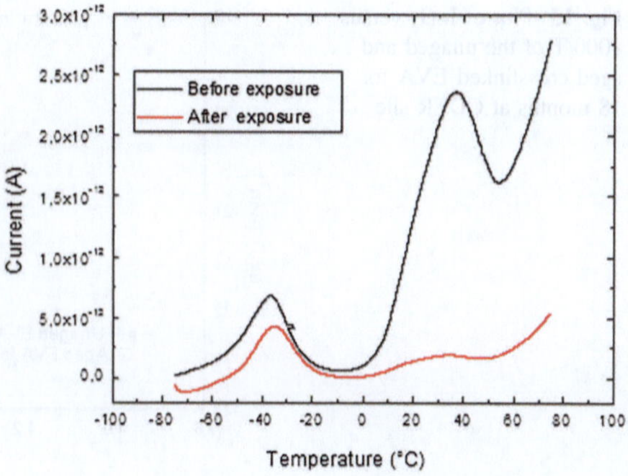

Fig. 11 TSDC spectra of unaged and aged crosslinked EVA under polarization temperature of 60 °C at URAER site

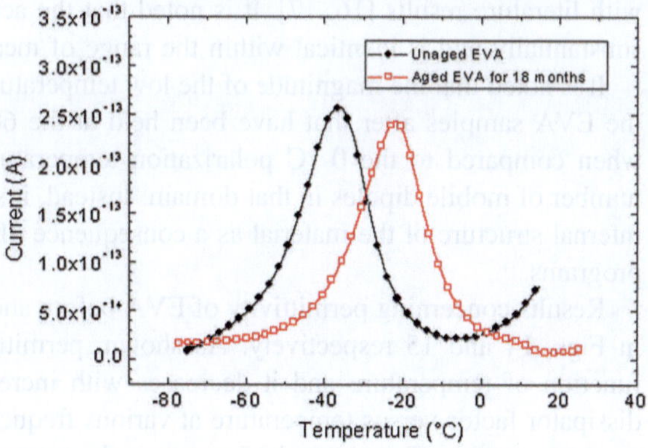

Fig. 12 TSDC curves of unaged and aged EVA for 18 months under polarization temperature of 0 °C at CDER site

Table 1 Peak positions at low temperature at $T_p = 60$ °C

Sample	Low temperature (°C)	Thermo-stimulated current (A)
Before exposure	−37	6.8×10^{-13}
After exposure	−35	4.2×10^{-13}

Table 2 Peak positions at high temperature at $T_p = 60$ °C

Sample	High temperature (°C)	Thermo-stimulated current (A)
Before exposure	36	23.5×10^{-13}
After exposure	32	2.3×10^{-13}

activation energy can be obtained from the plot of ln(J) versus $1000/T_m$ as seen in Fig. 13. The activation energy of the unaged and aged EVA for 18 months is calculated to be 33.5 and 33.8 kJ/mol, where the corresponding relaxation frequency is 4.9×10^{-3} and 5.5×10^{-3} Hz respectively, which is in good agreement

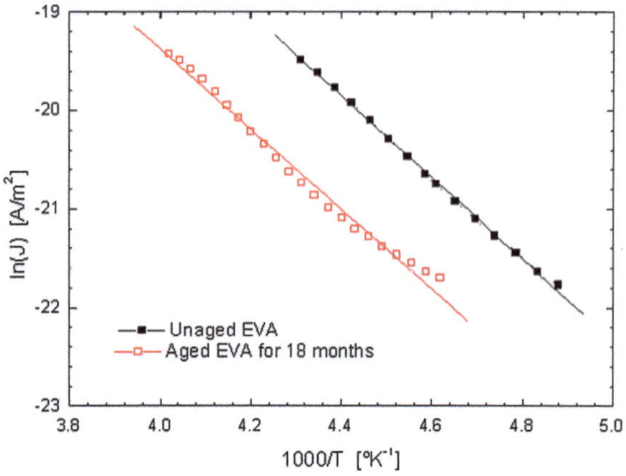

Fig. 13 Plot of ln(J) versus 1000/T of the unaged and aged crosslinked EVA for 18 months at CDER site

with literature results [16, 17]. It is noted that the activation energy is not affected substantially and is identical within the range of measurement.

It is noted that the magnitude of the low temperature relaxation peak is larger for the EVA samples after that have been held at the 60 °C polarization temperature when compared to the 0 °C polarization temperature, due to an increase in the number of mobile dipoles in that domain. Instead, it seems to reflect changes in the internal structure of the material as a consequence of the two different temperature programs.

Results concerning permittivity of EVA before and after exposure are illustrated in Figs. 14 and 15 respectively. As shown, permittivity of EVA is a non-linear function of temperature and it decreases with increasing frequency. The plot of dissipator factor versus temperature at various frequency before and after exposure as represented in Figs. 16 and 17 respectively, shows that EVA exhibits relaxation phenomenon. It is found that the relaxation behavior changes with frequency. At frequency of 1 Hz the glass transition temperature of the crosslinked EVA is found

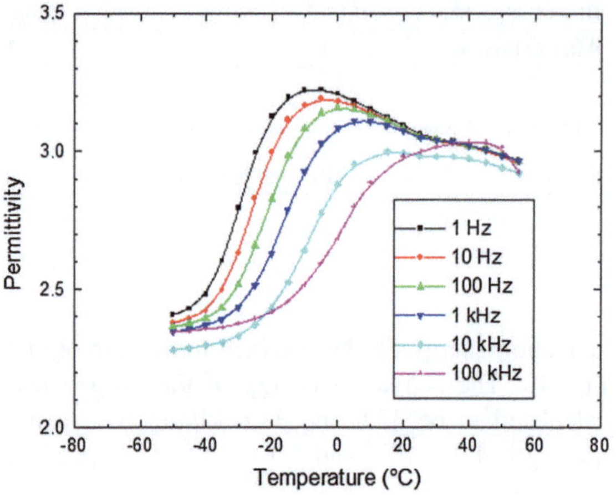

Fig. 14 Variation of the permittivity of cured EVA before exposure as function of temperature URAER site

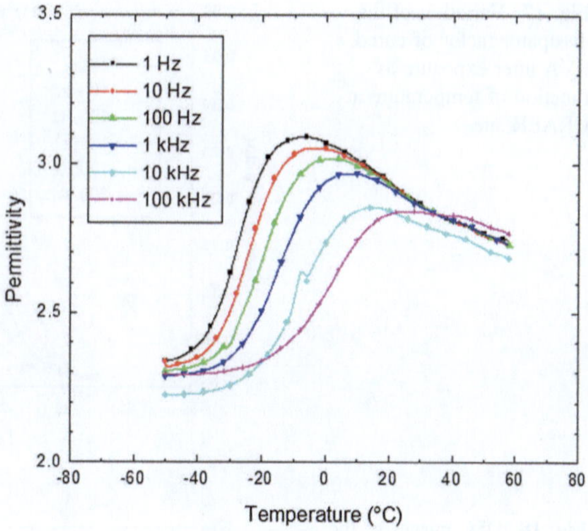

Fig. 15 Variation of the permittivity of cured EVA after exposure as function of temperature at URAER site

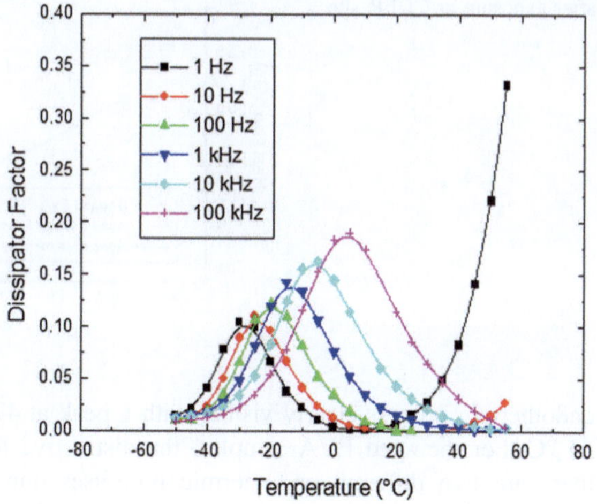

Fig. 16 Variation of the dissipator factor of cured EVA before exposure as function of temperature at URAER site

to be −28.7 °C. EVA after exposure is able to maintain dielectric constant and low dissipator factor over a wide temperature range. These low values are normally considered satisfactory for insulating material in the photovoltaic module encapsulation process.

4.2 Effect of Outdoor Exposure on Endothermic Peaks

Figure 18 shows the DSC experiments, in first heating mode, of the EVA samples before and after aging. On first heating, the unaged EVA sample show one

Fig. 17 Variation of the dissipator factor of cured EVA after exposure as function of temperature at URAER site

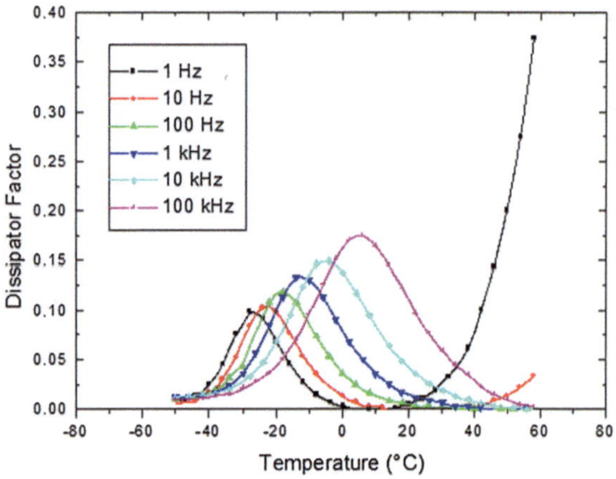

Fig. 18 DSC curves of the crosslinked EVA before and after exposure at CDER site

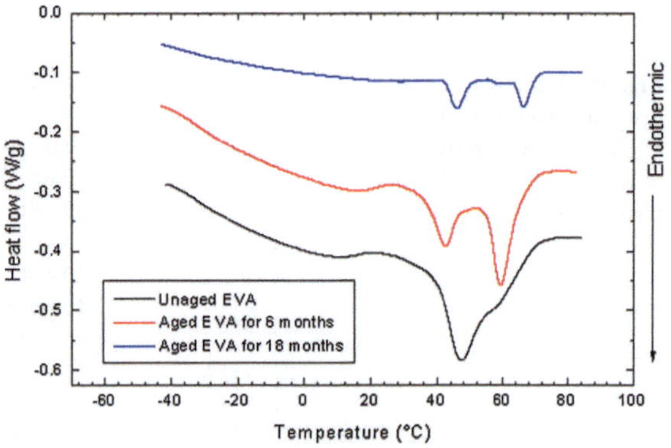

endothermic process clearly visible with a peak at 47.5 °C and with a shoulder at 55 °C. For the aged EVA samples, the distinctive feature of these results is that there are two different endothermic processes due to the recrystallization phenomenon. There is a low temperature event with peaks at 43 and 46 °C, and a higher temperature endothermic process clearly visible with peaks at 59 and 66 °C for EVA samples aged during 6 and 18 months respectively. This dual endotherm behavior in outdoor exposure is similar to that observed for EVA samples aged under accelerated test as high-UV conditions as described in IEC 61215. These peaks can be assigned to secondary crystallization due to the photo-degradation [17].

For EVA aged during 6 months, the integrated enthalpy for the low and high temperature peak is calculated to be 3.3 and 7.3 J/g respectively. This represents an increase of the low temperature endotherm by a factor of 2. Indeed, the magnitude of the integrated enthalpy for the low and high temperature peak decreases as the progress of the exposure time. There are examples of this dual endotherm behavior that have been published in the technical literature [18, 19]. The literature discussion explains this as the consequence of two populations of crystalline perfection. The low temperature endotherm is described as the melting of a population of imperfect, smaller crystallites, while the high temperature endotherm is the result of melting of larger, more regularly formed crystallites which occur in the area between the primary crystals during outdoor exposure. These additional crystals restrain the mobility of the polymer chains and therefore it can be assumed that the recrystallization is directly responsible for the decrease in thermal stimulated current as revealed by TSC technique. The imperfect crystals are the result of incorporation of branching and vinyl acetate comonomers into the polyethylene crystal lattice.

In order to examine the effect of outdoor exposure on crystallinity of EVA, the degree of crystallinity was calculated via the total enthalpy method [20] according to Eq. (8):

$$\chi_c(\%) = \frac{\Delta H_m}{\Delta H_{100}} 100\,\% \qquad (8)$$

where: χ_C is the degree of crystallinity, ΔH_m is the specific enthalpy of melting of the sample studied and ΔH_{100} is the specific enthalpy of melting for 100 % crystalline Polyethylene (288 J/g). Since the previous thermal history of a polymer affects the calculated degree of crystallinity, EVA crystallinity was evaluated after being subjected to a second heating designed to remove the effect of prior thermal history [21, 22]. The degree of crystallinity of EVA before and after environmental exposure for 6 months is calculated to be 5 and 2 % respectively.

It is noted that the degree of crystallinity does not change significantly after outdoor exposure. This result is in good agreement with X-ray diffraction. Indeed, Fig. 19 shows the X-ray diffraction patterns of the cured EVA before and after exposure. The X-ray patterns represent a broad peak than the crystalline polymer in general. Only a broad peak was observed at about 20° of its 2θ angle [23]. The high

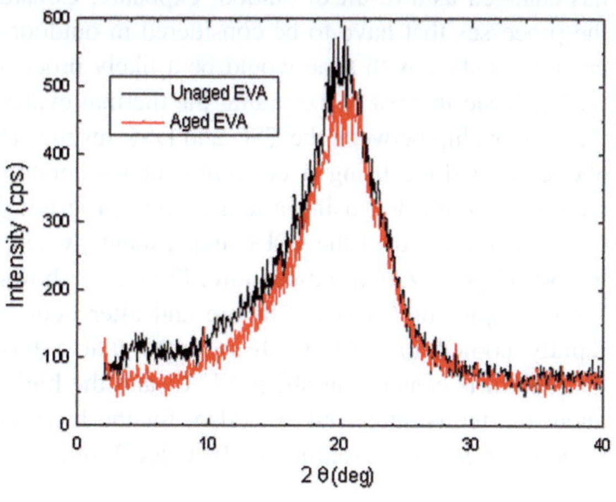

Fig. 19 X-ray diffraction of cured EVA before and after exposure for 6 months at CDER site

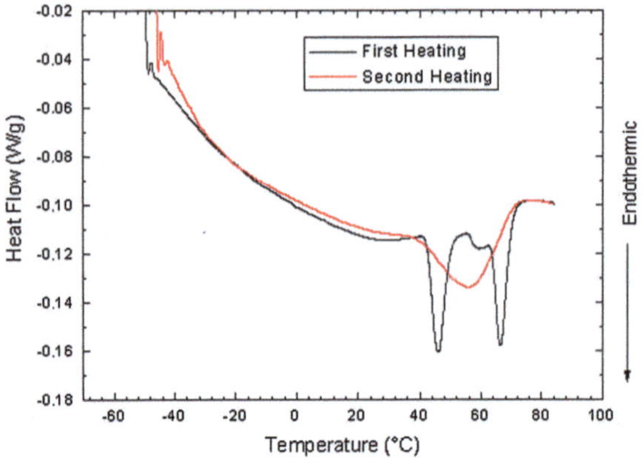

Fig. 20 DSC curves of the crosslinked EVA aged for 18 months at CDER site in heat-cool-heat mode with expanded heating cycles

angle peak is the reflection of amorphous polymer phase. The internal structure of the EVA exhibits a certain level of the organization, but this organization does not seem to have the three-dimensional characteristic periodicity of the typical crystal structure. It should be more likely to consider as a complex semi-crystalline system caused by the coexistence of two polymers, the Poly (Ethylene) and Poly (Vinyl-Acetate) with a block structure having different densities [24].

Figure 20 shows the heat-cool-heat data profile on the DSC instrument for crosslinked EVA sample aged during 18 months, where the heating cycles are expanded. The first heating shows the dual endotherm behavior previously noted. During the rapid cooling a large exothermic peak appears at about 36 °C due to exothermic crystallization. After cooling, the second heating is designed to remove the possible thermal history of the sample and also it ensures good contact with the DSC pan to enable good data acquisition to be taken. On the second heating, the dual exotherm is no longer observed; instead, there is a very broad endotherm with a peak at 55 °C and with a shoulder at 73 °C. It is noted that spans to higher temperature than the unaged EVA due to the fact that the structure of the material has changed as a result of outdoor exposure. Considering that annealing is one of the processes that have to be considered in outdoor exposure, the development of crystal structure with time would be a likely process to consider as important.

The basic interest in extending the thermal evaluation of EVA, was to examine the relationship between the TSC and DSC results. The DSC experiments described above were done using a conventional temperature program. DSC experiments were also conducted using a temperature program that simulated the temperature cycle used to conduct the TSDC experiments, where the EVA sample were held at the 60 °C polarization temperature. Figure 21 shows the DSC profiles of the aged EVA sample for 6 months before and after being held at 60 °C for 2 min and rapidly cooled to −50 °C. It is noted that a broad, shallow low temperature endotherm is centered at about 47 °C, and the high temperature peak is shifted to about the temperature, 64 °C. Also, for the high temperature peak the integrated peak is 7.1 J/g, after holding at 60 °C for 2 min, the integrated peak is 1.3 J/g. It is

Fig. 21 DSC curves of the crosslinked EVA before and after exposure for 6 months at CDER site after holding at 60 °C for 2 min

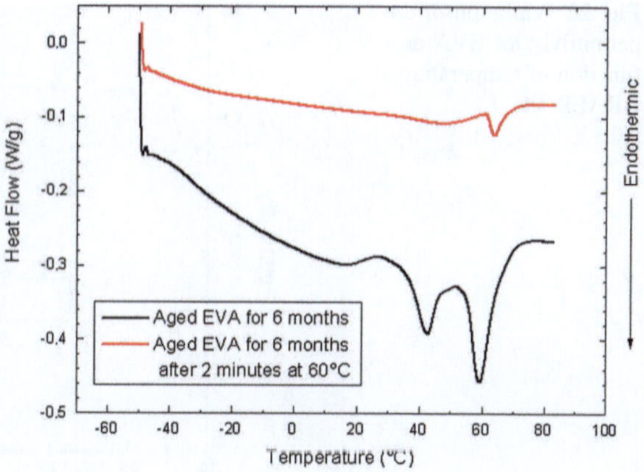

clear that in addition to being shifted to higher temperature, the high temperature peak is reduced in magnitude.

It appears that holding the specimen at 60 °C for 2 min melts out the entire low temperature, poorly formed crystal population. It also partially melts out a substantial fraction of the high temperature, well formed crystal population. A higher melting fraction of the well formed population remains and may even be annealed to a higher melting temperature. Rapid cooling to −50 °C does not allow the original crystallinity to reform; so that only a shallow, broad endotherm representing a small fraction of poorly formed crystals can be observed.

4.3 Effect of Outdoor Exposure on Dielectric Properties of EVA at Fixed Frequency

We have investigated the dielectric behaviour of EVA during outdoor exposure in the range 20–70 °C at fixed frequency of 50 Hz according to experimental set-up illustrated in Fig. 21. It is noted that, the dielectric constant of EVA is quite stable over the entire temperature range as illustrated in Fig. 22. In addition, Fig. 23 shows that the dissipator factor increases with increasing temperature from 30 to 70 °C. As shown in Fig. 24, the resistivity decreases with increasing temperature i.e. the EVA has a negative resistivity temperature coefficient.

The experimental measurement results show that EVA exhibits very high resistivity. This provides the high electrical insulation which is needed to guarantee circuit isolation from environmental damages. On the other hand, the cured EVA exhibits lower dielectric constant and dissipator factor than the original EVA. Low values of dissipator factor indicate low conversion of electrical energy to heat energy. EVA is able to maintain dielectric constant and low dissipator factor over a wide temperature range. These low values are normally considered satisfactory for insulating material in the PV module encapsulation process.

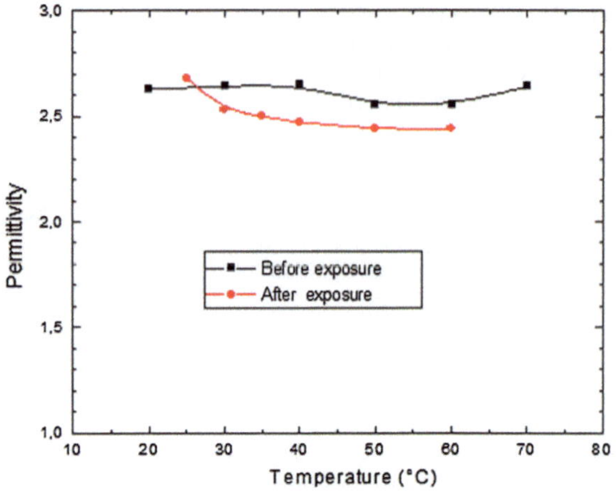

Fig. 22 Variation of the permittivity of EVA as function of temperature at URAER site

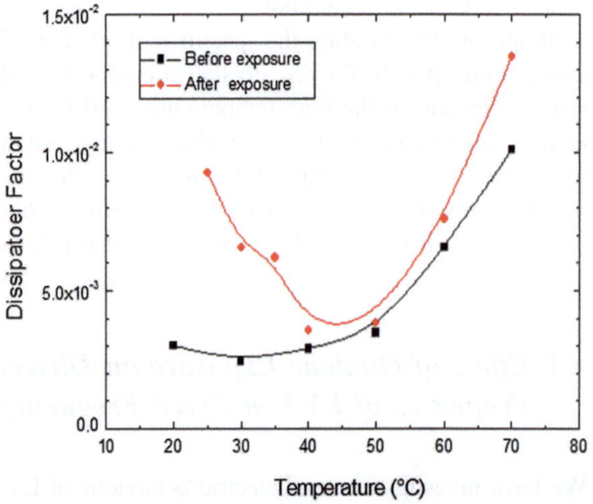

Fig. 23 Variation of the dissipater factor of EVA as function of temperature at URAER site

Experimentally, the temperature dependence of resistivity is given by Arrhenius model as described in Eq. (9):

$$\rho = \rho_\infty \cdot \exp\left(\frac{-E_a}{RT}\right) \tag{9}$$

where:

ρ: resistivity at temperature T; ρ_{∞}: resistivity at temperature T_∞; Ea (J/mol): Activation energy; T(°K): absolute temperature; R: Gas constant (8.314 J/mol °K).

A plot of Log ρ versus 1/T, will yield a straight line with a slope of −Ea/R as shown in Eq. (10):

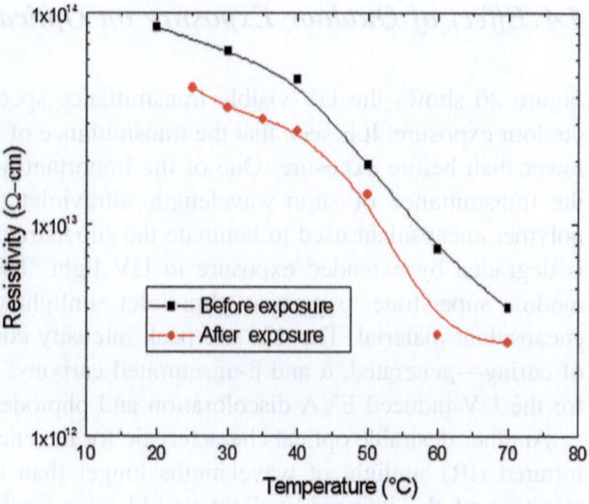

Fig. 24 Variation of the resistivity of EVA as function of temperature at URAER site

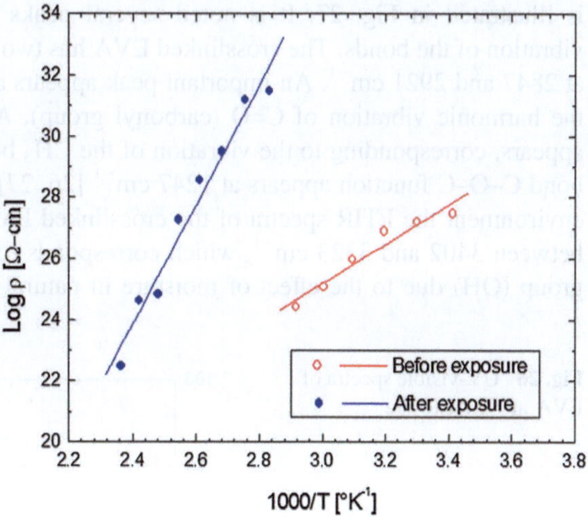

Fig. 25 Plot of ln(J) versus 1/T for cured EVA before and after exposure at URAER site

$$\mathrm{Log}\rho = \frac{-E_a}{R}\frac{1}{T} + \ln \rho_\infty \qquad (10)$$

The activation energy of the crosslinked EVA before and after exposure is calculated to be 52.8 and 162.2 kJ/mol respectively according to Fig. 25. After exposure in desert climate, the activation energy of EVA is very high due to environment weathering and time exposure duration.

4.4 Effect of Outdoor Exposure on Optical Properties

Figure 26 shows the UV-visible transmittance spectra for EVA before and after outdoor exposure. It is seen that the transmittance of EVA after outdoor exposure is lower than before exposure. One of the important properties of the encapsulant is the transmittance of short-wavelength ultraviolet (UV) sunlight. Typically, the polymer encapsulant used to laminate the superstrate to the solar cells in a module is degraded by extended exposure to UV light. Therefore, it is desirable for the module superstrate to prevent ultraviolet sunlight (<400 nm) from reaching the encapsulant material. The 420 nm peak intensity corresponds to the concentration of curing—generated, α and β-unsaturated carbonyl groups, which are responsible for the UV-induced EVA discoloration and photodegradation [25].

Another desirable optical characteristic for the encapsulant would be rejection of infrared (IR) sunlight of wavelengths longer than usable by the solar cells. The rejection of the infrared sunlight would raise module performance and increase module lifetime by reducing operating temperatures. However, a cost-effective method for rejecting the infrared heat has not been developed.

FTIR spectrum in ATR mode of the crosslinked EVA before and after exposure is illustrated in Fig. 27. It is noted several peaks related to different modes of vibration of the bonds. The crosslinked EVA has two peaks due to CH bond located at 2847 and 2921 cm^{-1}. An important peak appears at 1736 cm^{-1} corresponding to the harmonic vibration of C=O (carbonyl group). At 1372 cm^{-1} an intense peak appears, corresponding to the vibration of the CH_3 bond. The vibration of the ester bond C–O–C function appears at 1247 cm^{-1} [26, 27]. After exposure in the natural environment the FTIR spectra of the crosslinked EVA has a relatively broad peak between 3402 and 3323 cm^{-1}, which corresponds to the vibration of the hydroxyl group (OH) due to the effect of moisture in natural exposure. Nevertheless, other

Fig. 26 UV-visible spectra of EVA at URAER site

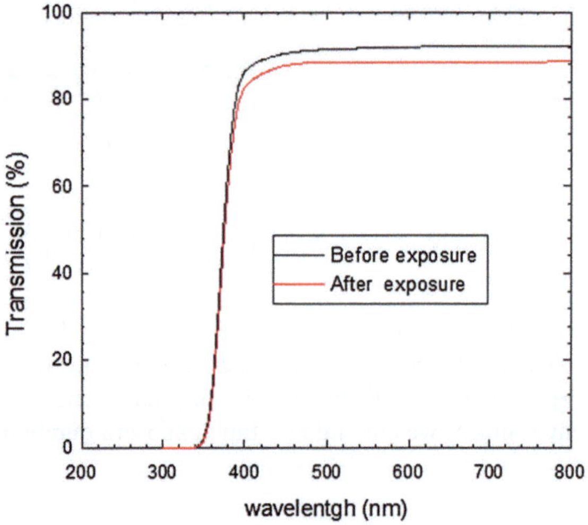

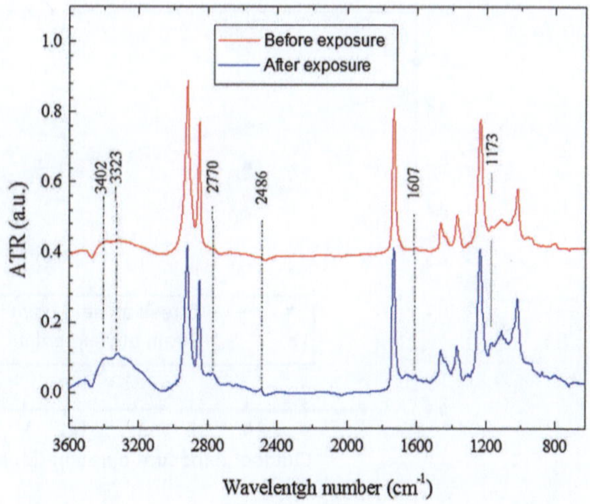

Fig. 27 UV-visible spectra of EVA at URAER site

low intensity peaks appear at around 2270 and 1173 cm^{-1} without major significant changes leading to the perfect overlapping of the two spectra.

4.5 Effect of Outdoor Exposure on Thermo-mechanical Properties

Regarding tensile testing, the unaged crosslinked EVA material exhibits a high ductile behavior with strain at break values of about 565 %, the stress at break is found to be 13.6 MPa as described in the technical guide. The ultimate mechanical properties of the crosslinked EVA before and after exposure are shown in Fig. 28. It is noted that the crosslinked EVA exhibits a significant change in strain and stress at yield point which are very sensitive to changes in molecular mass of the polymer and therefore sensitive to chemical aging. Also, this behavior can mainly be attributed to the recrystallization as revealed by DSC. For all investigated polymer films, a clear influence of the weathering stress factors like UV radiation, temperature and humidity on the ultimate mechanical properties of the crosslinked EVA was observed.

The DMTA curves represented by the storage modulus (E′) and loss tangent (tanδ) of the EVA before and after exposure are illustrated in Figs. 29 and 30 respectively. DMTA analysis revealed that the storage modulus before and after exposure increases from 536 to 1350 MPa respectively at Tg. In the same process, the loss tangent curve exhibits two relaxation peaks located at different temperatures due to the two distinct drops in modulus compared with the TSC thermogram, the low temperature drop is accompanied by a peak in the tanδ. This low temperature mechanical loss process is taken to correspond to the low temperature relaxation peak observed in the TSDC.

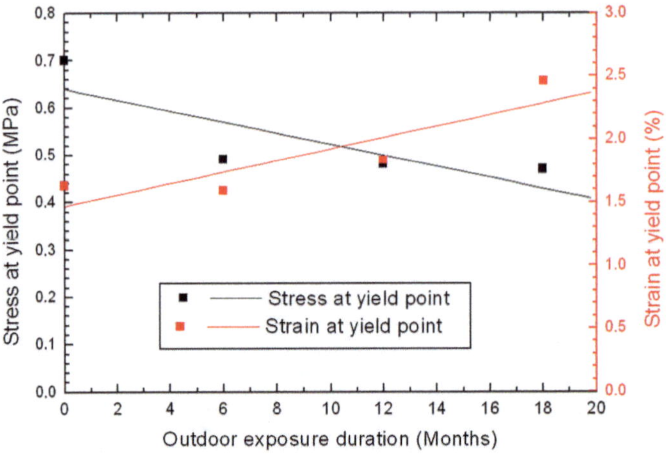

Fig. 28 Evolution of stress and strain at yield point of the crosslinked EVA before and after exposure at CDER site

Fig. 29 Storage modulus of the crosslinked EVA before and after exposure at CDER site

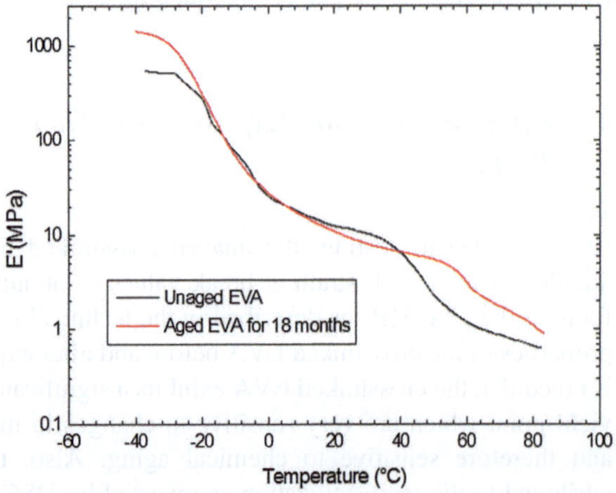

After long-term operation under desert climate at URAER site, some severely degraded PV modules were characterized [28, 29]. EVA samples taken from the front of degraded modules, continuously exposed to light and higher temperature, become opaque as shown in Fig. 31a. Their mechanical properties are greatly changed, such as reduced elongation compared with EVA samples taken from the back, not directly exposed to light and were at lower temperature than the front side as shown in Fig. 31b. It was surmised that cells in modules having large series resistance, or other defects that produced hot spots, could locally heat EVA leading to severe delamination [30].

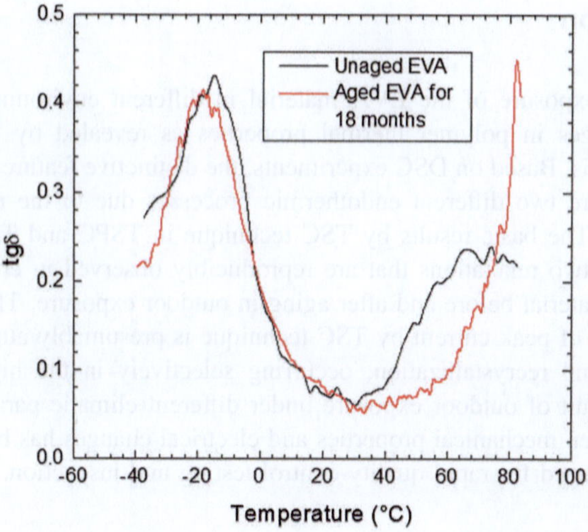

Fig. 30 Loss factor of the crosslinked EVA before and after exposure at CDER site

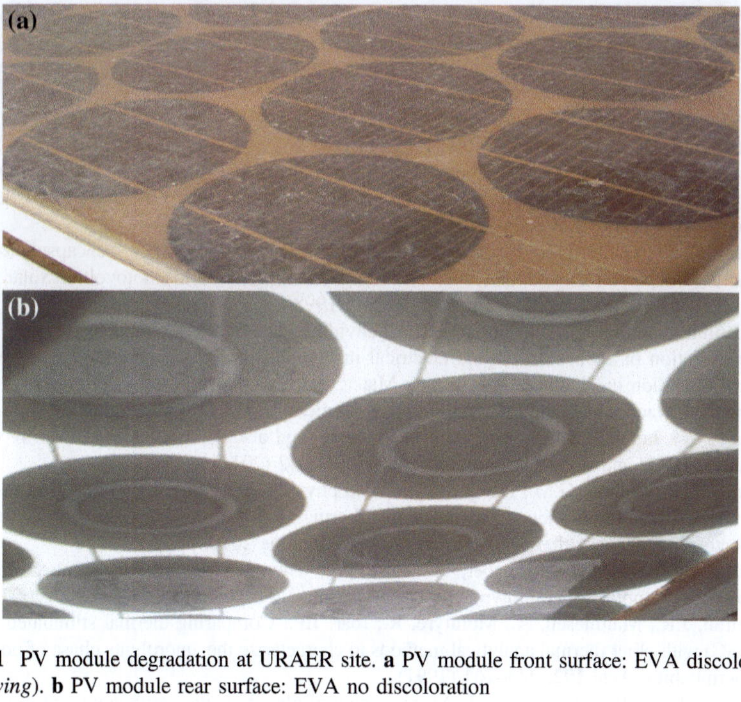

Fig. 31 PV module degradation at URAER site. **a** PV module front surface: EVA discoloration (*yellowing*). **b** PV module rear surface: EVA no discoloration

5 Conclusion

The outdoor exposure of the EVA material in different environmental exposure leads to changes in polymer thermal properties as revealed by TSC and DSC thermal analysis. Based on DSC experiments, the distinctive feature of these results is that there are two different endothermic processes due to the recrystallization phenomenon. The basic results by TSC technique in TSPC and TSDC modes, is that there are two relaxations that are reproducibly observed in crosslinked EVA encapsulant material before and after aging in outdoor exposure. The difference of the magnitude of peak current by TSC technique is presumably attributed to post-crosslinking and recrystallization, occurring selectively in the high temperature phase as a result of outdoor exposure under different climatic parameters. A correlation between mechanical properties and electrical changes has been performed, which can be used for rapid quality-control testing and inspection.

References

1. Schlothauer, J., Jungwirtha, S., Kohl, M., Roder, B.: Degradation of the encapsulant polymer in outdoor weathered photovoltaic modules: spatially resolved inspection of EVA ageing by fluorescence and correlation to electroluminescence. Sol. Energy Mater. Sol. Cells **102**, 75–85 (2012)
2. Wohlgemuth, J.H., Cunningham, D.W., Monus, P., Miller, J., Nguyen, A.: Long term reliability of photovoltaic modules. In: Proceedings of the 4th World Conference on PV Energy Conversion (2006)
3. Herrmann, W., Bogdanski, N., Reil, F., Köhl, M., Weiss, K.-A., Assmus, M., Heck, M.: PV module degradation caused by thermo-mechanical stress: real impacts of outdoor weathering versus accelerated testing in the laboratory. In: Proceedings of the SPIE 7773 (2010). http://dx.doi.org/10.1117/12. 859809
4. Dhere, N.G., Gadre, K.S.: Comparison of mechanical properties of EVA encapsulant in new and field-deployed PV modules. In: Proceedings of the 2nd World Photovoltaic Solar Energy Conference and Exhibition, Vienna, 6–10 July 1998
5. Dechthummarong, C., Wiengmoon, B., Chenvidhya, D., Jivacate, C., Kirtikara, K.: Physical deterioration of encapsulation and electrical insulation properties of PV modules after long-term operation in Thailand. Sol. Energy Mater. Sol. Cells **94**(9), 1437–1440 (2010)
6. Standard practice for outdoor weathering of plastics. ASTM D1435–05, 1985
7. Stark, W., Jaunich, M.: Investigation of ethylenevinyl acetate copolymer (EVA) by thermal analysis DSC and DMA. Polym. Test **30**(2), 236–242 (2011)
8. Oreski, G., Wallner, G.M.: Damp heat induced physical aging of PV encapsulation materials. In: 12th IEEE Intersociety Conference on Thermal and Thermo-Mechanical Phenomena in Electronic Systems (Itherm 2010), Las Vegas, 2–5 June 2010, pp. 1–6
9. Collins, G., Yoo, S.U., Recber, A., Jaffe, M.: Thermal analysis of complex relaxation processes in poly (desaminotyrosyl-tyrosine arylates). Polymer **48**(4), 975–988 (2007)
10. Saffell, J.R., Matthiesen, A., McIntyre, R., Ibar, J.P.: Comparing thermal stimulated current (TSC) with other thermal analytical methods to characterize the amorphous phase of polymers. Thermochim. Acta **192**, 243–264 (1991)
11. Agroui, K., Collins, G., Farenc, J.: Measurement of glass transition temperature of crosslinked EVA encapsulant by thermal analysis for photovoltaic application. Renew. Energy **43**, 218–223 (2012)

12. Czanderna, A.W., Pern, F.J.: Encapsulation of PV modules using ethylene vinyl acetate copolymer as a pottant: a critical review. Sol. Energy Mater. Sol. Cells **43**, 101–181 (1996)
13. Agroui, K., Segui, Y., Farenc, J., Benrekaa, N.: Characterisation of EVA encapsulant by glass transition temperature method for photovoltaic application. In: 1st International Conference on Energy (IEC 2000), University of El-Ain (UAE) 7–9 May 2000
14. Agroui, K.: Experimental aging study of EVA in outdoor exposure, internal report UDTS/2011
15. Agroui, K., Collins, G.: Thermal relaxations and transitions in EVA encapsulant material during photovoltaic module encapsulation process, in materials and processes for energy: communicating current research and technological developments. In: Méndez-Vilas, A. (ed.) Formatex Research Center, ISBN: 978-84-939843-7-3, pp. 150–157 (2013) http://www.formatex.info/energymaterialsbook/book/150-157.pdf
16. Agroui, K., Collins, G.: Characterization of EVA by thermally stimulated current technique. Sol. Energy Mater. Sol. Cells **80**(2), 33–45 (2003)
17. Bregulla, M., Köhl, M., Lampe, B., Oreski, G., Philipp, D., Wallner, G., Andersnd Weiß, K.: Degradation mechanisms of ethylene vinyl acetate copolymer: new studies including ultra fast cure foils. In: Proceedings of the 22nd European Photovoltaic Solar Energy Conference, Milan, 3–7 Sept 2007, pp. 2704–2707
18. Bistac, S., Kunemannand, P., Schultz, J.: Crystalline modifications of ethylene-vinyl acetate copolymers induced by a tensile drawing: effect of the molecular weight. Polymer **39**(20), 4875–4881 (1998)
19. Tsocheva, D., Tsanovand, T., Terlemezyan, L.: Structure of composite films containing polyaniline studied by DSC. J. Therm. Anal. Calorim. **66**(2), 415–422 (2001)
20. Reyes-Labarta, J.A., Olaya, M.M., Marcilla, A.: DSC and TGA study of the transitions involved in the thermal treatment of binary mixtures of PE and EVA copolymer with a crosslinking agent. Polymer **47**, 8194–8202 (2006)
21. McEvoy, R.L., Krause, S., Wut, P.: Surface characterization of ethylene vinyl acetate (EVA) and ethylene-acrylic acid (EAA) co-polymers using XPS and AFM. Polymer **39**(21), 5223–5239 (1998)
22. Brogly, M., Nardin, M., Schultz, J.: Effect of vinyl acetate content on crystallinity and second-order transitions in ethylene—vinyl acetate copolymers. J. Appl. Polym. Sci. **64**(10), 1903–1912 (1998)
23. Sung, Y.T., Kum, C.K., Lee, H.S., Kim, J.S., Yoon, H.G., Kim, W.N.: Effects of crystallinity and crosslinking on the thermal and rheological properties of ethylene vinyl acetate copolymer. Polymer **46**(25), 11844–11848 (2005)
24. Agroui, K., Belghachi, A., Collins, G., Farenc, J.: Quality control of EVA during photovoltaic module encapsulation process. Desalination **209**, 1–9 (2007)
25. Pern, F.J.: Ethylene-vinyl acetate (EVA) encapsulants for PV modules: degradation and discoloration mechanisms and formulation modifications for improved photostability. Sol. Energy Mater. Sol. Cells **252**, 195–216 (1997)
26. Allen, N.S., et al.: Aspects of the thermal oxidation of ethylene vinyl acetate copolymer. Polym. Degrad. Stab. **68**(3), 363–371 (2000)
27. Giurgincaa, M., Popab, L., Zaharescuc, T.: Thermo-oxidative degradation and radio-processing of ethylene vinyl acetate elastomers. Polym. Degrad. Stab. **82**(3), 463–466 (2003)
28. Agroui, K., Hadj Mahammed, I., Hadj Arab, A., Belghachi, A.: Characterization photovoltaic modules based on thin films solar cells in environmental operating conditions of Algerian Sahara. Proc. SPIE **7048**, 70480Q.1–70480Q.8 (2008)
29. Sadok, M., Mehdaoui, A.: Outdoor testing of photovoltaic arrays in the Saharan region. Renew. Energy **33**(12), 2516–2524 (2008)
30. Wiengmoon, B., Sangpongsanont, Y., Jivacate, C., Chenvidhya, D., Kirtikara, K.: Determination of PV module deterioration based on physical properties investigation of EVA. In: Proceedings of the 21st European Photovoltaic Solar Energy Conference, pp. 2514–2516 (2006)

Thermal Degradation of Bio-nanocomposites

Kieran A. Murray, John A. Killion, Ian Major and Luke M. Geever

Abstract Bio-nanocomposites have attracted a great deal of attention over the last number of years due to the excellent characteristics the material has to offer. With ever increasing demands of environmental controls, more sustainable materials like bio-nanocomposites are required to substitute the various petropolymers utilised nowadays. These bio-based polymers provide exceptional performance and have smart properties that have proven useful to the food packaging industry and a wide range of other applications. This chapter reviews the recent developments of bio-nanocomposites where the related biodegradable polymers include Polylactic acid (PLA), polycaprolactone (PCL), polyhydroxyvalerate (PHV), polyhydroxyalkanoates (PHAs), polyhydroxybutyrate (PHB), poly(3-hydroxybutyrate-co-3-hydroxyvalerate) (PHBV) and poly(d,l-lactide) (PDLLA). A concise history outlining the development of bio-nanocomposites materials is explored, while the importance of environmental conditions and in particular the rate of biodegradability is highlighted. Furthermore, this chapter addresses the steps of thermal degradation and the systematic approaches used to overcome these concerns. It discusses the behaviour of various nanoparticles on the thermal stability of biopolymers and other topics related to research challenges, future trends and applications.

K.A. Murray (✉) · J.A. Killion · I. Major · L.M. Geever
Applied Polymer Technologies, Athlone Institute of Technology,
Dublin Road, Athlone, Co. Westmeath, Ireland
e-mail: kmurray@research.ait.ie

J.A. Killion
e-mail: jkillion@research.ait.ie

I. Major
e-mail: imajor@ait.ie

L.M. Geever
e-mail: lgeever@ait.ie

K.A. Murray
Material Research Institute, Athlone Institute of Technology,
Dublin Road, Athlone, Co. Westmeath, Ireland

P.M. Visakh and Y. Arao (eds.), *Thermal Degradation of Polymer Blends, Composites and Nanocomposites*, Engineering Materials,
DOI 10.1007/978-3-319-03464-5_9

1 Introduction

Each year approximately 140 million tons of petroleum based synthetic polymers are developed worldwide, where large quantities of various polymers are introduced into the ecosystem as industrial waste products [1]. Due to the environmental concerns of synthetic polymers in terms of environmental pollution, depletion of fossil resources and greenhouse gas emissions, alternative polymers derived from renewable resources are necessary to replace these microbial resistant materials [2, 3], especially for use in short term packaging and disposable applications. These renewable biological resource polymers are usually called biopolymers, which are excellent vehicles for adding a wide range of additives such as antimicrobials, antioxidants, antifungal agents, colour and other nutrients [3, 4]. Under suitable conditions of temperature, moisture and oxygen availability, biodegradation can initiate fragmentation or disintegration of the materials with no toxic or environmentally harmful residues [5]. Biopolymers can be broken up into various categories which are based on the type of manufacturing process and the source of raw materials. These categories include natural biopolymers, synthetic biodegradable polymers and biopolymers developed by microbial fermentation [6].

Biopolymers such as aliphatic polyesters are now produced on a semi-commercial scale by numerous companies that make biodegradable plastics [7]. Aliphatic polyesters have repeating units that are bonded via ester linkages and these natural esters can be degraded by enzymes (esterases) that are ubiquitous in nature [8]. Polylactic acid (PLA), polyglycolate (PGA), poly(3-hydroxybutyrate), polycaprolactone (PCL), polyhydroxyvalerate (PHV) and their copolymers poly(butylene succinate) (PBSu), poly(ethylene succinate) (PESu), poly(propylene adipate) (PPAd), etc., are the most commonly used aliphatic polyesters for such applications as packaging materials, mulch films, tissue engineering, implants, drug delivery etc. [9]. However, aliphatic polyesters exhibit relatively poor thermal stability, mechanical strength, solvent resistance, gas permeability, hydrophobic character and slower resorption/degradation kinetics as compared to other polymers which have strongly limited their applications [10, 11]. Pandey et al. [12] and Rhim et al. [6] mentioned that the problems related to biopolymers are threefold: performance, processing and cost. These issues are similar to some degree, where the problems related to performance and processing are common to all biodegradable polymers irrespective of their origin.

In order to make biopolymers more attractive for commercial end-use applications, it is necessary to enhance the thermomechanical properties through the utilisation of nanocomposite technology. In the last decade, polymer nanocomposites have become a global research interest, due to the superior advances in polymer characteristics by the addition of nano-scale materials (preferably in the range 10–100 nm) into the polymer matrix [13]. Due to the nanoscopic dimensions and extreme aspect ratios of the nanofillers, this results in six interrelated characteristics, which defines the type of nanocomposite. These are (1) low-percolation threshold (~0.1–2 vol%), (2) particle to particle correlation (orientation and position) arising at low volume fractions ($\Phi c < 0.001$), (3) high density of particles per particle volume (10^6–10^8 particles/μm^3), (4) extensive interfacial area per volume of particles (10–50 nm at $\Phi \sim 1$–8 vol%), and (6) comparable size scales among the rigid nanoparticles inclusion, distance between particles, and the relaxation volume of polymer chains [13, 14]. One of the key advantages of nanofillers is the miscibility with the polymer matrix due to the low size, which provides unique synergism between the combined materials. Nanofillers can be classified into three different categories depending on the dimensions of the dispersed particles [15]:

(a) Nanolayers: these are a one dimensional layered material with typical dimensions of 1 nm thick and an aspect ratio following their two remaining dimensions of at least 25. They will usually stack on top of each other to form tactoids and these tactoids can contain up to hundreds of layers of platelets. In this category, layered silicate include sematic clays, layered double hydroxides and grapheme sheets.
(b) Nanotubes: these are two dimensional with a diameter below 100 nm and an aspect ratio of at least 100. This category includes nanocellulose subtracts carbon nanotubes silica and titanium nanotubes.
(c) Nanoparticles: these have three dimensions below 100 nm and are sometimes called isodimensional nanoparticles. The most popular nanofillers in this category include, silica particles, polyhedral oligomeric silesquioxane and metal oxides.

These nanocomposites can be further classified into the following areas:
(1) clayed based nanocomposites, (2) nanocellulose based nanocomposites, (3) carbonaceous based nanocomposites, (4) metal/metallic hydroxide based nanocomposites and silicon based nanocomposites. Table 1 displays each of the

Table 1 Biopolymers and inorganic or antimicrobial materials typically used for the preparation of bio-nanocomposites [6]

Biodegradable polymers	Antimicrobial materials
Starch or thermoplastic starch (TPS)	Clay (e.g. Montmorillonite (MMT))
Chemically modified cellulose (e.g., cellulose acetate (CA) and cellulose acetate butyrate (CAB))	Organically modified nanoclay (quaternary ammonium modified MMT, Ag-zeolite)
Polylactic acid (PLA)	Metal ions (e.g. silver, copper, gold, platinum)
Polycaprolactone (PCL)	Metal oxide (e.g. TiO_2, ZnO, MgO)
Polyhydroxyalkanoate (PHA)	Natural biopolymers (e.g. chitosan)
Polyhydroxybutyrate (PHB)	Natural antimicrobial agents (e.g. nisin, thymol, carvacrol, isothiocyanate, antibiotics)
Poly(butylene succinate) (PBS)	Enzymes (peroxidase, lysozyme)
	Synthetic antimicrobial agents (quaternary ammonium salts, EDTA, propionic acid, benzoic acid, sorbic acid)

biopolymers and inorganic or antimicrobial materials typically used for the preparation of bio-nanocomposites.

Clay based nanocomposites have received remarkable interest since they were first introduced [16]. Based on the pioneered worked performed by the Toyota's research team [17], widespread research has been carried out by various industrial and academic groups [6, 18, 19]. Among the various nanocomposites available, clay based nanocomposites would be regarded as the most popular for the improvement of various biopolymer characteristics, such as the mechanical and thermal properties. Montmorillonite is a semantic clay which has weak bonding between its layers; where each layer consists of two sheets of silica tetrahedral with an edge shared octahedral sheet of alumina as demonstrated (layer thickness of around 1 nm) Fig. 1 [20]. As a result of this isomorphic substitution of alumina between the layers, each unit cell has a negative charge between 0.5 and 1.3. These layers are held together by a layer of charge compensating ions such as Li^+, Na^+, K^+ or Ca^{2+} which are allocated within the galleries of the silicate layers [21].

Layered silicates have positive ions present on the surface which makes them hydrophilic and therefore incompatible with many polymers as they are generally hydrophobic [22]. To make bio-nanocomposites, it is necessary to modify the layered silicates by replacing the interlayer cations with cations bearing long alkyl chains, such as alkylphosphonium or alkylammonium [23]. Alkyl ammonium cations are employed to lower the surface energy and enhance the wetting characteristics with the polymer [24]. By incorporating polymers or monomers into the interlayer, this can lead to the formation of a bio-nanocomposite. Different structures can exist for polymer/clay bio-nanocomposites and they are as follows (see Fig. 2):

(a) Phase separated (Microcomposite): When a polymer is not able to intercalate with the layered silicate. Their properties remain in the same range as traditional microcomposites.
(b) Intercalated: When polymer chains penetrate in between the layered silicate which in turn increases the gallery height.

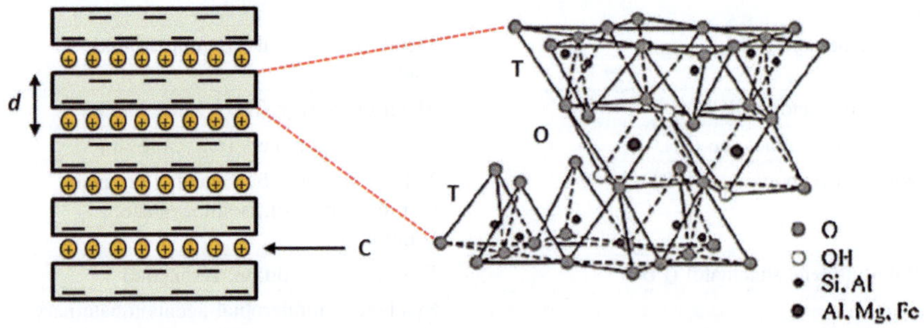

Fig. 1 Representation for the crystal structure of montmorillonite [14]. Copyright 2011. Reproduced by permission of Elsevier Science Ltd.

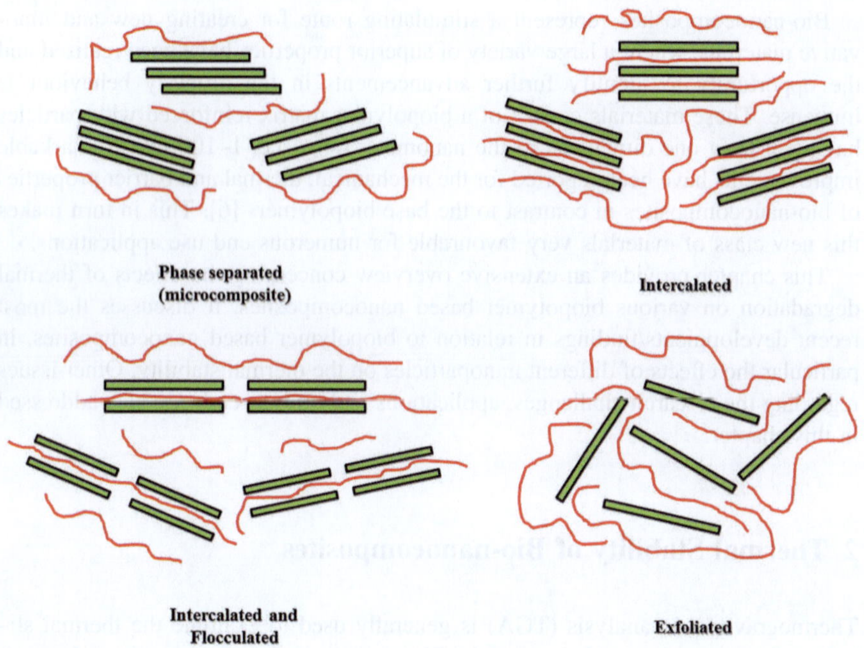

Fig. 2 Schematic representation of different structure of and morphologies of polymer/clay nanocomposites

(c) Intercalated and flocculated: Theoretically this is like intercalated nanocomposites, however; the layered silicates are sometimes flocculated by the hydroxylated edge–edge interaction of the layered silicates [13].

(d) Exfoliated: When the silicate layers are completely and homogeneously delaminated and dispersed in the polymers matrix. The best materials characteristics are achieved for exfoliated nanocomposites for most applications [25].

The effectiveness of inorganic nanoparticles as additives such as nanoclays, cellulous nanowhiskers, carbon nanotubes and ultrafine layered titanate have been used to enhance the performance of many biopolymers [13, 21, 26]. In particular, organically modified layered silicate (OMLS) nanocomposites and polymers have received a great deal of interest due to their substantial property improvement in contrast to the unmodified polymer resin. These include greater thermal and environmental stability, flammability resistance, solvent uptake and rate of biodegradability of bio-nanocomposites [26, 27]. For OMLS, property enhancements are generally obtained following the addition of low silicate content (5 wt%) compared to conventional filler loaded systems. This in turn aids in the development of a lighter composite, making it a more competitive material for specific applications. These property enhancements in contrast to the conventional composites is mainly as a result of stronger interfacial interactions between the polymer matrix and the OMLS [28].

Bio-nanocomposites represent a stimulating route for creating new and innovative materials, where a large variety of superior properties have been realised and the opportunity to identify further advancements in the property behaviour is immense. These materials consist of a biopolymer matrix reinforced with particles having at least one dimension in the nanometer range i.e. 1–100 nm. Remarkable improvements have been reported for the mechanical, thermal and barrier properties of bio-nanocomposites in contrast to the base biopolymers [6]. This in turn makes this new class of materials very favourable for numerous end use applications.

This chapter provides an extensive overview concerning the effects of thermal degradation on various biopolymer based nanocomposites. It discusses the most recent developments/findings in relation to biopolymer based nanocomposites, in particular the effects of different nanoparticles on the thermal stability. Other issues regarding the research challenges, applications and future trends are also addressed in this chapter.

2 Thermal Stability of Bio-nanocomposites

Thermogravimetric analysis (TGA) is generally used to examine the thermal stability of bio-nanocomposite materials. This technique involves continuous monitoring of the weight of a sample in a controlled environment (inert using nitrogen, oxidising using oxygen or reducing using 8–10 % hydrogen in nitrogen) as a function of temperature and/or time [29, 30]. TGA uses a controlled heating element to heat the samples while an electrobalance measures the weight on a continuous basis. During the process, the weight loss due to the formation of volatile products is plotted as a function of temperature. A mechanism for the compositional analysis of the material is generated by weight loss over the specific temperature ranges and the environments employed [31].

Aliphatic polyesters appear as a solution to the emerging environmental concerns that have arisen in recent years, however; such limitations as poor thermal and mechanical resistance limits their access to the industrial sector. By enhancing the thermomechanical properties via blending, copolymerisation and filling techniques, this could overcome such weaknesses [32]. Bio-nanocomposites represent an emerging group of hybrid materials that can provide excellent material characteristics through the incorporation of various nanofillers into the biopolymer matrix [33, 34]. Good thermal stability is one of many important properties that are essential for various industrial applications [32] and this can be achieved by a low filler concentration such as clay. Clay nanofiller acts as a superior insulator and mass transport barrier to the volatile products generated during decomposition [26]. For this reason, TGA is employed to identify the clay content of bio-nanocomposites, as clay minerals such as montmorillonite (MMT) are thermally stable up to a temperature of 900 °C [31]. There are several factors that govern the thermal stability of nanocomposite materials such as nanofiller content, intrinsic thermal resistance of polymer matrix, chemical constitution of organic modifier and chemical character of polar compatibilisers and access of oxygen to the composite material during heating [13].

The first work to report an enhancement in the thermal stability of bio-nanocomposites, combined PLA and organically modified fluorohectorite (FH) or MMT clay [35]. Bandyopadhyay et al. [35] identified that the PLA was intercalated between layers of FH or MMT clay and this provided thermal degradation resistance towards conditions that would otherwise completely degrade pure PLA. Based on the work performed by previous authors [36–38], it was observed that nanofillers significantly increase the thermal stability of PLA in contrast to microfillers under a thermooxidative environment. There have been many reports concerning about the improved thermal stability of PLA based nanocomposites prepared with numerous organically modified layers silicates (OMLS) [39–41]. Paul et al. [42] found an increase in the thermal stability of PLA nanocomposites with a maximum clay loading of 5 %. Above this loading the thermal stability decreased due to the relative extent of exfoliation/delamination in the function in the amount of OMLS.

However, other conclusions were reached by Chang et al. [40], where hexadecyl-amine-montmorillonite (C16-MMT), dodecyltrimethyl ammonium bromide-montmorillonite (DTA-MMT), and Cloisite 25A PLA nanocomposites from solution were prepared. They observed a linear decrease in the thermal stability under nitrogen when increasing the amounts of organoclay from 2 to 8 % at a 2 % weight loss when C16-MMT and Cloiste 25A hybrids were compared to virgin PLA. In addition, the temperature at 2 % weight loss in the hybrid with DTA-MMT, nearly had consistent values, irrespective of the clay loading [43].

Thermal stability of PLA containing 2 wt% Cloisite 30B and three different chain extenders: polycarbodiimide, tris (nonyl phenyl) phosphite and Joncryl ADR 4368 were investigated through rheometry and TGA which are presented in Figs. 3 and 4 [9]. From the rheological data it was revealed that PLA in the presence of Cloisite 30B had significant degradation, however; following the incorporation of chain extenders, a profound effect was noticed for an increase in the molecular

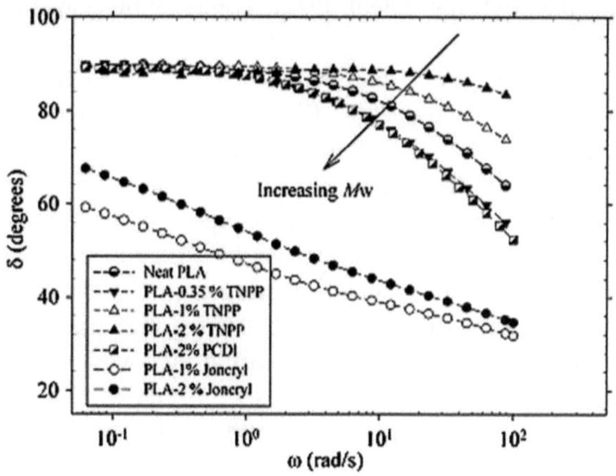

Fig. 3 Effect of molecular structure on loss angle of neat PLA and PLA containing different chain extenders (T = 190 °C) [44]. Copyright 2012. Reproduced by permission of Elsevier Science Ltd.

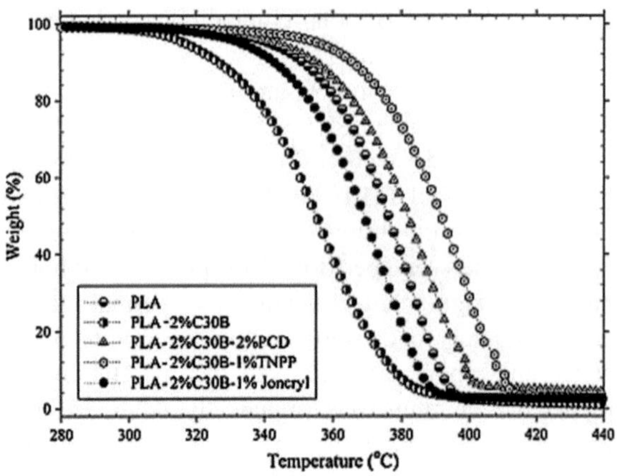

Fig. 4 Effect of clay and different chain extenders on thermal degradation of PLA nanocomposites [44]. Copyright 2012. Reproduced by permission of Elsevier Science Ltd.

weight and the control of degradation (see Fig. 3). Based on the results from TGA (Fig. 4), the addition of clay into the PLA matrix decreased the thermal stability, whereas the incorporation of a chain extender resulted in an increase in the onset temperature for thermal degradation. Joncryl was the most efficient chain extender which strongly influence the rheological properties and the control of degradation over a wide range of processing temperatures [44].

Thermal degradation of PCL in an inert atmosphere can transpire through the rupture of polyester chains via ester pyrolysis reaction with the release of CO_2, H_2O and a carboxylic acid [45]. According to Chrissafis et al. [46], organomodified MMT and nanosilica accelerated the decomposition of PCL, however; unmodified MMT and multi-walled carbon nanotubes inhibited the thermal degradation of PCL. Based on the results from the calculated activation energies through using the Ozawa, Flynn and Wall (OFW) and Friedman methods, it was determined that virgin PCL degraded by two different mechanisms. The first one corresponded to a small mass loss, whereas the second was related to the main decomposition mechanism, where a substantial mass loss took place. In addition, the nanocomposites did not affect the decomposition mechanism but only the activation energies.

From the TGA data obtained for virgin PCL and different loadings of dodecyl sulphate modified CoAl-LDH (OCoAl-LDH) in an nitrogen atmosphere, it was identified that degradation of nanocomposites occurs at lower temperatures in contrast to the virgin PCL. Furthermore, the thermal degradation temperature decreases gradually with the increase of OCoAl-LDH concentration [47]. Chen et al. [22] reported that PCL/multi-walled carbon nanotube (MWCNT) composites prepared by ultrasonically mixing the PCL and as-fabricated MWCNT in a tetrahydrofuran solution lead to better thermal stability. The activation energy of the composites was smaller than virgin PCL which can be attributed to the MWCNT loading. MWCNT loading causes a decrease in the degradation rate and an increase in the residual weight for PCL/MWCNT nanocomposites. An enhancement in the thermal stability is due to the CNTs having exceptional thermal conductivity but also due to the formation and stabilisation of MWCNTs bonded macroradicals.

It appears that no consistent conclusions are available concerning the thermal stability of nanocomposites in contrast to the parent polymer. However, Carrasco et al. [48], stated that the thermostability is determined by the nature and content of nanoclay, but in particular the method used to produce the nanocomposite as a less depolymerising process would provide better stability.

3 Thermal Degradation of Bio-nanocomposites

The different types of polymer degradation include: thermal, oxidative, biological, hydrolytic, chemical, mechanical, photoinduced and radioactive degradation. Thermal degradation of polymers can be defined as "molecular deterioration as a result of overheating" [49]. When a polymer is heated at a high temperature, i.e. above its melting temperature, chain scission will occur; the polymer chain backbone will react with other molecules in turn, changing the properties of the material. Recent studies of bio-nanocomposites have shown that the rate of thermal degradation is controlled by factors which include: molecular weight, purity, crystallinity, processing temperature, water permeability, additives, bacteria and fillers [50].

4 Different Steps in Thermal Degradation of Bio-nanocomposites

4.1 *Thermal Degradation*

The thermal degradation and stability of bio-nanocomposites have been extensively investigated via different processing techniques [51, 52]. These studies have shown that the degradation reaction occurs during the mixing of the polymer melt in the barrel of the extruder or injection moulding machine. Throughout this section, the thermal degradation steps will focus on both PLA and polyhydroxyalkanoates (PHA) based bio-nanocomposites.

The thermal degradation process can affect the chemical structure, stereo configuration, relative hydrophobicity, crystallinity, fillers and conformational flexibility contribute of the polymers. In the case of PLA, it is well known that this polymer easily degrades upon thermal processing due to the formation of various reactions. These reaction mechanisms include hydrolysis by trace amounts of water, zipper like depolymerisation, oxidation, random main-chain scission, intermolecular transesterification to monomer and oligomeric esters and intermolecular transesterification resulting in monomer and oligomer lactide of low molecular weight formation. Due to these aforementioned reactions, the molecular weight and melt viscosity of PLA can be significantly altered, resulting in a bio-nanocomposite with a reduction in mechanical and thermal properties. For example, a study by Araujo et al. [53] compared the molecular weight distribution using gel permeation chromatography (GPC) for non-degraded and degraded polylactic acid based bio-nanocomposites. Figure 5 demonstrates that the molecular weight of the samples was reduced after exposure to thermal degradation for a set period of time.

Unlike PLA, biodegradation of polyhydroxyalkanoates (PHAs) is a one stage process; where no hydrolysis is required to initiate degradation. PHAs are extremely prone to enzymatic degradation [54]. There are two main groups of PHA's, polyhydroxybutyrate (PHB) and polyhydroxyvalerate (PHV). Both PHB and PHV undergo a degradation process from the surface of the polymer resulting in chain scission, therefore; causing the polymer to erode. Hakkarainen et al. [54] reports that in co-polymers of PHB and PHV, the amorphous regions were the first to be attacked by micro-organisms, indicating a greater tendency for PHB to degrade. The rate of biodegradation in PHA is dependent on many factors including moisture level, nutrient level, temperature and the surface area exposed to the compost. The biodegradation of both PHB and PHBV were assessed by Rosa et al. [55]. Tests

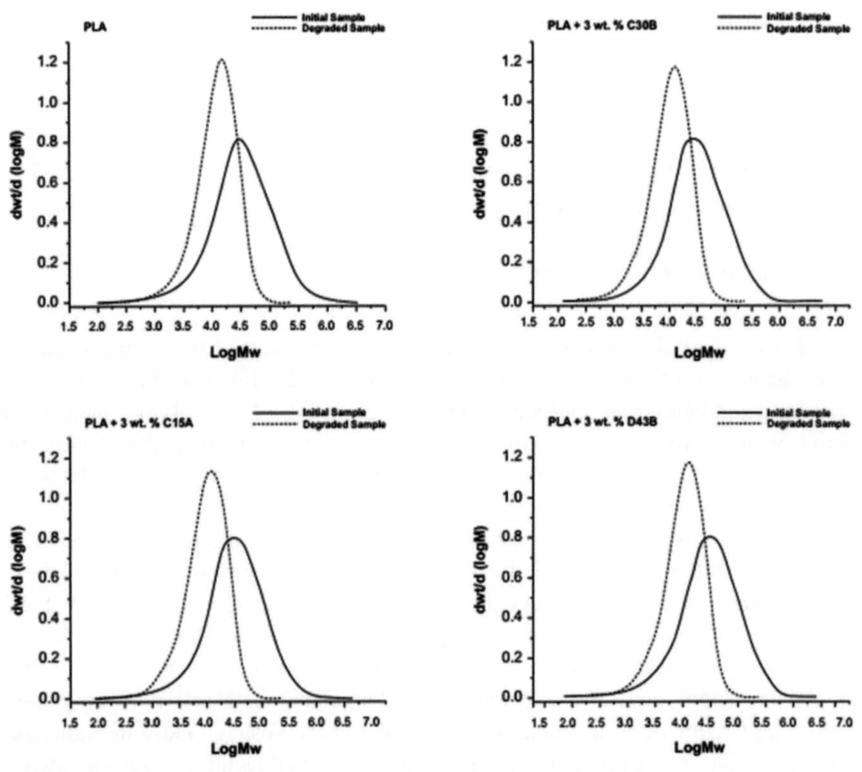

Fig. 5 Molecular weight distribution of initial and degraded samples [53]. Copyright 2014. Reproduced by permission of Elsevier Science Ltd.

were carried out at 46 and 24 °C in a soil composting medium. Both materials were assessed by examining loss of mass, tensile strength and polymer roughness. Temperatures were shown to increase the rate of the biodegradation process with both samples completely degraded after 104 days at 46 °C, while only 51 % of PHB and 56 % of PHBV were degraded after 321 days of composting at 24 °C [56]. Studies have also demonstrated that PHAs degrade in aquatic environments within 254 days, at temperatures below 6 °C [57].

4.2 Hydrolysis

Hydrolysis is a chemical reaction during which molecules of water are split into hydrogen cations and hydroxide anions. In the case of polyester based polymers, the molecular weight decrease is principally instigated by the hydrolysis of the ester linkage which randomly occurs throughout the polymer. The basic mechanism of hydrolysis for PLA is shown in Fig. 6. The amount of hydrolysis or degradation is controlled by the temperature, amount of water, acid or base catalyst and morphology of the polymer [58].

As mentioned above, temperature is a critical parameter in controlling the hydrolysis reaction. Degradation studies have been carried out at temperatures to simulate in vivo conditions ($T = 37$ °C), soil or compost conditions, ($T = 25$–60 °C) and polymer processing conditions ($T = 200$ °C) in which the biopolymers are hydrolysed to give low molecular weight water-soluble oligomers [59]. The hydrolytic degradation of bio-nanocomposites has been reported to take place mainly in the bulk of the material rather than on the surface [60] and has been assumed to be an autocatalytic hydrolysis of biopolymers which occurs homogeneously along sample cross-sections.

Figure 7 shows the TGA degradation profiles for polylactic closite bio-nanocomposites in which the samples were exposed for 8 weeks at both 37 and 58 °C. In summary, the results showed an overall increase in thermal degradation of the bio-nanocomposites with an increase in temperature. In comparing the effect of nanocomposites on the degradation profile, hydrolysis degradation was a lot more prevalent at the high temperature [59].

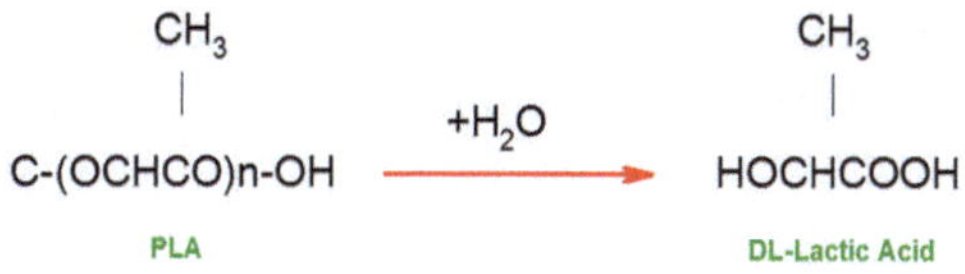

Fig. 6 Hydrolysis of polylactic acid

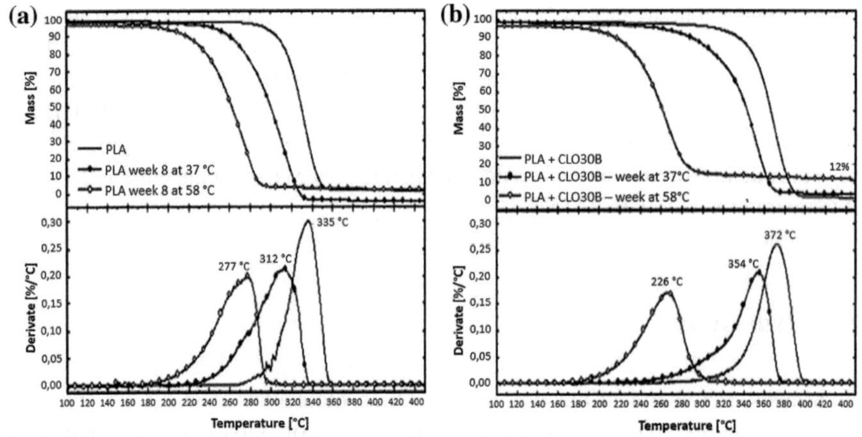

Fig. 7 TGA profile of hydrolysis degradation of PLA nanocomposites at different temperatures [59]. Copyright 2011. Reproduced by permission of Elsevier Science Ltd.

4.3 Chain-Scission

The chain-scission reaction is another common degradation pathway leading to a gradual reduction of the average molecular weight and subsequently diminished physical properties. Signori et al. [61] studied the effect of processing conditions (150–200 °C) on the thermal degradation of PLA and poly(butylene adipate-co-terephthalate) [61]. Results showed that the melt volume rate (MVR) was found to increase with the increase in processing temperature. This suggests that the increase in flow resulted in a reduction of molecular weight, i.e. chain scission. Chain scission can occur at both the middle of the repeat unit and at the chain ends [62].

4.4 Transesterification

The transesterification mechanism occurs when the carbonyl carbon of the starting ester ($RCOOR^1$) undergoes nucleophilic attack by the incoming alkoxide (R^2O^-) to give a tetrahedral intermediate, which either reverts to the starting material, or proceeds to the transesterified product ($RCOOR^2$). This is not an attractable occurrence as it increases the polydispersity and reduces the molecular weight, which affects the mechanical properties of the material.

There are two different types of transesterfication: intramolecular and intermolecular. Intramolecular transesterfication for PLA bio-nanocomposites is found to dominate at high temperatures (>200 °C) and results in the formation and degradation of cyclic polylactide oligomers. In comparison, intermolecular transesterfication affects the sequence of different polymeric segments [44]. As a result of such reactions, the molecular weight and the mechanical properties decrease. Considering that the rate of thermal degradation increases often after clay loading, controlling the thermal degradation of PLA nanocomposites is a major challenge.

5 Effect of Reinforcement on Thermal Stability on Bio-nanocomposites

Carrasco et al. [50] conducted a wide ranging study into the thermal stability of polylactic acid. The researchers studied the effects of different processing and thermal conditioning steps and of multiple polymer processing steps on thermal stability. The unprocessed PLA was more thermally stable than the processed and thermally conditioned PLA. The raw material decomposed over a narrow range (331–375 °C). All the processed and annealed samples began to decompose sooner (325 °C) and would be fully decomposed 5 mins sooner than the raw PLA. The difference in thermal stability was shown to be related to the presence of lower molecular weight PLA chains in the processed samples formed due to thermal decomposition at elevated temperatures in melt processing. A follow on study by Carrasco et al. [48] investigated organo-modified montmorillonite reinforced PLA nanocomposites (0.5 and 2.5 % loading) where the 0.5 % was exfoliated and the 2.5 % was intercalated and exfoliated. The nanoclay had only limited reinforcement effect on the PLA (no change in Young's modulus), however; it did promote plastic deformation. Processed nanocomposite samples decomposed sooner than the processed PLA, however; the nanocomposite samples took longer to fully decompose. The authors attributed the protective effects of nanoclay to the tortuous path created by the exfoliated clay hindering diffusion of volatile degradation products. The 2.5 % loaded sample was much less thermally stable due to the destabilising effect of ammonium cation present in the nanoclay. Indeed, there has only been limited success with nanoclay and PLA in the literature, due mainly to the difficulty in achieving exfoliation of the clay through melt processing. Nieddu et al. [63] investigated the effect of different nanoclays on the properties of PLA, achieving an intercalated structure for all but one of the nanoclays (Nanofil). The PLA nanocomposites had higher moduli compared to the neat PLA with maximum increase (47 %) achieved for a 10 % Somasif® MEE loading. These reinforced PLA samples showed increased thermal stability; the reinforced samples had both a 20 °C higher onset decomposition temperature and an 18 °C higher total decomposition temperature.

Starch containing organo-modified sepiolite provided for a considerable increase (164 %) in Young's modulus for a 6 % loaded sample [64]. This large reinforcement effect had no bearing on thermal stability; instead increasing organo-modified nanoclay loading actually reduced the total decomposition temperature by 9 °C at up to 6 % loading. Reduction in thermal stability had a linear relationship to increasing the cationic starch content that had been used to modify the sepiolite. Unmodified sepiolite provided for the opposite trend and increasing nanoparticle loading produced a 3 and 8 °C increase in the total decomposition temperature for the 3 and 6 % sepiolite loading, respectively. The authors speculated that the increase in thermal stability was due to the interplay of two actions—the retardation of volatile degradation compounds by extensive molecular interactions, via physisorption and/or chemisorption, occurring at the polymer/sepolite interface and the sepiolite forming a protective inorganic layer at the surface of the starch. The unmodified sepiolite did reinforce the starch but to a much less of an extent than the cationic starch modified sepiolite, with Young's modulus only increasing 77 % for the 6 % loading. Another study investigated the montmorillonite (MMT) based starch nanocomposites. The authors reported that MMT had a reinforcement effect which increased with increasing loading, with 198 % increase in Young's modulus observed for the 5 % loading. Concurrently, MMT also enhanced the thermal stability of starch. The onset decomposition temperature of starch was 223 °C and the addition of 5 % MMT increased this temperature by 19 °C (242 °C), indicating an improvement in the stability. However, this increase in thermal stability is probably, as stated previously, a factor of both the tortuous path created by the exfoliated nanoclay and the thermal barrier presented by the nanoclay layers.

Bruzaud and Bourmaud [65] investigated a poly(3-hydroxybutyrate-co-3-hydroxyvalerate) (PHBV) polymer reinforced with an organo-modified montmorillonite. The nanoclay provided a 65, 107 and 165 % increase in Young's modulus for 1, 2.5 and 5 % loading, respectively. Thermal stability increased with increasing nanoclay loading, a 30 °C improvement was obtained with a 5 % loading of nanoclay in the 50 % decomposition temperature. The nanoclay acts as a heat barrier retarding the diffusion of oxygen and volatile degradation products. Cellulose nanowhiskers reinforced PHBV showed the opposite trend with reinforcement [66]. Young's modulus increased with an increase in loading, where the 5 % loading increased the stiffness by 77 %. The thermal stability of the neat PHBV was higher compared to the sample loaded with 5 % cellulose nanofiber. Addition of the nanoparticle reduced the total decomposition temperature by 5 °C. In essence this reduction in thermal stability is due to the cellulose based nanoparticles acting as a thermal conductor as opposed to the nanoclay which acted as a thermal barrier. Taken as a review of the work in the literature, it must be concluded that thermal stability is less dependent on reinforcement, but is in fact much more dependent on properties of the actual nanoparticle during decomposition.

6 Effect of Nanoparticle on Thermal Stability on Bio-nanocomposites

In the previous section, it was outlined how the thermal stability of bio-nanocomposites is more related to the type and to the modification of the nanoparticle dispersed within the biopolymer, than to do with any reinforcement effect attributable to the nanoparticle. Different nanoparticles have different effects on the thermal stability of biopolymers. Nanoclay can greatly enhance the thermal stability of a number of different biopolymers—PLA [48, 63], PCL [67], starch [68, 69] and PHBV [65]. As stated above this stabilisation effect is related to the nanolayers acting as a thermal barrier slowing the diffusion of volatile compounds through the polymer matrix.

For cellulose based nanoparticles there is a mixed effect on the thermal stability of biopolymers. A study by de Paula et al. [70] investigated the effects of incorporating cellulose nanowhiskers in poly (d,l-lactide). These nanoparticles significantly increased the thermal stability of biopolymers, where the onset decomposition temperature increased by 12 and 26 °C for 1 and 5 % loadings, respectively. Cellulose nanowhiskers were shown to have negligible effect on the thermal stability of PLA [71]. Wang et al. [72] investigated the cellulose nanowhisker nanocomposites based on a novel poly (propylene carbonate) (PPC) synthesised by copolymerisation of propylene oxide and carbon dioxide. The addition of cellulose nanowhiskers significantly increased thermal stability of the biodegradable polymer. From up to 5 % loading of the nanoparticle, the onset decomposition temperature of the PPC increased by 20 °C. After this point additional cellulose nanowhiskers had less of a stabilisation effect on the PPC and 10 % loading had no effect on the onset decomposition temperature. Identical trends were observed for the total decomposition temperature.

Carbon nanotubes (CNTs) do not greatly affect the thermal stability of biopolymers. Single wall CNTs did change the thermal decomposition temperatures of PCL [73]. In the same study, silver nanoparticles significantly destabilised PCL reducing both the onset and total decomposition temperatures by 50 °C. The thermal stability hybrid film containing both CNTs and silver nanoparticles was similar to the binary film containing only silver nanoparticles, reflecting that the CNTs did not act to stabilise the system. A study incorporating multi-wall CNTs in polylactic acid also reported negative effects on thermal stability [74]. In air, the onset decomposition temperature of the bio-nanocomposite material significantly decreased with increased loading, while under nitrogen the decrease in the onset decomposition temperature only occurred for in excess of 2 % loading. The destabilisation mechanism was speculated to be different under the two conditions. In the presence of oxygen CNTs have a catalysing effect for decomposition, while under nitrogen and in excess of 2 % the CNTs for a percolating network acts as a conducting pathway for assisting decomposition reactions.

7 Recent Studies of Bio-nanocomposites

Recently bio-nanocomposites have attracted considerable attention since a minute amount of filler, i.e. nanoclay can significantly enhance the mechanical properties, dimensional stability, flame retardancy, gas barrier properties, chemical resistance, surface appearance, electrical conductivity and thermal stability [75, 76]. These bio-nanocomposites have been extensively studied in areas such as food packaging, automotive and tissue engineering. These studies report on the measuring of the bio-nanocomposites thermal stability and/or thermal degradation by thermogravimetric analysis (TGA), rheological studies, x-ray diffraction, gel permeation

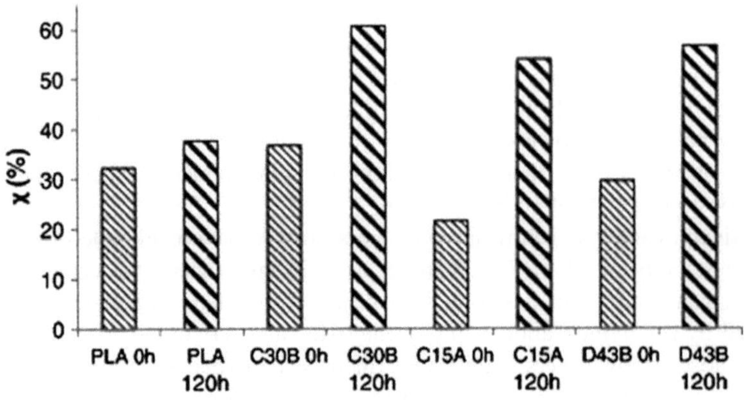

Fig. 8 Crystallinity degree of PLA bio-nanocomposites before and after thermal degradation [53]. Copyright 2014. Reproduced by permission of Elsevier Science Ltd.

chromatography (GPC), etc. Examples of recent related work on bio-nanocomposites based on various aliphatic polyesters are given below.

Araújo et al. [53] investigated the effect of clay organic modifier on the thermal-stability of PLA based bio-nanocomposites. From the results illustrated in Fig. 8, the incorporation of composites (C30B, C15A and D43B) stemmed in an increase in crystallinity for all samples. The reason for the higher degree of crystallinity was due to its increase in hydrophobicity of the nanoclays, which enhances the chain scission of PLA chains within the bio-nanocomposites. This property improves the dispersion of composites, since PLA is a hydrophobic-polymer due to the presence of CH_3 side groups.

Authors have reported the thermal degradation behaviour of polycaprolactone (PCL) bio-nanocomposites [47, 77]. A study by Chrissafis et al. [46] investigated the thermal behaviour of modified and unmodified nanocomposites on PCL. They revealed that both unmodified montmorillonite and multiwalled carbon nanotubes inhibited the thermal degradation of the bio-nanocomposites. On the other hand, organically modified montmorillonite and nanosilica increased the rate of degradation of PCL bio-nanocomposites (Fig. 9).

Liu et al. [78] investigated the thermal/oxidative degradation properties of PDLLA bio-nanocomposites. They discovered that the degradation process of the bio-nanocomposites was due to random chain scission mechanisms which included two or three stages [78]. In stage one the reaction was controlled by the oligomers containing carboxylic acid and hydroxyl groups. Subsequent to stage one, the oligomers were consumed and oxygen had a promoting effect on the thermo-oxidative properties (stage two). In the final stage it was observed when the bio-nanocomposite was degraded under nitrogen over 200 °C, it resulted in the appearance of carboxylic acids. This shows the importance of removing the low molecular oligomers and the significance of polymer protection from the atmosphere when bio-nanocomposites are in a melt.

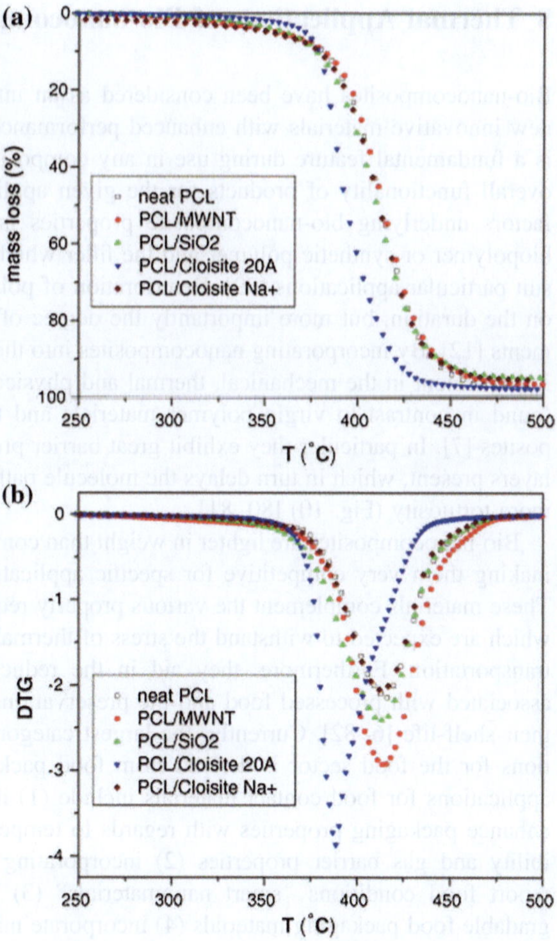

Fig. 9 Thermogravimetric analysis of silica, modified and unmodified montmorillonite [46]. Copyright 2007. Reproduced by permission of Elsevier Science Ltd.

In the literature some authors suggest that clays significantly improves the PLA thermal stability under thermo-oxidative conditions, i.e. nitrogen environment [36]. However, a portion of the literature suggests the thermal stability under nitrogen increased linearly as the quantity of clay reduced from 2 to 8 wt% [53]. Pluta et al. [36] studied the thermal properties of PLA bio-nanocomposites and microcomposites by TGA under helium and oxygen environments. Samples were prepared by melt blending and quench processing. Under both environments, nano and micro composites did not alter the polymer thermal degradation. However, when comparing virgin PLA, microcomposites and nanocomposites, the thermal stability for the nanocomposite considerably improved under thermo-oxidative conditions.

8 Thermal Applications of Bio-nanocomposites

Bio-nanocomposites have been considered as an interesting route for developing new innovative materials with enhanced performance characteristics. Performance is a fundamental feature during use in any composite material which decides the overall functionality of products for the given applications. One of the principle factors underlying bio-nanocomposite properties is the interaction between the biopolymer or synthetic polymer and the filler which can be tuned or perfected to suit particular applications [79]. Deterioration of polymeric materials is dependent on the duration, but more importantly the degree of exposure to various environments [12]. By incorporating nanocomposites into the polymer matrix, a significant improvement in the mechanical, thermal and physicochemical properties has been found in contrast to virgin polymer materials and their conventional microcomposites [7]. In particular they exhibit great barrier properties as a result of the clay layers present, which in turn delays the molecule pathway giving the diffusion path more tortuosity (Fig. 10) [80, 81].

Bio-nanocomposites are lighter in weight than conventional composite materials, making them very competitive for specific applications such as food packaging. These materials complement the various property requirements for food packaging which are expected to withstand the stress of thermal food processing, storage and transportation. Furthermore, they aid in the reduction of the packaging waste associated with processed food and the preservation of packaged foods extending their shelf-life [6, 82]. Currently, the largest category of nanotechnology applications for the food sector is derived from food packaging materials. The primary applications for food contact materials include (1) incorporating nanomaterials to enhance packaging properties with regards to temperature/moisture stability, flexibility and gas barrier properties (2) incorporating nanosensors to monitor and report food conditions "smart nanomaterials" (3) "bio-nanocomposites" biodegradable food packaging materials (4) incorporate nanoparticles with antimicrobial or oxygen scavenging properties [83].

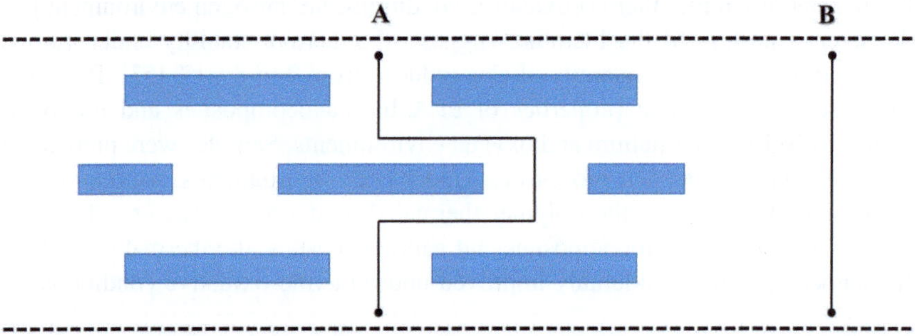

Fig. 10 Diagram demonstrating the tortuosity of a diffusing penetrant introduced on an exfoliating solid, layered in a polymer matrix. (*A*) Filled polymer and (*B*) unfilled polymer [82]. Copyright 2007. Reproduced by permission of Elsevier Science Ltd.

As bio-nanocomposites offer superior mechanical and thermal properties and are biocompatible and/or biodegradable, this renders them as one of the most versatile materials available today. For this reason, bio-nanocomposite materials are of immense interest to biomedical technologies such as controlled drug delivery, tissue engineering, bone replacement/repair and dental applications [84]. In addition, these unique nanomaterials offer excellent prospects in the development of electrochemical biosensors. The high sensitivity of these devices, together with their compatibility with modern microfabrication technologies, minimal power requirements, disposability, portability and independence of optical pathway or sample turbidity makes them exceptional for clinical diagnostics [84].

The relationship of nanofillers with PLA can improve a number of the PLA properties, however; they do not satisfy all requirements when high performances are needed in durable applications like electronics [81, 85]. Research efforts have been performed to overcome this problem by combining nanofillers with one or more nanofillers, microfillers or polymers and/or using multifunctional nanofillers [21, 86]. Murariu et al. [87] reported the synergistic combination between $CaSO_4$ filler under its β-anhydrite AII form as microfillers and some organomodified MMT as nanofillers in order to achieve flame retardant properties for specific end-use applications for resulting PLA based materials. PCL is a biodegradable polyester material that exhibits high elongation at break and low modulus, while its commercial availability makes it very attractive for commodity applications. However; due to its low melting point, PCL must be blended with other materials in order to meet the requirements of conventional applications. Cabedo et al. [88] found that by blending PCL with PLA, this lead to improvements in the mechanical properties and thermal stability without decreasing the barrier properties. In addition, they found by melt mixing PCL/PLA with kaolinite, this further improved the mechanical properties, but in particular the gas barrier characteristics. Although bio-nanocomposite materials have very promising future prospects, the low levels of production and the high costs restrains them from being used in a wide range of applications [13].

9 Future Trends

The future of bio-nanocomposites will be very much tied up in the application areas to which these materials are best suited. By far the greatest utility of these materials is within packaging. Packaging plays an essential role in society by meeting the needs of consumers for convenience and portability, as well as protecting and preserving products throughout the supply chain. Conventional plastics have been ideal to achieve these aims due to having limited interaction with food; not supporting microorganisms; being light-weight and strong enough to survive the distribution chain; and being easily formed into a wide range of structures, shapes and designs which are effective, convenient and aesthetically pleasing. The global packaging material and machinery industry generates \$500 billion yearly, accounting for between 1 and 2 % of gross domestic product in industrialised nations, according to the World Packaging Organisation [89]. There are 100,000 packaging manufacturing companies in operation, representing 5 million jobs. However, due to environmental regulations and also consumer demands there is an increasing need for more recyclable and/or biodegradable packaging. Although conventional plastic packaging is now expediently streamed for recycling and well able to tolerate extensive recycling, it is non-biodegradable and is almost exclusively sourced from non-renewable fossil fuels. Biopolymers would be the perfect replacement for conventional plastics as they are derived from sustainable resources. Over the last decade, improvements in the performance of biodegradable packaging make them a match of conventional plastic packaging in some niche areas. However, more technological development is required to extend the utility of biopolymers in packaging applications. Nanotechnology is one of the key methods to provide increased product functions as the addition of nanoparticles can provide packaging with improved performance and smarter properties. Nanoclay technology has been deployed in biopolymer based packaging to improve the mechanical performance and the barrier properties to oxygen, water vapour and carbon dioxide.

Active and smart packaging solutions are rapidly being developed for conventional plastic packaging based on nanotechnology. Most active packaging technologies are aimed at extending the shelf-life of food. The shelf-life of food is determined by such environmental factors as relative humidity, pH, temperature, UV light and gas composition. These factors can alter taste, quality and safety of the food through chemical modification, biochemical change or microbiological spoilage. Nanotechnology can be incorporated into plastic packaging to overcome these shortcomings and improve food safety. The majority of active packaging techniques are concerned with substances that absorb oxygen, ethylene, moisture, carbon dioxide, flavours/odours and those which release carbon dioxide, antimicrobial agents, antioxidants and flavours. Therefore, the future of bio-nanocomposites will be shaped by two complimentary needs—the requirement to match the properties of conventional plastic packaging and the demand for smarter packaging. The effect of incorporating novel nanoparticles on the thermal stability of biopolymers will need to be extensively studied.

10 Conclusion

Even though bio-nanocomposites have strong future prospects, they present low levels of production and high costs limit them from a wide range of applications. Based on the discussions throughout this chapter, there have been different and sometimes contradictory results concerning the effects of nanoparticles on the biopolymer thermal degradation and stability. The largest discrepancy was found for clay nanoparticles like montmorillonite, which are generally used for property improvement of biopolymers. Depending on the chemical structure of the biopolymer and the interaction of it with the various nanoparticles, this can decide the effect it has on the thermal stability and the different stages of thermal degradation.

This is due to the difficulty of the clay plates dispersing in the polymer matrix where the untreated clays form microcomposites instead of nanocomposites. Improvements in the thermal stability can be as a result of (1) the absorption of gas formed in the clay plates (2) the improvement in barrier properties (3) the large surface volume (4) low permeability and a reduction in the rate of evolution of the formed volatile products and (5) the development of high performance carbonaceous silicate char on the surface of the nanoparticles, therefore insulating the underline material while reducing the escape of volatiles formed during decomposition [14]. When a low concentration of clay nanoparticles is added to the polymer matrix they are well dispersed and the barrier effect is predominant. However, if the concentration is increased, degradation rapidly rises and becomes impressible, hence, leading to a decrease in the thermal stability of the nanocomposite [14]. Carbon nanotubes also demonstrated an increase in the thermal stability of polymers due to their excellent thermal conductivity and the development and stabilisation of multi-walled carbon nanotube-bonded macroradicals. Overall, this chapter has shown that the addition of nanoparticles can provide thermal stabilisation of polymers during decomposition, except for montmorillonite clays. Successful thermal stability enhancement generally takes place where a low concentration (4–5 wt%) of nanoparticles is used. Above this threshold, the thermal stability can deteriorate significantly due the formation of aggregates which forms macrocomposites instead of nanocomposites. At the moment the level of improvement for biopolymers is not enough to compete with petroleum based polymers. Further enhancements of bio-nanocomposites is required, for example, developing optimum formulations for individual biopolymers and processing techniques to promote the desirability to meet a diverse range of end use applications and also to reduce the cost of bio-nanocomposites.

References

1. Shah, A.A., Hasan, F., Hameed, A., Ahmed, S.: Biological degradation of plastics: a comprehensive review. Biotechnol. Adv. **26**, 246–265 (2008)
2. Jenck, J.F., Agterberg, F., Droescher, M.J.: Products and processes for a sustainable chemical industry: a review of achievements and prospects. Green Chem. **6**, 544–556 (2004)
3. Kümmerer, K.: Sustainable from the very beginning: rational design of molecules by life cycle engineering as an important approach for green pharmacy and green chemistry. Green Chem. **9**, 899–907 (2007)
4. Clarinval, A.M., Halleux, J.: Classification of biodegradable polymers, pp. 3–31. CRC Press, Boca Raton (2005)
5. Chandra, R., Rustgi, R.: Biodegradable polymers. Prog. Polym. Sci. **23**, 1273–1335 (1998)
6. Rhim, J.W., Park, H.M., Ha, C.S.: Bio-nanocomposites for food packaging applications. Prog. Polym. Sci. **38**, 1629–1652 (2013)
7. Bikiaris, D.N.: Nanocomposites of aliphatic polyesters: an overview of the effect of different nanofillers on enzymatic hydrolysis and biodegradation of polyesters. Polym. Degrad. Stab. **98**, 1908–1928 (2013)
8. Shimao, M.: Biodegradation of plastics. Curr. Opin. Biotechnol. **12**, 242–247 (2001)

9. Gandini, A.: Polymers from renewable resources: a challenge for the future of macromolecular materials. Macromolecules **41**, 9491–9504 (2008)
10. Koh, H.C., Park, J.S., Jeong, M.A., Hwang, H.Y., Hong, Y.T., Ha, S.Y., et al.: Preparation and gas permeation properties of biodegradable polymer/layered silicate nanocomposite membranes. Desalination **233**, 201–209 (2008)
11. Trznadel, M.: Biodegradable polymer materials. Int Polym Sci Technol. **22**, 58–65 (1995)
12. Pandey, J.K., Raghunatha R.K., Pratheep K.A., Singh, R.P.: An overview on the degradability of polymer nanocomposites. Polym Degrad Stab. **88**, pp. 234−50 (2005)
13. Kumar, A.P., Depan, D., Singh Tomer, N., Singh, R.P.: Nanoscale particles for polymer degradation and stabilization-trends and future perspectives. Prog in Polym Sci (Oxford) **34**, 479–515 (2009)
14. Chrissafis, K., Bikiaris, D.: Can nanoparticles really enhance thermal stability of polymers? Part I: an overview on thermal decomposition of addition polymers. Thermochim. Acta **523**, 1–24 (2011)
15. Herron, N., Thorn, D.L.: Nanoparticles: uses and relationships to molecular cluster compounds. Adv. Mater. **10**, 1173–1184 (1998)
16. Carter, L.W., Hendricks, J.G., Bolley, D.S.: Elastomer reinforced with a modified clay. Google patents (1950)
17. Deguchi, R., Nishio, T., Okada, A.: Polyamide composite material and method for preparing the same. Google patents (1992)
18. Wang, Y., Chen, F.-B., Li, Y.-C., Wu, K.-C.: Melt processing of polypropylene/clay nanocomposites modified with maleated polypropylene compatibilizers. Compos. B Eng. **35**, 111–124 (2004)
19. Hasegawa, N., Kawasumi, M., Kato, M., Usuki, A., Okada, A.: Preparation and mechanical properties of polypropylene-clay hybrids using a maleic anhydride-modified polypropylene oligomer. J. Appl. Polym. Sci. **67**, 87–92 (1998)
20. Liang, Z.M., Yin, J.: Poly(etherimide)/montmorillonite nanocomposites prepared by melt intercalation. J. Appl. Polym. Sci. **90**, 1857–1863 (2003)
21. Raquez, J.M., Habibi, Y., Murariu, M., Dubois, P.: Polylactide (PLA)-based nanocomposites. Prog. Polym. Sci. **38**, 1504–1542 (2013)
22. Bafna, A., Beaucage, G., Mirabella, F., Mehta, S.: 3D hierarchical orientation in polymer–clay nanocomposite films. Polymer **44**, 1103–1115 (2003)
23. Alexandre, M., Dubois, P.: Polymer-layered silicate nanocomposites: preparation, properties and uses of a new class of materials. Mater Sci Eng. R Reports **28**, 1–63 (2000)
24. Giannelis, E.P., Krishnamoorti, R., Manias, E.: Polymer-silicate nanocomposites: model systems for confined polymers and polymer brushes. Adv. Polym. Sci. **138**, 108−147 (1999)
25. Dennis, H.R., Hunter, D.L., Chang, D., Kim, S., White, J.L., Cho, J.W., et al.: Effect of melt processing conditions on the extent of exfoliation in organoclay-based nanocomposites. Polym. **42**, 9513–9522 (2001)
26. Sinha Ray, S., Bousmina, M.: Biodegradable polymers and their layered silicate nanocomposites: in greening the 21st century materials world. Prog. Mater Sci. **50**, 962–1079 (2005)
27. Sinha Ray, S., Okamoto, M.: Polymer/layered silicate nanocomposites: a review from preparation to processing. Prog. Polym. Sci. **28**, 1539–1641 (2003)
28. Chen, J.-S., Poliks, M.D., Ober, C.K., Zhang, Y., Wiesner, U., Giannelis, E.: Study of the interlayer expansion mechanism and thermal–mechanical properties of surface-initiated epoxy nanocomposites. Polymer **43**, 4895–4904 (2002)
29. Prime, R.B., Bair, H.E., Vyazovkin, S., Gallagher, P.K., Riga, A.: Thermogravimetric analysis (TGA). In: Menczel, D.J., Prime, B.R. (eds.) Thermal Analysis of Polymers, p. 241. Wiley, Hoboken (2009)
30. Earnest, C.M., Compositional Analysis by Thermogravimetry: ASTM International (1988)
31. Kumar, P.: Development of Bio-nanocomposite Films with Enhance Mechanical and Barrier Properties using Extrusion Processing. North Carolina State University, Raleigh (2009)

32. Bikiaris, D.: Can nanoparticles really enhance thermal stability of polymers? Part II: An overview on thermal decomposition of polycondensation polymers. Thermochim. Acta **523**, 25–45 (2011)
33. Yang, K.K., Wang, X.L., YZ, W.: Progress in nanocomposite of biodegradable polymer. J Ind Eng Chem. **13**, 485–500 (2007)
34. Mohanty, A.K., Wibowo, A., Misra, M., Drzal, L.T.: Development of renewable resource–based cellulose acetate bioplastic: effect of process engineering on the performance of cellulosic plastics. Polym. Eng. Sci. **43**, 1151–1161 (2003)
35. Bandyopadhyay, S., Chen, R., Giannelis, E.P.: Biodegradable organic-inorganic hybrids based on poly(L-lactide). Polym. Mater. Sci. Eng. **81**, 159–160 (1999)
36. Pluta, M., Galeski, A., Alexandre, M., Paul, M.A., Dubois, P.: Polylactide/montmorillonite nanocomposites and microcomposites prepared by melt blending: structure and some physical properties. J. Appl. Polym. Sci. **86**, 1497–1506 (2002)
37. Chen, C.X., Yoon, J.S.: Morphology and thermal properties of poly(L-lactide)/poly(butylene succinate-co-butylene adipate) compounded with twice functionalized clay. J. Polym. Sci. Part B: Polym. Phys. **43**, 478–487 (2005)
38. Marras, S.I., Zuburtikudis, I., Panayiotou, C.: Nanostructure vs. microstructure: morphological and thermomechanical characterization of poly(l-lactic acid)/layered silicate hybrids. Eur. Polymer J. **43**, 2191–2206 (2007)
39. Chang, J.H., An, Y.U., Cho, D., Giannelis, E.P.: Poly(lactic acid) nanocomposites: comparison of their properties with montmorillonite and synthetic mica (II). Polymer **44**, 3715–3720 (2003)
40. Chang, J.H., An, Y.U., Sur, G.S.: Poly(lactic acid) nanocomposites with various organoclays. I. Thermomechanical properties, morphology and gas permeability. J. Polym. Sci., Part B: Polym. Phys. **41**, 94–103 (2003)
41. Paul, M.A., Alexandre, M., Degée, P., Calberg, C., Jérôme, R., Dubois, P.: Exfoliated polylactide/clay nanocomposites by in-situ coordination-insertion polymerization. Macromol. Rapid Commun. **24**, 561–566 (2003)
42. Paul, M.A., Alexandre, M., Degée, P., Henrist, C., Rulmont, A., Dubois, P.: New nanocomposite materials based on plasticized poly(L-lactide) and organo-modified montmorillonites: thermal and morphological study. Polymer **44**, 443–450 (2003)
43. Zhou, Q., Xanthos, M.: Nanosize and microsize clay effects on the kinetics of the thermal degradation of polylactides. Polym. Degrad. Stab. **94**, 327–338 (2009)
44. Najafi, N., Heuzey, M.C., Carreau, P.J., Wood-Adams, P.M.: Control of thermal degradation of polylactide (PLA)-clay nanocomposites using chain extenders. Polym. Degrad. Stab. **97**, 554–565 (2012)
45. Sivalingam, G., Madras, G.: Thermal degradation of binary physical mixtures and copolymers of poly(ε-caprolactone), poly(D, L-lactide), poly(glycolide). Polym. Degrad. Stab. **84**, 393–398 (2004)
46. Chrissafis, K., Antoniadis, G., Paraskevopoulos, K.M., Vassiliou, A., Bikiaris, D.N.: Comparative study of the effect of different nanoparticles on the mechanical properties and thermal degradation mechanism of in situ prepared poly(ε-caprolactone) nanocomposites. Compos. Sci. Technol. **67**, 2165–2174 (2007)
47. Peng, H., Han, Y., Liu, T., Tjiu, W.C., He, C.: Morphology and thermal degradation behavior of highly exfoliated CoAl-layered double hydroxide/polycaprolactone nanocomposites prepared by simple solution intercalation. Thermochim. Acta **502**, 1–7 (2010)
48. Carrasco, F., Gámez-Pérez, J., Santana, O.O., Maspoch, M.L.: Processing of poly(lactic acid)/organomontmorillonite nanocomposites: microstructure, thermal stability and kinetics of the thermal decomposition. Chem. Eng. J. **178**, 451–460 (2011)
49. Reich, L.: Elements of Polymer Degradation. McGraw-Hill, New York (1971)
50. Carrasco, F., Pagès, P., Gámez-Pérez, J., Santana, O.O., Maspoch, M.L.: Processing of poly (lactic acid): characterization of chemical structure, thermal stability and mechanical properties. Polym. Degrad. Stab. **95**, 116–125 (2010)

51. Moreira, F.K.V., Pedro, D.C.A., Glenn, G.M., Marconcini, J.M., Mattoso, L.H.C.: Brucite nanoplates reinforced starch bionanocomposites. Carbohydr. Polym. **92**, 1743–1751 (2013)
52. Espino-Pérez, E., Bras, J., Ducruet, V., Guinault, A., Dufresne, A., Domenek, S.: Influence of chemical surface modification of cellulose nanowhiskers on thermal, mechanical, and barrier properties of poly(lactide) based bionanocomposites. Eur. Polymer J. **49**, 3144–3154 (2013)
53. Araújo, A., Botelho, G., Oliveira, M., Machado, A.V.: Influence of clay organic modifier on the thermal-stability of PLA based nanocomposites. Appl. Clay Sci. **88–89**, 144–150 (2014)
54. Hakkarainen, M. Aliphatic Polyesters: abiotic and biotic degradation and degradation products. In: Albertsson, A.-C (ed.) Degradable Aliphatic Polyester, pp. 113–138. Springer, Heidelberg (2002)
55. Rosa, D.S., Lotto, N.T., Lopes, D.R., Guedes, C.G.F.: The use of roughness for evaluating the biodegradation of poly-β-(hydroxybutyrate) and poly-β-(hydroxybutyrate-co-β-valerate). Polym. Testing **23**, 3–8 (2004)
56. Lotto, N.T., Calil, M.R., Guedes, C.G.F., Rosa, D.S.: The effect of temperature on the biodegradation test. Mater. Sci. Eng. C **24**, 659–662 (2004)
57. Reddy, C.G.: R. Rashmi, Kalia, VC. Polyhydroxyalkanoates: an overview. Bioresour. Technol. **87**, 137–146 (2003)
58. de Jong, S.J., Arias, E.R., Rijkers, D.T.S., van Nostrum, C.F., Kettenes-van den Bosch, J.J., Hennink, W.E.: New insights into the hydrolytic degradation of poly(lactic acid): participation of the alcohol terminus. Polymer **42**, 2795–2802 (2001)
59. Fukushima, K., Tabuani, D., Dottori, M., Armentano, I., Kenny, J.M., Camino, G.: Effect of temperature and nanoparticle type on hydrolytic degradation of poly(lactic acid) nanocomposites. Polym. Degrad. Stab. **96**, 2120–2129 (2011)
60. Zhou, Q., Xanthos, M.: Nanoclay and crystallinity effects on the hydrolytic degradation of polylactides. Polym. Degrad. Stab. **93**, 1450–1459 (2008)
61. Signori, F., Coltelli, M.-B., Bronco, S.: Thermal degradation of poly(lactic acid) (PLA) and poly(butylene adipate-co-terephthalate) (PBAT) and their blends upon melt processing. Polym. Degrad. Stab. **94**, 74–82 (2009)
62. Gleadall, A., Pan, J., Kruft, M.-A., Kellomäki, M.: Degradation mechanisms of bioresorbable polyesters. Part 1. Effects of random scission, end scission and autocatalysis. Acta Biomater. **10**, 2223–2232 (2014)
63. Nieddu, E., Mazzucco, L., Gentile, P., Benko, T., Balbo, V., Mandrile, R., et al.: Preparation and biodegradation of clay composites of PLA. React. Funct. Polym. **69**, 371–379 (2009)
64. Chivrac, F., Pollet, E., Schmutz, M., Avérous, L.: Starch nano-biocomposites based on needle-like sepiolite clays. Carbohydr. Polym. **80**, 145–153 (2010)
65. Bruzaud, S., Bourmaud, A.: Thermal degradation and (nano)mechanical behavior of layered silicate reinforced poly(3-hydroxybutyrate-co-3-hydroxyvalerate) nanocomposites. Polym. Testing **26**, 652–659 (2007)
66. Ten, E., Turtle, J., Bahr, D., Jiang, L., Wolcott, M.: Thermal and mechanical properties of poly (3-hydroxybutyrate-co-3-hydroxyvalerate)/cellulose nanowhiskers composites. Polymer **51**, 2652–2660 (2010)
67. Pantoustier, N., Alexandre, M., Degée, P., Calberg, C., Jérôme, R., Henrist, C., et al.: Poly(η-caprolactone) layered silicate nanocomposites: effect of clay surface modifiers on the melt intercalation process. e-Polymers 77 (2001)
68. Cyras, V.P., Manfredi, L.B., Ton-That, M.-T., Vázquez, A.: Physical and mechanical properties of thermoplastic starch/montmorillonite nanocomposite films. Carbohydr. Polym. **73**, 55–63 (2008)
69. Schlemmer, D., Angélica, R.S., Sales, M.J.A.: Morphological and thermomechanical characterization of thermoplastic starch/montmorillonite nanocomposites. Compos. Struct. **92**, 2066–2070 (2010)
70. Luiz de Paula E, Mano V, Pereira FV. : Influence of cellulose nanowhiskers on the hydrolytic degradation behavior of poly (d,l-lactide). Polym Degrad Stab. **96**, 1631−1638 (2011)

71. Hossain, K.Z., Ahmed, I., Parsons, A., Scotchford, C., Walker, G., Thielemans, W., et al.: Physico-chemical and mechanical properties of nanocomposites prepared using cellulose nanowhiskers and poly(lactic acid). J. Mater. Sci. **47**, 2675–2686 (2012)
72. Wang, D., Yu, J., Zhang, J., He, J., Zhang, J.: Transparent bionanocomposites with improved properties from poly (propylene carbonate) (PPC) and cellulose nanowhiskers (CNWs). Compos. Sci. Technol. **85**, 83–89 (2013)
73. Fortunati, E., D'Angelo, F., Martino, S., Orlacchio, A., Kenny, J.M., Armentano, I.: Carbon nanotubes and silver nanoparticles for multifunctional conductive biopolymer composites. Carbon **49**, 2370–2379 (2011)
74. Hapuarachchi, T.D., Peijs, T.: Multiwalled carbon nanotubes and sepiolite nanoclays as flame retardants for polylactide and its natural fibre reinforced composites. Compos. A Appl. Sci. Manuf. **41**, 954–963 (2010)
75. Sadegh-Hassani, F., Mohammadi Nafchi, A.: Preparation and characterization of bionanocomposites films based on potato starch/halloysite nanoclay. Int. J. Biol. Macromol. **67**, pp. 458-462 (2014)
76. Ojijo, V., Ray, S.S.: Nano-biocomposites based on synthetic aliphatic polyesters and nanoclay. Prog. Mater. Sci. **62**, 1–57 (2014)
77. Nerantzaki, M., Papageorgiou, G.Z., Bikiaris D.N.: Effect of nanofiller's type on the thermal properties and enzymatic degradation of poly(ε-caprolactone). Polym. Degrad. Stab. **108**, 257–268 (2014)
78. Liu, X., Zou, Y., Li, W., Cao, G., Chen, W.: Kinetics of thermo-oxidative and thermal degradation of poly(d, l-lactide) (PDLLA) at processing temperature. Polym. Degrad. Stab. **91**, 3259–3265 (2006)
79. Hule, R.A., Pochan, D.J.: Polymer nanocomposites for biomedical applications. MRS Bull. **32**, 354–358 (2007)
80. Bharadwaj, R.K.: Modeling the barrier properties of polymer-layered silicate nanocomposites. Macromolecules **34**, 9189–9192 (2001)
81. Sorrentino, A., Gorrasi, G., Vittoria, V.: Potential perspectives of bio-nanocomposites for food packaging applications. Trends Food Sci. Technol. **18**, 84–95 (2007)
82. Emamifar, A., Kadivar, M., Shahedi, M., Soleimanian-Zad, S.: Evaluation of nanocomposite packaging containing Ag and ZnO on shelf life of fresh orange juice. Innovative Food Sci. Emerg. Technol. **11**, 742–748 (2010)
83. Chaudhry, Q., Scotter, M., Blackburn, J., Ross, B., Boxall, A., Castle, L., et al.: Applications and implications of nanotechnologies for the food sector. Food Addit. Contam. Part A Chem. Anal. Control Exposure Risk Asses. **25**, 241–258 (2008)
84. Tiwari, A.: Frontiers in bio-nanocomposites. Advanced. Mater. Lett. **2**, 377 (2011)
85. Madhavan Nampoothiri, K., Nair, N.R., John, R.P.: An overview of the recent developments in polylactide (PLA) research. Bioresour. Technol. **101**, 8493–8501 (2010)
86. Li, H.Y., Chang, C.M., Hsu, K.Y., Liu, Y.L.: Poly(lactide)-functionalized and Fe 3O 4 nanoparticle-decorated multiwalled carbon nanotubes for preparation of electrically-conductive and magnetic poly(lactide) films and electrospun nanofibers. J. Mater. Chem. **22**, 4855–4860 (2012)
87. Murariu, M., Bonnaud, L., Yoann, P., Fontaine, G., Bourbigot, S., Dubois, P.: New trends in polylactide (PLA)-based materials: "Green" PLA-Calcium sulfate (nano) composites tailored with flame retardant properties. Polym. Degrad. Stab. **95**, 374–381 (2010)
88. Cabedo, L., Feijoo, J.L., Villanueva, M.P., Lagarén, J.M., Giménez, E.: Optimization of biodegradable nanocomposites based on aPLA/PCL blends for food packaging applications. Macromol. Symp. **233**, 191–197 (2006)
89. ReportLinker. Packaging Industry: Market Research Reports, Statistics and Analysis. ReportLinker (2014)

MIX
Papier aus verantwortungsvollen Quellen
Paper from responsible sources
FSC® C105338

If you have any concerns about our products,
you can contact us on
ProductSafety@springernature.com

In case Publisher is established outside the EU,
the EU authorized representative is:
Springer Nature Customer Service Center GmbH
Europaplatz 3, 69115 Heidelberg, Germany

Printed by Libri Plureos GmbH
in Hamburg, Germany